Springer Series on

Wave Phenomena 14

Edited by L. B. Felsen

Springer Series on

Wave Phenomena

Volume 1 **Mechanics of Continua and Wave Dynamics**
By L. M. Brekhovskikh, V. Goncharov

Volume 2 **Rayleigh-Wave Theory and Application**
Editors: E. A. Ash, E. G. S. Paige

Volume 3 **Electromagnetic Surface Excitations**
Editors: R. F. Wallis, G. I. Stegeman

Volume 4 **Short-Wavelength Diffraction Theory**
Asymptotic Methods
By V. M. Babič, V. S. Buldyrev

Volume 5 **Acoustics of Layered Media I**
Plane and Quasi-Plane Waves
By L. M. Brekhovskikh, O. A. Godin

Volume 6 **Geometrical Optics of Inhomogeneous Media**
By Yu. A. Kravtsov, Yu. I. Orlov

Volume 7 **Recent Developments in Surface Acoustic Waves**
Editors: D. F. Parker, G. A. Maugin

Volume 8 **Fundamentals of Ocean Acoustics** 2nd Edition
By L. M. Brekhovskikh, Yu. P. Lysanov

Volume 9 **Nonlinear Optics in Solids**
Editor: O. Keller

Volume 10 **Acoustics of Layered Media II**
Point Sources and Bounded Beams
By L. M. Brekhovskikh, O. A. Godin

Volume 11 **Resonance Acoustic Spectroscopy**
By N. Veksler

Volume 12 **Scalar Wave Theory**
Green's Functions and Applications
By J. A. DeSanto

Volume 13 **Modern Problems in Radar Target Imaging**
Editors: W.-M. Boerner, F. Molinet, H. Überall

Koichi Furutsu

Random Media and Boundaries

Unified Theory, Two-Scale Method, and Applications

With 22 Figures

Springer-Verlag
Berlin Heidelberg New York
London Paris Tokyo
Hong Kong Barcelona
Budapest

Dr. Koichi Furutsu

Nakato 4-15-3, Musashi-Murayama
Tokyo 208, Japan

Series Editors:

Professor Leonid M. Brekhovskikh, Academician

Russian Academy of Sciences, P. P. Shirshov Institute of Oceanology,
23 Krasikowa St., 117218 Moscow, Russia

Professor Leopold B. Felsen, Ph.D.

Polytechnic University, Route 110, Farmingdale, NY 11735, USA

Professor Hermann A. Haus, Ph.D.

Department of Electrical Engineering & Computer Sciences, MIT,
Cambridge, MA 02139, USA

Managing Editor: Dr.-Ing. Helmut K.V. Lotsch

Springer-Verlag, Tiergartenstrasse 17,
W-6900 Heidelberg, Fed. Rep. of Germany

Statistical Theory of Waves in Random Media by Koichi Furutsu
Copyright © 1982 by Koichi Furutsu
Originally published in Japanese by Iwanami Shoten, Publishers, Tokyo 1982
Translated by Koichi Furutsu

ISBN-13: 978-3-642-84809-4 e-ISBN-13: 978-3-642-84807-0
DOI: 10.1007/978-3-642-84807-0

Library of Congress Cataloging-in-Publication Data
Furutsu, Kōichi. [Randamu baishitsunai no hadō denpan. English] Random media and boundaries: unified theory, two-scale method, and applications / Koichi Furutsu. p. cm. -- (Springer series on wave phenomena; 14)
Translation of: Randamu baishitsunai no hadō denpan. Includes bibliographical references and index.
1. Waves--Mathematics. 2. Scattering (Physics) 3. Random fields. 4. Boundary value problems. I. Title. II.Series.
QC 157. F8713 1993 530. 1'24--dc20 92-23248

Softcover reprint of the hardcover 2nd edition 1993

54/3140-5 4 3 2 1 0 – Printed on acid-free paper

Preface

This book is based on the author's works on random media and boundaries, and is written as an introduction and basis for the theory of transport in bounded space, a fixed scatterer embedded in a random system, and optical wave propagaiton in a turbulent medium (as a typical case of the forward scattering approximation).

Random media have attracted much attention, not only in atmospheric wave propagation, but also in solid-state physics in connection with wave transport through disordered media. Independently, rough boundaries have also been investigated to solve boundary-value problems of the transport equation, as well as to predict the angular distribution of waves scattered from a rough surface of a particular material and structure. Independent of whether media or boundaries are random, the governing equation can be written in the form of the Bethe-Salpeter equation, and the problem is therefore effectively reduced to finding a good (mostly optical) approximation to the solution. Also, for a system composed of a random medium and rough boundaries, the Bethe-Salpeter equation for the entire system can be constructed so that the medium and the boundaries are treated on exactly the same footing. This enables us to obtain several expressions for the solution by interchanging the roles of the medium and the boundaries, and hence a variety of expressions to choose from, depending on the particular situation and on the information required.

In this book the basic equations are all treated nonperturbatively in terms of a scattering matrix introduced separately for each of the media and the boundaries, and also for combinations constructed according to several addition formulas of scattering matrices, which can be purely coherent or incoherent in character. The unified Bethe-Salpeter equation, say, for a system composed of a random medium and rough boundaries, can also be directly applied to a fixed scatterer imbedded in a random system, leading naturally to an effective cross section that includes the shadowing effect and others.

Another area of interest has been the statistical properties of optical wave propagation through a turbulent medium, in which case all the equations can be formulated based on the forward scattering approximation. Here the problems include the intensity probability distribution function, as well as the intensity fluctuation, so that high-order moments of the wave functions are required to be evaluated based on the moment equations. Here, for the intensity moments of the first few orders, a two-scale method is available, which is based on the Lagrange variational principle in dynamics as applied to exact (functional) integral representations for the solutions of the moment equations. For the higher-order moments, the cluster approximation as introduced in solid-state physics enables those moments to be effectively expressed in terms of the first low order moments.

Tokyo, July 1992 K.Furutsu

Contents

1 **Operator Representation of a Random Medium** 1
1.1 Single Random Quantity 1
1.2 N Discrete Random Quantities 5
1.3 Random Function in Space 6
1.4 Multi-Component Random Medium 8

2 **Waves in a Homogeneously Random Medium** 9
2.1 Operator Representation of Statistical Green's Functions in a Medium of Independent Particles 9
2.2 Green's Functions of First and Second Orders 11
2.3 Bethe-Salpeter Equation in a General Random Medium 15
2.3.1 Optical Condition 16
2.3.2 Optical Expressions and Transport Equation 18
2.4 Random Layer with Free Boundaries 22
2.4.1 Boundary Condition 23
2.4.2 Addition of Scattering Matrices 23
2.5 Eigenfunction Expansions and Diffusion Approximation 27
2.5.1 Symmetries of the Eigenfunctions and Eigenvalues 30
2.5.2 Diffusion Eigenfunctions 32
2.5.3 Mode Expansions in a Homogeneous Random Medium 36
2.5.4 Orthogonality and Power Carried by the Mode Waves 40

3 **Random Rough Boundaries** 43
3.1 Rough Surface (One-Sided Boundary) 43
3.1.1 Reflection Coefficient 48
3.1.2 Scattering Matrix of a Small Boss on a Smooth Boundary 50
3.1.3 Optical Conditions 51
3.1.4 Surface Wave and Power Equation 52
3.1.5 Derivation of Surface Impedance 54
3.1.6 Integral Equation for the Reflection Coefficient 57
3.1.7 Tangent Plane Method – Case of a Large-Scale Rough Surface 58

3.2 Statistical Green's Functions of First and Second Orders ... 66
3.2.1 Incoherent Scattering Matrix ... 69
3.2.2 Optical Condition ... 69
3.2.3 M and K for a Surface of Randomly Distributed Bosses ... 72
3.2.4 Continuation to the Outside Space ... 73
3.2.5 Rough Surface as an Effective Random Medium ... 75
3.2.6 Optical Expressions and Cross Sections per Unit Area ... 76
3.2.7 Optical Relations ... 81
3.2.8 Examples ... 83
3.3 Transmissible (Two-Sided) Rough Boundary ... 85
3.3.1 Basic Equations ... 86
3.3.2 Evaluation of the Surface Green's Function ... 91
3.3.3 Power Equations ... 97
3.3.4 Statistical Green's Functions of First and Second Orders ... 99
3.3.5 Continuation to the Outside Spaces ... 103
3.3.6 Scattering Cross Sections and Optical Relations ... 107
3.3.7 Case of a Slightly Random Boundary ... 111

4 System of Random Media and Rough Boundaries ... 113
4.1 Bethe-Salpeter Equation for the Entire System and Scattering Matrices ... 113
4.1.1 Statistical Green's Functions ... 117
4.1.2 Optical Relations and a Dispersive Medium ... 119
4.1.3 Case of Three Random Layers ... 122
4.1.4 Scattering Matrices and Solutions ... 123
4.2 Effective Boundary Scattering Matrices in a Random Medium and Construction of Solutions ... 127
4.2.1 Case of Three Random Layers with Two Rough Boundaries ... 132
4.2.2 $I^{(12+23)}$ and Boundary Scattering Matrices ... 135
4.2.3 Another Expression for $I^{(q+12+23)}$ ... 136
4.2.4 Addition Formulas of Scattering Matrices Utilized ... 139
4.2.5 Optical Relations of Random Layers ... 143

5 Optical Cross Sections of a Random Layer ... 147
5.1 Construction of the Cross Sections ... 147
5.1.1 Case Involving No Boundary Scattering ... 150
5.1.2 Case Involving Boundary Scattering ... 152
5.1.3 Optical Relations and Reciprocity ... 154
5.2 Application of the Diffusion Approximation ... 155
5.2.1 Boundary Condition of the Diffusion Equation ... 157
5.2.2 Boundary-Value Solution of the Diffusion Equation ... 160
5.2.3 Case of a Random Layer with Smooth Boundaries ... 164

5.2.4 Reciprocity ... 166
5.2.5 Boundary Condition when Media are Random on Both Sides of the Boundary ... 168

6 **Fixed Scatterer** ... 171
6.1 Basic Equations ... 171
6.2 Power Equations and Optical Relations ... 179
6.3 Optical Cross Section and Shadowing Effect ... 180
6.4 Observation of a Fixed Scatterer Embedded in a Semi-Infinite Random Layer ... 182

7 **Forward Scattering Approximation** ... 185
7.1 Moment Equations of a Light Wave in a Turbulent Medium 185
7.1.1 Turbulent Medium of Kolmogorov Spectrum ... 192
7.2 Solutions of the Moment Equations ... 194
7.2.1 Basic Equations ... 194
7.2.2 Phase Screen ... 200
7.2.3 Formal Generalization to a Turbulent Continuum ... 202
7.2.4 Exact Solutions of All Orders in a Special Case ... 205
7.3 High Order Intensity Moments and the Cluster Approximation ... 207
7.3.1 Intensity Fluctuation ... 208
7.3.2 Cluster Approximation ... 212
7.4 Probability Distribution Function of Intensity ... 216
7.4.1 Rice-Nakagami Distribution ... 217
7.4.2 Integral Representation of the Probability Density Function for Given $\langle I^n \rangle$, $n = 1, 2, 3$ 219
7.4.3 $\langle I^\nu \rangle$ as the Characteristic Function of $\ln(I)$ ... 221
7.4.4 Intensity Distribution for a Beam Wave Purely in the State of Wandering ... 224
7.4.5 Intensity Distribution Function in a Turbulent Medium ... 225
7.4.6 Comparison of Theories to Experimental Results ... 230
7.5 Two-Scale Method ... 235
7.5.1 Exact Integral Representation of $M_{22}(1)$ ... 236
7.5.2 Two-Scale Representation as an Effective Approximation ... 242
7.5.3 Two-Frequency Intensity Correlation ... 249
7.5.4 Third Order Intensity Correlation Function ... 253
7.5.5 Summary and Discussion ... 261

References ... 263

Bibliography ... 266

Subject Index ... 267

1. Operator Representation of a Random Medium

The entire statistical information of a random medium, say $q(\hat{\boldsymbol{x}})$, as a random function in space, can be obtained through its characteristic functional, say Z_q, and the situation is also the same for a wave in the random medium, which is a functional of the random medium $q(\hat{\boldsymbol{x}})$, say $f[q]$. The problem is therefore basically reduced to finding the characteristic functional of the wave for a given Z_q. We begin this subject by learning how the statistical average of the given functional, $f[q]$, defined explicitly or implicitly through its governing equation, can be obtained in terms of Z_q. In this section, the medium will be represented by an operator, $\hat{q}(\hat{\boldsymbol{x}})$, so that the average value of $f[q]$ can be obtained as an expectation value of the same functional of the operator, $f[\hat{q}]$. The medium will first be regarded as a single random quantity in Sect. 1.1, as N discrete random quantities in Sect. 1.2, and as a random function in space in Sect. 1.3.

1.1 Single Random Quantity

The characteristic function, $Z_q(p)$, of a single random quantity, q, is defined, for an arbitrary variable, p, by

$$Z_q(p) = \langle \exp(pq) \rangle, \qquad Z_q(0) = 1 \ , \tag{1.1.1}$$

where $\langle \ldots \rangle$ means the statistical averaging. Hence,

$$\left(\frac{\partial}{\partial p}\right)^n Z_q(p) \mid_{p=0} = \langle q^n \rangle, \qquad n = 0, 1, 2, \ldots \ , \tag{1.1.2}$$

and, therefore, for any function, $f(q)$,

$$f(\partial/\partial p) Z_q(p) \mid_{p=0} = \langle f(q) \rangle \ , \tag{1.1.3}$$

when the power series expansion of q is possible. Even when this is not the case, use of the Fourier representation,

$$f(q) = (2\pi)^{-1} \int_{-\infty}^{\infty} dp \exp(ipq) \tilde{f}(p) \ , \tag{1.1.4}$$

yields

$$\langle f(q)\rangle = (2\pi)^{-1}\int_{-\infty}^{\infty} dp\, Z_q(\mathrm{i}p)\tilde{f}(p) \; . \tag{1.1.5}$$

Equation (1.1.3) is certainly convenient when $f(q)$ is explicitly given, but this is not the case when it is implicitly given, e.g., through its governing equation. For such a case, the following operator representation of q is much more convenient to use.

For an arbitrary variable, c, use of definition (1.1.1) yields

$$\begin{aligned} Z_q(\partial/\partial c)f(c) &= \langle \exp[q\partial/\partial c]f(c)\rangle \\ &= \left\langle \sum_{n=0}^{\infty} n!^{-1}q^n\left(\frac{\partial}{\partial c}\right)^n f(c)\right\rangle = \langle f(c+q)\rangle \;, \end{aligned} \tag{1.1.6}$$

which shows that $\langle f(q)\rangle$ can be given by the left-hand side with the limit $c=0$. Here, the evaluation of the left-hand side becomes simple by introduction of an operator, $\hat{q}$, defined by

$$\hat{q} = Z_q(\partial/\partial c)c[Z_q(\partial/\partial c)]^{-1} \; . \tag{1.1.7}$$

Hence, with $Z_q = Z_q(\partial/\partial c)$,

$$\begin{aligned} &Z_q c^2 Z_q^{-1} = (Z_q c Z_q^{-1})(Z_q c Z_q^{-1}) = \hat{q}^2, \\ &Z_q c^n Z_q^{-1} = \hat{q}^n, \qquad n = 0,1,2,\dots \;, \end{aligned} \tag{1.1.8}$$

and, therefore, generally,

$$Z_q f(c) Z_q^{-1} = f(Z_q c Z_q^{-1}) = f(\hat{q}) \;, \tag{1.1.9}$$

which enables (1.1.6) to be written as

$$\langle f(c+q)\rangle = f(\hat{q})Z_q = f(\hat{q}) \;, \tag{1.1.10}$$

since, in the last derivation, there is nothing for $Z_q(\partial/\partial c)$ to operate on.

From (1.1.10) we learn that the statistical average of any function, $f(q)$, can be found simply by the replacement of q by the operator, $\hat{q}$, and therefore that

$$\langle (c+q)f(c+q)\rangle = \hat{q}f(\hat{q}) = \hat{q}\langle f(c+q)\rangle \;, \tag{1.1.11}$$

which is particularly convenient to find $\langle qf(q)\rangle$ for given $\langle f(q)\rangle$.

Here, obtaining an explicit expression of $\hat{q}$ is facilitated by introducing the cumulant of $Z_q(p)$, defined by

$$\theta(p) = \ln[Z_q(p)] \; . \tag{1.1.12}$$

Hence, in terms of the conventional notation for a commutator $[\widehat{A},\widehat{B}] = \widehat{A}\widehat{B} - \widehat{B}\widehat{A} = -[\widehat{B},\widehat{A}]$, (1.1.7) gives, as will be proved below,

$$\begin{aligned}\widehat{q} &= \exp[\theta(\partial/\partial c)]\; c \exp[-\theta(\partial/\partial c)] \\ &= c + [\theta, c] + \frac{1}{2!}[\theta,[\theta,c]] + \frac{1}{3!}[\theta,[\theta,[\theta,c]]] + \dots \;,\end{aligned} \tag{1.1.13}$$

being a power series expansion with respect to the operator, $\theta(\partial/\partial c)$.

The proof is given, on redefining $\widehat{q}$ by

$$\widehat{q} = e^{\lambda\theta} c e^{-\lambda\theta} \;,$$

with arbitrary constant λ, so that

$$\frac{\partial}{\partial\lambda}\widehat{q} = \theta\widehat{q} - \widehat{q}\theta = [\theta,\widehat{q}] \;,$$

$$\left(\frac{\partial}{\partial\lambda}\right)^2 \widehat{q} = \left[\theta, \frac{\partial}{\partial\lambda}\widehat{q}\right] = \left[\theta,[\theta,\widehat{q}]\right], \quad \text{etc.} \;,$$

and $\widehat{q} = c$ for $\lambda = 0$; hence (1.1.13) is obtained as the power series expansion of the new $\widehat{q}$, with respect to $\lambda = 1$ [1.1].

Here, in (1.1.13),

$$[\theta(\partial/\partial c), c] = \theta'(\partial/\partial c) \;, \tag{1.1.14}$$

where

$$\theta'(p) \equiv \frac{\partial}{\partial p}\theta(p) = \frac{\partial}{\partial p}\ln Z_q(p) \;. \tag{1.1.15}$$

Therefore, on the right-hand side of (1.1.13), the non-vanishing terms are only the first two terms, yielding

$$\widehat{q} = c + \theta'(\partial/\partial c) \;. \tag{1.1.16}$$

The proof of (1.1.14) is straightforward by expanding $\theta(\partial/\partial c)$ in a series of $\partial/\partial c$, and using the commutation relation,

$$[(\partial/\partial c)^n, c] = n(\partial/\partial c)^{n-1} \;.$$

Using (1.1.1) and (1.1.12), the power series expansion of $\theta(p)$ (cumulant series) yields, when $\langle q\rangle = 0$,

$$\theta(p) \sim \frac{1}{2}\langle q^2\rangle p^2, \qquad \theta'(p) \sim \langle q^2\rangle p \;, \tag{1.1.17}$$

to the lowest order of p. Hence by using (1.1.16),

$$\widehat{q} \sim c + \langle q^2\rangle \frac{\partial}{\partial c} \;. \tag{1.1.18}$$

Here, when $\hat{q}$ is given by (1.1.18), statistics of q is called Gaussian, and utilization of the formula (1.1.11) becomes particularly simple.

As an example, let us evaluate the characteristic function $Z_q(p)$ of (1.1.1) according to formula (1.1.10, 18), i.e.,

$$Z_q(p) = \exp(p\hat{q}) = \exp\left[p\left(c + \langle q^2\rangle \frac{\partial}{\partial c}\right)\right]\bigg|_{c=0} . \tag{1.1.19}$$

This can be achieved most simply by use of lemma [1.1],

$$\exp(\hat{A} + \hat{B}) = \exp(\hat{A})\exp(\hat{B})\exp(2^{-1}[\hat{B}, \hat{A}]) , \tag{1.1.20}$$

which holds for arbitrary operators $\hat{A}$ and $\hat{B}$ when the commutator $[\hat{A}, \hat{B}]$ is commutable with both $\hat{A}$ and $\hat{B}$, as will be proved later. Hence, when $\hat{A} = pc$ and $\hat{B} = \langle q^2\rangle p\,\partial/\partial c$, then, $[\hat{B}, \hat{A}] = \langle q^2\rangle p^2$ is constant, yielding, in the limit $c = 0$,

$$Z_q(p) = \exp(2^{-1}\langle q^2\rangle p^2) , \tag{1.1.21}$$

in view of $\exp(\hat{B}) = 1$ with its right-hand side factor independent of c.

The proof of (1.1.20) is given by introducing an operator, $\hat{f}(\lambda)$, defined by

$$\exp\left[\lambda(\hat{A} + \hat{B})\right] = \exp(\lambda\hat{A})\hat{f}(\lambda) ,$$

where λ is an ordinary number. Hence, differentiating both sides with respect to λ, the resulting equation can be converted into

$$\frac{\partial}{\partial\lambda}\hat{f}(\lambda) = \exp(-\lambda\hat{A})\hat{B}\exp(\lambda\hat{A})\hat{f}(\lambda) ,$$

which provides a differential equation for the unknown $\hat{f}(\lambda)$. Here, since $[\hat{A}, \hat{B}]$ is commutable with both $\hat{A}$ and $\hat{B}$ by the assumption,

$$\exp(-\lambda\hat{A})\hat{B}\exp(\lambda\hat{A}) = \hat{B} - \lambda[\hat{A}, \hat{B}] ,$$

in exactly the same way as in (1.1.13). Hence, upon the substitution, the solution is found to be

$$\hat{f}(\lambda) = \exp\left(\lambda\hat{B} - \frac{1}{2}\lambda^2[\hat{A}, \hat{B}]\right), \qquad \hat{f}(0) = 1 ,$$

yielding (1.1.20) after setting $\lambda = 1$.

1.2 N Discrete Random Quantities

When a random quantity is made up of N components, say $\{q\} = \{q_1, q_2, \ldots, q_N\}$, the characteristic function, $Z_q\{p\}$, where $\{p\} = \{p_1, p_2, \ldots, p_N\}$, is defined by

$$Z_q\{p\} = \left\langle \exp\left(\sum_{j=1}^{N} p_j q_j\right)\right\rangle, \qquad Z_q\{0\} = 1 \ , \tag{1.2.1}$$

in just the same way as in (1.1.1). Hence, it follows that almost every equation of Sect. 1.1 holds as is, with a minor dimensional change, yielding particularly from (1.1.10, 11),

$$\langle f\{c+q\}\rangle = f\{\widehat{q}\} \ , \tag{1.2.2}$$

$$\langle (c+q)_j f\{c+q\}\rangle = \widehat{q}_j f\{\widehat{q}\} = \widehat{q}_j \langle f\{c+q\}\rangle \ . \tag{1.2.3}$$

Here, from (1.1.15, 16) with $\{c\} = \{c_1, c_2, \ldots, c_N\}$,

$$\widehat{q}_j = c_j + \theta_j\{\partial/\partial c\}, \qquad j = 1, 2, \ldots, N \ , \tag{1.2.4}$$

where

$$\theta_j\{p\} = \frac{\partial}{\partial p_j}\theta\{p\} \equiv \frac{\partial}{\partial p_j}\ln Z_q\{p\} \ . \tag{1.2.5}$$

Hence, as in (1.1.17), if the power series of $\theta\{p\}$ can be approximated by the first non-vanishing term, i.e.,

$$\theta\{p\} \sim \sum_{i,j=1}^{N} \frac{1}{2}\langle q_i, q_j\rangle p_i p_j, \qquad \langle q_j\rangle = 0 \ , \tag{1.2.6}$$

then, by (1.2.4, 5)

$$\theta_j\{p\} \sim \sum_{i=1}^{N} \langle q_j q_i\rangle p_i \ , \tag{1.2.7}$$

$$\widehat{q}_j \sim c_j + \sum_{i=1}^{N} \langle q_j q_i\rangle \frac{\partial}{\partial c_i} \ , \tag{1.2.8}$$

giving the expression when $\{q\}$ is Gaussian.

Here, using formula (1.2.2) with (1.2.8), the characteristic function (1.2.1) is evaluated as

$$\begin{aligned} Z_q\{p\} &= \exp\left(\sum_{j=1}^{N} p_j \widehat{q}_j\right)\Bigg|_{c=0} \\ &= \exp\left(\sum_{i,j=1}^{N} 2^{-1}\langle q_i q_j\rangle p_i p_j\right) \ , \end{aligned} \tag{1.2.9}$$

with the aid of lemma (1.1.20), and the cumulant (1.2.6) is thus reproduced.

1.3 Random Function in Space

The previous situation remains unchanged even when the random quantity is a function, $q(\hat{\boldsymbol{x}})$, of a space coordinate $\hat{\boldsymbol{x}} = (x_1, x_2, x_3)$, with the element of volume, $d\hat{\boldsymbol{x}} = dx_1 dx_2 dx_3$. Thus, for an arbitrary function, $p(\hat{\boldsymbol{x}})$, the characteristic functional, $Z_q[p]$, is defined, instead of (1.2.1), by

$$Z_q[p] = \left\langle \exp\left[\int d\hat{\boldsymbol{x}}\, p(\hat{\boldsymbol{x}})q(\hat{\boldsymbol{x}})\right]\right\rangle . \tag{1.3.1}$$

Here, since $p(\hat{\boldsymbol{x}})$ is a continuous function of $\hat{\boldsymbol{x}}$, a particular consideration is necessary when differentiating (1.3.1), for example, with respect to $p(\hat{\boldsymbol{x}})$. Let $\delta p(\hat{\boldsymbol{x}})$ be an arbitrary variation of $p(\hat{\boldsymbol{x}})$ for $\hat{\boldsymbol{x}}$ over the entire range of space, and let the variation of $Z_q[p]$, due to the change of $p(\hat{\boldsymbol{x}})$ to $p(\hat{\boldsymbol{x}}) + \delta p(\hat{\boldsymbol{x}})$, be $\delta Z_q[p] = Z_q[p + \delta p] - Z_q[p]$. Then, to the first order of $\delta p(\hat{\boldsymbol{x}})$, $\delta Z_q[p]$ can always be written in the form

$$\delta Z_q[p] = \int d\hat{\boldsymbol{x}}'\, \frac{\delta Z_q[p]}{\delta p(\hat{\boldsymbol{x}}')}\, \delta p(\hat{\boldsymbol{x}}') . \tag{1.3.2}$$

Here, from (1.3.1),

$$\frac{\delta Z_q[p]}{\delta p(\hat{\boldsymbol{x}}')} = \left\langle \exp\left[\int d\hat{\boldsymbol{x}}\, p(\hat{\boldsymbol{x}})q(\hat{\boldsymbol{x}})\right] q(\hat{\boldsymbol{x}}')\right\rangle . \tag{1.3.3}$$

Equation (1.3.2) also gives the definition of the functional derivative, $\delta Z_q[p]/\delta p(\hat{\boldsymbol{x}})$, of any functional, $Z_q[p]$, and therefore, for example, the functional differentiation of (1.3.3) yields

$$\begin{aligned}&[\delta/\delta p(\hat{\boldsymbol{x}}_1)]\,[\delta/\delta p(\hat{\boldsymbol{x}}_2)]\, Z_q[p] \\ &= \left\langle \exp\left[\int d\hat{\boldsymbol{x}}\, p(\hat{\boldsymbol{x}})q(\hat{\boldsymbol{x}})\right] q(\hat{\boldsymbol{x}}_1)q(\hat{\boldsymbol{x}}_2)\right\rangle .\end{aligned} \tag{1.3.4}$$

More generally,

$$f[\delta/\delta p]Z_q[p] = \left\langle \exp\left[\int d\hat{\boldsymbol{x}}\, p(\hat{\boldsymbol{x}})q(\hat{\boldsymbol{x}})\right] f[q]\right\rangle \longrightarrow \langle f[q]\rangle, \quad p = 0 , \tag{1.3.5}$$

which obviously corresponds to (1.1.3), suggesting that all the equations in Sect. 1.2 will be reproduced with the replacement of $\partial/\partial c_j$ with $\delta/\delta c(\hat{\boldsymbol{x}})$, and $\sum_j$ with $\int d\hat{\boldsymbol{x}}$.

Thus, in the same way as (1.1.7), let $\hat{q}(\hat{\boldsymbol{x}})$ be a spatial operator, defined by

$$\hat{q}(\hat{\boldsymbol{x}}) = Z_q[\delta/\delta c]c(\hat{\boldsymbol{x}})\{Z_q[\delta/\delta c]\}^{-1} . \tag{1.3.6}$$

Then, on reference to (1.2.4, 5), we obtain

$$\widehat{q}(\widehat{\boldsymbol{x}}) = c(\widehat{\boldsymbol{x}}) + \theta(\widehat{\boldsymbol{x}}, \delta/\delta c) \ , \tag{1.3.7}$$

where

$$\theta(\widehat{\boldsymbol{x}}, p) = \frac{\delta}{\delta p(\widehat{\boldsymbol{x}})}\theta[p] \equiv \frac{\delta}{\delta p(\widehat{\boldsymbol{x}})} \ln Z_q[p] \ , \tag{1.3.8}$$

and $\theta(\widehat{\boldsymbol{x}}, \delta/\delta c)$ is the functional of the functional differential operator, $\delta/\delta c(\widehat{\boldsymbol{x}})$. Here, from (1.2.2, 3), the following rules hold [1.2],

$$\langle f[c+q] \rangle = f[\widehat{q}] \ , \tag{1.3.9}$$

$$\langle [c(\widehat{\boldsymbol{x}}) + q(\widehat{\boldsymbol{x}})] f[c+q] \rangle = \widehat{q}(\widehat{\boldsymbol{x}}) \langle f[c+q] \rangle \ . \tag{1.3.10}$$

On the other hand, when $\langle q(\widehat{\boldsymbol{x}}) \rangle = 0$, the power series expansion of $\theta[p]$ becomes, using (1.3.8, 1),

$$\begin{aligned}\theta[p] &= \frac{1}{2} \int d\widehat{\boldsymbol{x}}_1 \, d\widehat{\boldsymbol{x}}_2 \, \langle q(\widehat{\boldsymbol{x}}_1) q(\widehat{\boldsymbol{x}}_2) \rangle p(\widehat{\boldsymbol{x}}_1) p(\widehat{\boldsymbol{x}}_2) \\ &+ \frac{1}{6} \int d\widehat{\boldsymbol{x}}_1 \, d\widehat{\boldsymbol{x}}_2 \, d\widehat{\boldsymbol{x}}_3 \, \langle q(\widehat{\boldsymbol{x}}_1) q(\widehat{\boldsymbol{x}}_2) q(\widehat{\boldsymbol{x}}_3) \rangle p(\widehat{\boldsymbol{x}}_1) p(\widehat{\boldsymbol{x}}_2) p(\widehat{\boldsymbol{x}}_3) + \dots \ . \end{aligned} \tag{1.3.11}$$

Hence, by (1.3.7, 8),

$$\begin{aligned}\widehat{q}(\widehat{\boldsymbol{x}}) &= c(\widehat{\boldsymbol{x}}) + \int d\widehat{\boldsymbol{x}}_1 \, \langle q(\widehat{\boldsymbol{x}}) q(\widehat{\boldsymbol{x}}_1) \rangle \frac{\delta}{\delta c(\widehat{\boldsymbol{x}}_1)} \\ &+ \frac{1}{2} \int d\widehat{\boldsymbol{x}}_1 \, d\widehat{\boldsymbol{x}}_2 \, \langle q(\widehat{\boldsymbol{x}}) q(\widehat{\boldsymbol{x}}_1) q(\widehat{\boldsymbol{x}}_2) \rangle \frac{\delta}{\delta c(\widehat{\boldsymbol{x}}_1)} \frac{\delta}{\delta c(\widehat{\boldsymbol{x}}_2)} + \dots \ , \end{aligned} \tag{1.3.12}$$

which, when the third and higher order terms are negligible on the right-hand side, gives the operator representation of a random medium subject to the Gaussian statistics, i.e.,

$$\widehat{q}(\widehat{\boldsymbol{x}}) \sim c(\widehat{\boldsymbol{x}}) + \int d\widehat{\boldsymbol{x}}' \langle q(\widehat{\boldsymbol{x}}) q(\widehat{\boldsymbol{x}}') \rangle \frac{\delta}{\delta c(\widehat{\boldsymbol{x}}')} \ . \tag{1.3.13}$$

In the last case, the characteristic functional, $Z_q[p]$, is again reproduced according to formula (1.3.9), by

$$\begin{aligned}Z_q[p] &= \exp \left[\int d\widehat{\boldsymbol{x}} \, p(\widehat{\boldsymbol{x}}) \widehat{q}(\widehat{\boldsymbol{x}}) \right] \Big|_{c=0} \\ &= \exp \left[2^{-1} \int d\widehat{\boldsymbol{x}}_1 \, d\widehat{\boldsymbol{x}}_2 \, \langle q(\widehat{\boldsymbol{x}}_1) q(\widehat{\boldsymbol{x}}_2) \rangle p(\widehat{\boldsymbol{x}}_1) p(\widehat{\boldsymbol{x}}_2) \right] \ , \end{aligned} \tag{1.3.14}$$

with the aid of lemma (1.1.20), and the commutation relation,

$$[\delta/\delta c(\hat{\boldsymbol{x}}), c(\hat{\boldsymbol{x}}')] = \delta(\hat{\boldsymbol{x}} - \hat{\boldsymbol{x}}') \ . \tag{1.3.15}$$

It may be noticed here that the approximation (1.3.13) can be used in the right-hand side of (1.3.10) even when the third and higher order terms of the series (1.3.12) are not zero, if the magnitude of $q(\hat{\boldsymbol{x}})$ is small enough so that the change of $f[q]$ (wave function) is negligible for the possible fluctuation of $q(\hat{\boldsymbol{x}})$ over the range of its correlation distance. This is actually the case of light wave propagation in a turbulent medium (Chap. 7).

On the other hand, when $q(\hat{\boldsymbol{x}})$ changes by a large magnitude, as in a medium of random particles, so that $f[q]$ also changes by a large amount with every fluctuaiton of $q(\hat{\boldsymbol{x}})$, locally, the approximation of (1.3.13) (keeping only the linear term of $\delta/\delta c(\hat{\boldsymbol{x}})$) does not make sense any longer, and $\theta(\hat{\boldsymbol{x}}, \delta/\delta c)$ must be constructed with all order terms of $\delta/\delta c(\hat{\boldsymbol{x}})$ involved (Sect. 2.1).

Finally, the operators, $\hat{q}(\hat{\boldsymbol{x}})$, are commutable to each other, i.e., for arbitrary points $\hat{\boldsymbol{x}}$ and $\hat{\boldsymbol{x}}'$, the definition (1.3.6) shows that

$$[\hat{q}(\hat{\boldsymbol{x}}), \hat{q}(\hat{\boldsymbol{x}}')] = Z_q [c(\hat{\boldsymbol{x}}), c(\hat{\boldsymbol{x}}')] Z_q^{-1} = 0 \ , \tag{1.3.16}$$

and, therefore, $\hat{q}(\hat{\boldsymbol{x}})$ can be treated in exactly the same way as an ordinary (commutable) function, $q(\hat{\boldsymbol{x}})$, is.

1.4 Multi–Component Random Medium

When the medium is composed of several independent components, say,

$$q(\hat{\boldsymbol{x}}) = q_1(\hat{\boldsymbol{x}}) + q_2(\hat{\boldsymbol{x}}) \ , \tag{1.4.1}$$

the characteristic functional of the medium is given by

$$\begin{aligned} Z_q[p] &= \left\langle \exp\left\{ \int d\hat{\boldsymbol{x}}\, p(\hat{\boldsymbol{x}}) [q_1(\hat{\boldsymbol{x}}) + q_2(\hat{\boldsymbol{x}})] \right\} \right\rangle \\ &= Z_{q_1}[p] Z_{q_2}[p] \ , \end{aligned} \tag{1.4.2}$$

where $Z_{q_j}[p], \ \ j = 1, 2$, is the characteristic functional of $q_j(\hat{\boldsymbol{x}})$ alone. Hence, according to (1.3.8),

$$\theta(\hat{\boldsymbol{x}}, p) = \theta_1(\hat{\boldsymbol{x}}, p) + \theta_2(\hat{\boldsymbol{x}}, p) \ , \tag{1.4.3a}$$

$$\theta_j(\hat{\boldsymbol{x}}, p) = \frac{\delta}{\delta p(\hat{\boldsymbol{x}})} \ln Z_{q_j}[p], \qquad j = 1, 2 \ , \tag{1.4.3b}$$

which shows that the contributions from the independent medium components are simply added up to construct $\theta(\hat{\boldsymbol{x}}, p)$ of (1.3.7), as

$$\hat{q}(\hat{\boldsymbol{x}}) = c(\hat{\boldsymbol{x}}) + \theta_1(\hat{\boldsymbol{x}}, \delta/\delta c) + \theta_2(\hat{\boldsymbol{x}}, \delta/\delta c) \ . \tag{1.4.4}$$

2. Waves in a Homogeneously Random Medium

Here we typically consider a medium of independent particles, in which the medium, $q(\hat{x})$, changes by a large magnitude in space so that the wave is seriously affected by every fluctuation of the medium, locally, in contrast to the case of turbulent air, in which the wave is negligibly affected by the medium fluctuation for the propagation over a distance of the order of the medium correlation distance.

2.1 Operator Representation of Statistical Green's Functions in a Medium of Independent Particles

We first suppose that N non-dissipative particles are involved in a space of volume, V, and that the contribution from one particle to the medium is described by a function, $q_a(\hat{x})$, with the center at the coordinates $\hat{x}_a$. Hence,

$$q(\hat{x}) = q^*(\hat{x}) = \sum_a q_a(\hat{x}) \ . \tag{2.1.1}$$

In order to find, according to (1.3.7, 8), the operator representation, $\hat{q}(\hat{x})$, of the medium, we observe that when the particles are independently distributed (even with a possibility of overlapping), use of (2.1.1) in (1.3.1) yields the characteristic functional of the present medium as [2.1],

$$Z_q[p] = \left[V^{-1} \int_V d\hat{x}_a \left\langle \exp\left[\int_V d\hat{x}\, p(\hat{x}) q_a(\hat{x}) \right] \right\rangle' \right]^N \ , \tag{2.1.2}$$

where bracket $\langle \ldots \rangle'$ means the average over all possible properties of the involved particles (excluding the average over the center coordinates, $\hat{x}_a$), e.g., their sizes, shapes, orientations, refractive indices, etc. Hence, as V and $N \to \infty$, and keeping the density of particles, $n = N/V$, constant, (2.1.2) tends to

$$Z_q[p] = \exp \left[\!\left[n \int d\hat{x}_a \left\{ \left\langle \exp\left[\int d\hat{x}\, p(\hat{x}) q_a(\hat{x}) \right] \right\rangle' - 1 \right\} \right]\!\right] \ , \tag{2.1.3}$$

which, according to (1.3.8), gives

$$\theta(\hat{\boldsymbol{x}}, p) = n \int d\hat{\boldsymbol{x}}_a \left\langle q_a(\hat{\boldsymbol{x}}) \exp\left[\int d\hat{\boldsymbol{x}}'\, p(\hat{\boldsymbol{x}}') q_a(\hat{\boldsymbol{x}}')\right]\right\rangle' . \tag{2.1.4}$$

Thus, in terms of the notation,

$$\langle \ldots \rangle_a = n \int d\hat{\boldsymbol{x}}_a \, \langle \ldots \rangle' , \tag{2.1.5}$$

the operator, $\hat{q}(\hat{\boldsymbol{x}})$, is finally given by

$$\hat{q}(\hat{\boldsymbol{x}}) = c(\hat{\boldsymbol{x}}) + \theta(\hat{\boldsymbol{x}}, \delta/\delta c) , \tag{2.1.6a}$$

with

$$\theta(\hat{\boldsymbol{x}}, \delta/\delta c) = \left\langle q_a(\hat{\boldsymbol{x}}) \exp\left[\int d\hat{\boldsymbol{x}}'\, q_a(\hat{\boldsymbol{x}}') \frac{\delta}{\delta c(\hat{\boldsymbol{x}}')}\right]\right\rangle_a . \tag{2.1.6b}$$

A scalar wave function, $\psi(\hat{\boldsymbol{x}})e^{i\omega t}$, where $\omega > 0$ and t is time, is hereafter considered with the equation of the form

$$[L - q(\hat{\boldsymbol{x}})]\, \psi(\hat{\boldsymbol{x}}) = j(\hat{\boldsymbol{x}}) , \tag{2.1.7a}$$

$$[L - q(\hat{\boldsymbol{x}})]\, \psi^*(\hat{\boldsymbol{x}}) = j^*(\hat{\boldsymbol{x}}) . \tag{2.1.7b}$$

Here, the last is the complex-conjugate wave equation,

$$L = -\left(\frac{\partial}{\partial \hat{\boldsymbol{x}}}\right)^2 - k_0^2, \qquad \mathrm{Im}\,\{k_0\} = -0 , \tag{2.1.7c}$$

$q(\hat{\boldsymbol{x}})$ is the random part of the medium, and $j(\hat{\boldsymbol{x}})$ is a source term. The power flux vector, $\hat{\boldsymbol{w}}(\hat{\boldsymbol{x}})$, is defined by

$$\hat{\boldsymbol{w}}(\hat{\boldsymbol{x}}) = (2i)^{-1} \psi^* \left(\frac{\overleftarrow{\partial}}{\partial \hat{\boldsymbol{x}}} - \frac{\overrightarrow{\partial}}{\partial \hat{\boldsymbol{x}}}\right) \psi(\hat{\boldsymbol{x}}) , \tag{2.1.7d}$$

so that the equation of continuity becomes

$$\frac{\partial}{\partial \hat{\boldsymbol{x}}} \cdot \hat{\boldsymbol{w}} = (2i)^{-1} (\psi^* j - \psi j^*) . \tag{2.1.7e}$$

The Green's function of the wave equation (2.1.7a), say $G_q(\hat{\boldsymbol{x}}|\hat{\boldsymbol{x}}')$, is defined by

$$[L - q(\hat{\boldsymbol{x}})]\, G_q(\hat{\boldsymbol{x}}|\hat{\boldsymbol{x}}') = \delta(\hat{\boldsymbol{x}} - \hat{\boldsymbol{x}}') , \tag{2.1.8a}$$

or, in coordinate matrix form, by

$$[L - q]G_q = 1 \ , \tag{2.1.8b}$$

and is hence a functional of $q(\hat{\boldsymbol{x}})$. We introduce here a symbolical Green's function, $\widehat{G}_q(\hat{\boldsymbol{x}}|\hat{\boldsymbol{x}}')$, which is the same functional of the operator, $\widehat{q}(\hat{\boldsymbol{x}})$, as $G_q(\hat{\boldsymbol{x}}|\hat{\boldsymbol{x}}')$ is of the deterministic, $q(\hat{\boldsymbol{x}})$; so that the first order Green's functions, $G(\hat{\boldsymbol{x}}|\hat{\boldsymbol{x}}') = \langle G_q(\hat{\boldsymbol{x}}|\hat{\boldsymbol{x}}')\rangle$ and $G^*(\hat{\boldsymbol{x}}|\hat{\boldsymbol{x}}') = \langle G_q^*(\hat{\boldsymbol{x}}|\hat{\boldsymbol{x}}')\rangle$, are given according to formula (1.3.9) and with a constant, $Z_o = 1$, by

$$G(\hat{\boldsymbol{x}}|\hat{\boldsymbol{x}}') = \widehat{G}_q(\hat{\boldsymbol{x}}|\hat{\boldsymbol{x}}')Z_o \ , \tag{2.1.9a}$$

$$G^*(\hat{\boldsymbol{x}}|\hat{\boldsymbol{x}}') = \widehat{G}_q^*(\hat{\boldsymbol{x}}|\hat{\boldsymbol{x}}')Z_o \ . \tag{2.1.9b}$$

In the same way, the second order Green's function is defined by

$$\begin{aligned} I(\hat{\boldsymbol{x}}_1;\hat{\boldsymbol{x}}_2|\hat{\boldsymbol{x}}_1';\hat{\boldsymbol{x}}_2') &\equiv \langle G_q^*(\hat{\boldsymbol{x}}_1|\hat{\boldsymbol{x}}_1')G_q(\hat{\boldsymbol{x}}_2|\hat{\boldsymbol{x}}_2')\rangle \\ &= \widehat{G}_q^*(\hat{\boldsymbol{x}}_1|\hat{\boldsymbol{x}}_1')\widehat{G}_q(\hat{\boldsymbol{x}}_2|\hat{\boldsymbol{x}}_2')Z_o \ , \end{aligned} \tag{2.1.10a}$$

or, in the matrix form,

$$I(1;2) \equiv \langle G_q^*(1)G_q(2)\rangle \tag{2.1.10b}$$

$$= \widehat{G}_q^*(1)\widehat{G}_q(2)Z_o = \widehat{G}_q^*(1)G(2) \ . \tag{2.1.10c}$$

Here, in the last derivation, use has been made of (2.1.9a), and the auxiliary function, $c(\hat{\boldsymbol{x}})$, is to vanish in the final results.

2.2 Green's Functions of First and Second Orders

The Green's function of first order, $G(\hat{\boldsymbol{x}}|\hat{\boldsymbol{x}}')$, is the solution of

$$[L - \widehat{q}(\hat{\boldsymbol{x}})]G(\hat{\boldsymbol{x}}|\hat{\boldsymbol{x}}') = \delta(\hat{\boldsymbol{x}} - \hat{\boldsymbol{x}}') \ , \tag{2.2.1}$$

as directly follows from (2.1.9a, 8a), or by averaging the deterministic equation (2.1.8a) with formula (1.3.10). To evaluate the left-hand side by using $\widehat{q}(\hat{\boldsymbol{x}})$ of (2.1.6a,b), we observe that

$$\begin{aligned} &\theta(\hat{\boldsymbol{x}}, \delta/\delta c)G(\hat{\boldsymbol{x}}|\hat{\boldsymbol{x}}') = \theta(\hat{\boldsymbol{x}}, \delta/\delta c)\widehat{G}_q(\hat{\boldsymbol{x}}|\hat{\boldsymbol{x}}')Z_o \\ &= \left\langle q_a(\hat{\boldsymbol{x}}) \exp\left[\int d\hat{\boldsymbol{x}}''\, q_a(\hat{\boldsymbol{x}}'')\frac{\delta}{\delta c(\hat{\boldsymbol{x}}'')}\right] \widehat{G}_q(\hat{\boldsymbol{x}}|\hat{\boldsymbol{x}}')\right\rangle_a Z_o \ , \end{aligned} \tag{2.2.2}$$

which becomes

$$\langle q_a(\hat{\boldsymbol{x}})\widehat{G}_{q+a}(\hat{\boldsymbol{x}}|\hat{\boldsymbol{x}}')\rangle_a Z_o \ , \tag{2.2.3}$$

in terms of the notation,

$$\widehat{G}_{q+a} = \widehat{G}_q \big|_{\widehat{q} \to \widehat{q} + q_a} , \tag{2.2.4}$$

in consequence of $\widehat{q}(\widehat{\boldsymbol{x}})$ having the term $c(\widehat{\boldsymbol{x}})$ linear in (2.1.6a), and $\exp\left[\int d\widehat{\boldsymbol{x}}''\, q_a(\widehat{\boldsymbol{x}}'')\delta/\delta c(\widehat{\boldsymbol{x}}'')\right]$ as a functional displacement operator of $c(\widehat{\boldsymbol{x}})$ by the amount, $q_a(\widehat{\boldsymbol{x}})$ [cf. (1.1.6)].

Here we introduce a notation, $G_{a+b}(\widehat{\boldsymbol{x}}|\widehat{\boldsymbol{x}}')$, or a coordinate matrix, G_{a+b}, for the Green's function in a medium, $a+b$, so that it is the solution of

$$(L - a - b)G_{a+b} = 1 . \tag{2.2.5}$$

By using a similar equation for G_b, the equation can be rewritten as

$$G_{a+b} = G_b[1 + aG_{a+b}] \tag{2.2.6}$$

$$= G_b + G_b T_a^b G_b , \tag{2.2.7}$$

in terms of the scattering matrix, T_a^b, of a, defined by

$$aG_{a+b} = T_a^b G_b, \qquad G_{a+b}\, a = G_b T_a^b , \tag{2.2.8}$$

and given by

$$T_a^b = a(1 + G_b T_a^b) = (1 - aG_b)^{-1} a \tag{2.2.9a}$$

$$= a + aG_b a + aG_b a G_b a + \dots . \tag{2.2.9b}$$

Here, T_a^b means a resultant effect caused by the scatterer, a, in the medium, b. In the case of a non-dissipative scatterer, $a^* = a$, it is subject to a (optical) condition from

$$a = \left[1 + (T_a^b)^* G_b^*\right]^{-1} (T_a^b)^* = T_a^b (1 + G_b T_a^b)^{-1} , \tag{2.2.10a}$$

hence,

$$(T_a^b)^* - T_a^b = (T_a^b)^* (G_b^* - G_b) T_a^b . \tag{2.2.10b}$$

Now the result (2.2.2, 3) can be written using relation (2.2.8) with $a \to q_a$ and $b \to \widehat{q}$, as

$$\theta(\widehat{\boldsymbol{x}}, \delta/\delta c) G(\widehat{\boldsymbol{x}}|\widehat{\boldsymbol{x}}') = \int d\widehat{\boldsymbol{x}}'' \langle \widehat{T}_a^q (\widehat{\boldsymbol{x}}|\widehat{\boldsymbol{x}}'')\rangle_a \widehat{G}_q(\widehat{\boldsymbol{x}}''|\widehat{\boldsymbol{x}}') Z_o$$

$$= \int d\widehat{\boldsymbol{x}}'' \langle \widehat{T}_a^q (\widehat{\boldsymbol{x}}|\widehat{\boldsymbol{x}}'')\rangle_a G(\widehat{\boldsymbol{x}}''|\widehat{\boldsymbol{x}}') , \tag{2.2.11}$$

with the notation,

$$\widehat{T}_a^q = T_a^b \big|_{a = q_a,\; b = \widehat{q}} . \tag{2.2.12}$$

Here, $\widehat{T}_a^q$ is a functional of $\widehat{q}(\widehat{\boldsymbol{x}})$ and is, therefore, still an operator; it means the scattering matrix of a particular particle, q_a, embedded in a random medium, $\widehat{q}$.

In (2.2.11), the correlation of $\widehat{T}_a^q$ and G is usually very small because $\widehat{T}_a^q$ is affected only by the medium, $\widehat{q}$, in the immediate neighborhood of the particle, q_a, as may be expected directly from definition (2.2.9b); whereas G is affected by the entire $\widehat{q}$ distributed along the wave path, therefore, seldom being affected by the same $\widehat{q}$ as involved in $\widehat{T}_a^q$, effectively. Mathematically, the correlation of $\langle\widehat{T}_a^q\rangle_a$ and G in (2.2.11) is negligible when G can be regarded as a constant independent of $c(\widehat{\boldsymbol{x}})$ when multiplied with the operator, $\langle\widehat{T}_a^q\rangle_a$.

Thus we can write,

$$\widehat{q}(\widehat{\boldsymbol{x}})G(\widehat{\boldsymbol{x}}|\widehat{\boldsymbol{x}}')\,\Big|_{c=0} = \int d\widehat{\boldsymbol{x}}''\, M(\widehat{\boldsymbol{x}}|\widehat{\boldsymbol{x}}'')G(\widehat{\boldsymbol{x}}''|\widehat{\boldsymbol{x}}') \;, \tag{2.2.13}$$

where, from (2.2.11) and (2.1.5),

$$M(\widehat{\boldsymbol{x}}|\widehat{\boldsymbol{x}}') = \left\langle\widehat{T}_a^q(\widehat{\boldsymbol{x}}|\widehat{\boldsymbol{x}}')\right\rangle_a \tag{2.2.14a}$$

$$\simeq n\int d\widehat{\boldsymbol{x}}_a\, \langle T_a^M(\widehat{\boldsymbol{x}}|\widehat{\boldsymbol{x}}')\rangle' \;. \tag{2.2.14b}$$

In the last derivation, the operator, $\widehat{q}$, has been replaced by a definite matrix, M, based on the observation that $\widehat{q}$ is equivalent, in view of (2.2.13), to M whenever operating on G, and hence, also, $(\widehat{q})^n G \sim M^n G$; and (2.2.14b) provides an equation to determine the still unknown matrix, M, self–consistently. The matrix, M, will be referred to as the "effective medium", hereafter.

To obtain the corresponding equation for the second order Green's function, we first introduce the coordinate matrices, $G^*(1)$ and $G(2)$, defined by the matrix elements, $G^*(\widehat{\boldsymbol{x}}_1|\widehat{\boldsymbol{x}}_1')$ and $G(\widehat{\boldsymbol{x}}_2|\widehat{\boldsymbol{x}}_2')$, respectively, and write their governing equations from (2.2.13) as

$$\left[L(1) - M^*(1)\right]G^*(1) = \delta(1) \;, \tag{2.2.15a}$$

$$\left[L(2) - M(2)\right]G(2) = \delta(2) \;, \tag{2.2.15b}$$

where $\delta(1)$ and $\delta(2)$ denote the unit matrices; then multiply (2.1.10b, 10c) to the left with $L(1) - \widehat{q}(1)$ to obtain

$$\left[L(1) - \widehat{q}(1)\right]I(1;2) = \delta(1)\widehat{G}_q(2)Z_o = \delta(1)G(2) \;. \tag{2.2.16}$$

Here, to evaluate the last term on the left-hand side , we can utilize the same method as when deriving (2.2.2, 3); hence, with the aid of relations (2.2.7, 8),

$$\begin{aligned}\theta(1,\delta/\delta c)I(1;2) &= \left\langle q_a(1)\widehat{G}^*_{q+a}(1)\widehat{G}_{q+a}(2)\right\rangle_a Z_o\\ &= \left\langle \widehat{T}_a^{q*}(1)\widehat{G}^*_q(1)\widehat{G}_q(2)\left[1+\widehat{T}_a^q(2)\widehat{G}_q(2)\right]\right\rangle_a Z_o\\ &= \left[\left\langle\widehat{T}_a^{q*}(1)\right\rangle_a + \widehat{G}_q(2)\left\langle\widehat{T}_a^{q*}(1)\widehat{T}_a^q(2)\right\rangle_a\right]I(1;2) \;,\end{aligned} \tag{2.2.17}$$

which can be approximated by replacing $\widehat{q}$ with M in the same fashion as in (2.2.14), by

$$\left[\langle T_a^{M*}(1)\rangle_a + G_M(2)\langle T_a^{M*}(1)T_a^M(2)\rangle_a\right] I(1;2) \ , \tag{2.2.18}$$

in terms of the notation, $G_M(2) = G(2)$.

Hence, upon introduction of a matrix, $K(1;2)$, defined by

$$K(1;2) = \langle T_a^{M*}(1)T_a^M(2)\rangle_a = n\int d\widehat{\boldsymbol{x}}_a \langle T_a^{M*}(1)T_a^M(2)\rangle' \ , \tag{2.2.19}$$

(2.2.16) is written, after setting $c(\widehat{\boldsymbol{x}}) = 0$, as

$$[L(1) - M^*(1) - G(2)K(1;2)]I(1;2) = \delta(1)G(2) \ , \tag{2.2.20}$$

with $M^*(1)$ given by the complex-conjugation of (2.2.14b); and further evolves, on multiplying to the left with $G^*(1)$ from (2.2.15a), as

$$I(1;2) = G^*(1)G(2)[1 + K(1;2)I(1;2)] \ . \tag{2.2.21}$$

Note that (2.2.21) is in the form of the Bethe-Salpeter (BS) equation.

In the same way, we obtain the equations for the other two Green's functions of second order, as

$$I(1,3) = G^*(1)G^*(3)[1 + K(1,3)I(1,3)] \ , \tag{2.2.22a}$$

$$I(2,4) = G(2)G(4)[1 + K(2,4)I(2,4)] \ , \tag{2.2.22b}$$

with

$$K(1,3) = n\int d\widehat{\boldsymbol{x}}_a \langle T_a^{M*}(1)T_a^{M*}(3)\rangle' \ , \tag{2.2.23a}$$

$$K(2,4) = n\int d\widehat{\boldsymbol{x}}_a \langle T_a^M(2)T_a^M(4)\rangle' \ . \tag{2.2.23b}$$

Here, and also hereafter, the subscripts of odd number are attached to the coordinates of quantities of the complex-conjugate wave function, and the subscripts of even number are attached to those of quantities of the original wave function.

The scattering matrix, defined by (2.2.9), is directly connected to the conventional scattering amplitude as follows: Let the medium, b, be free space so that

$$G_b(\widehat{\boldsymbol{x}}|\widehat{\boldsymbol{x}}') = G_o(\widehat{\boldsymbol{x}} - \widehat{\boldsymbol{x}}') = \left|4\pi(\widehat{\boldsymbol{x}} - \widehat{\boldsymbol{x}}')\right|^{-1} \exp\left(-k_o|\widehat{\boldsymbol{x}} - \widehat{\boldsymbol{x}}'|\right) \ .$$

Then, (2.2.7) is explicitly written by

$$G_{a+b}(\hat{x}|\hat{x}') = G_o(\hat{x}-\hat{x}') + \int d\hat{x}''\, d\hat{x}'''G_o(\hat{x}-\hat{x}'')T_a^b(\hat{x}''-\hat{x}_a|\hat{x}'''-\hat{x}_a)G_o(\hat{x}'''-\hat{x}') ,$$

with $\hat{x}_a$ as the center coordinates of the scatterer. Here, in the asymptotic region where both the points, $\hat{x}$ and $\hat{x}'$, are separated enough from the scatterer, the free space Green's functions in the integrand can be regarded as plane waves. Hence, the integral can be written in the form

$$|\hat{x}-\hat{x}_a|^{-1}\exp\Big(-ik_0|\hat{x}-\hat{x}_a|\Big)A(\hat{\Omega}|\hat{\Omega}')G_o(\hat{x}_a-\hat{x}') .$$

Here, with the Fourier transform of T_a^b defined by

$$\tilde{T}_a^b(\hat{\lambda}|\hat{\lambda}') = \int d\hat{x}\, d\hat{x}' \exp\Big[i(\hat{\lambda}\cdot\hat{x}-\hat{\lambda}'\cdot\hat{x}')\Big]T_a^b(\hat{x}|\hat{x}') ,$$

then

$$A(\hat{\Omega}|\hat{\Omega}') = (4\pi)^{-1}\tilde{T}_a^b(k_0\hat{\Omega}|k_0\hat{\Omega}') ,$$

where $\hat{\Omega}$ and $\hat{\Omega}'$ are the unit vectors in the directions of $\hat{x}-\hat{x}_a$ and $\hat{x}_a-\hat{x}'$, respectively, is the conventional scattering amplitude.

2.3 Bethe-Salpeter Equation in a General Random Medium

Equation (2.2.15) for the Green's functions of the first order holds true with the effective medium, M, defined, on changing the notation, $G_q \to g$, by

$$MG = \langle qg\rangle = \hat{g}G, \qquad g \equiv G_q . \tag{2.3.1}$$

The same is true also for the BS equation (2.2.21) with the (incoherent) factor, $K(1;2)$, defined as follows: The deterministic, g, as the solution of (2.1.8a), can be written in terms of G of (2.2.15), as

$$g = G(1+\Delta qg) , \tag{2.3.2}$$

$$\Delta q = q - M, \qquad \langle\Delta qg\rangle = 0 . \tag{2.3.3}$$

Hence, with the complex-conjugate expression

$$g^* = G^*(1+\Delta q^* g^*), \qquad \Delta q^* = q - M^* , \tag{2.3.4}$$

the substitution into the right-hand side of (2.1.10b) reproduces the BS equation by defining the factor, $K(1;2)$, according to

$$K(1;2)I(1;2) = \big\langle\Delta q^*(1)\Delta q(2)g^*(1)g(2)\big\rangle, \tag{2.3.5}$$

which means the incoherent source term. We now introduce a coherent propagator, $U(1;2)$, defined by

$$U(1;2) = G^*(1)G(2) \ , \tag{2.3.6}$$

and write the BS equation (2.2.21) as

$$I(1;2) = U(1;2)[1 + K(1;2)I(1;2)] \ , \tag{2.3.7}$$

with the solution

$$I = U + USU \ , \tag{2.3.8}$$

in terms of an incoherent scattering matrix, S of K, defined by

$$SU = KI, \qquad US = IK \ , \tag{2.3.9}$$

and given, therefore, as the solution of

$$S = K(1 + US) \ . \tag{2.3.10}$$

Thus the solution in matrix form is

$$S = (1 - KU)^{-1}K = K(1 - UK)^{-1} \tag{2.3.11a}$$
$$= K + KUK + KUKUK + \dots \ . \tag{2.3.11b}$$

Equation (2.3.10) can also be written as

$$S = K + KIK \ , \tag{2.3.12}$$

which is the same operator of K and I, as I by (2.3.8) is of U and S.

2.3.1 Optical Condition

The matrices, M and K, are not quite independent of each other, being subjected to a relation that ensures power conservation in the scattering at every point in space. It is straightforward to show, using (2.3.2, 4) and definition (2.3.5), that

$$\langle q(1)g^*(1)g(2)\rangle = [M^*(1) + G(2)K(1;2)]I(1;2) \ , \tag{2.3.13a}$$
$$\langle q(2)g^*(1)g(2)\rangle = [M(2) + G^*(1)K(1;2)]I(1;2) \ , \tag{2.3.13b}$$

wherein whenever the two column coordinates, $\hat{\boldsymbol{x}}_1$ and $\hat{\boldsymbol{x}}_2$, are set equal so that $\hat{\boldsymbol{x}}_1 = \hat{\boldsymbol{x}}_2 = \hat{\boldsymbol{x}}$, left–hand sides of the two matrix equations become the same and, therefore, so become the right-hand sides. This relation can be written in simple form by introducing a coordinate matrix, $\delta(\hat{\boldsymbol{x}}|1;2)$, defined by the matrix elements,

$$\delta(\hat{\boldsymbol{x}}|\hat{\boldsymbol{x}}_1;\hat{\boldsymbol{x}}_2) = \delta(\hat{\boldsymbol{x}}-\hat{\boldsymbol{x}}_1)\delta(\hat{\boldsymbol{x}}-\hat{\boldsymbol{x}}_2) \ , \tag{2.3.14a}$$

so that the product, $\delta(\hat{\boldsymbol{x}}|1;2)A^*(1)B(2) \equiv A^*B(\hat{\boldsymbol{x}}|1;2) \equiv A^*B(\hat{\boldsymbol{x}})$, represents

$$\begin{aligned}\int & d\hat{\boldsymbol{x}}_1\, d\hat{\boldsymbol{x}}_2\, \delta(\hat{\boldsymbol{x}}-\hat{\boldsymbol{x}}_1)\delta(\hat{\boldsymbol{x}}-\hat{\boldsymbol{x}}_2)A^*(\hat{\boldsymbol{x}}_1|\hat{\boldsymbol{x}}_1')B(\hat{\boldsymbol{x}}_2|\hat{\boldsymbol{x}}_2') \\ &= A^*(\hat{\boldsymbol{x}}|\hat{\boldsymbol{x}}_1')B(\hat{\boldsymbol{x}}|\hat{\boldsymbol{x}}_2') \equiv A^*B(\hat{\boldsymbol{x}}|\hat{\boldsymbol{x}}_1;\hat{\boldsymbol{x}}_2) \equiv A^*B(\hat{\boldsymbol{x}}) \ . \end{aligned} \tag{2.3.14b}$$

Hence, we can write the relation, from (2.3.13), as

$$\delta(\hat{\boldsymbol{x}}|1;2)\{M^*(1) - M(2) - [G^*(1) - G(2)]K(1;2)\} = 0 \ , \tag{2.3.15}$$

or, in terms of the new matrices,

$$\Gamma(1;2) = (2\mathrm{i})^{-1}[M^*(1) - M(2)] \ , \tag{2.3.16a}$$

$$\Delta G(1;2) = (2\mathrm{i})^{-1}[G^*(1) - G(2)] \ , \tag{2.3.16b}$$

simply as

$$\Delta G(\hat{\boldsymbol{x}})K = \Gamma(\hat{\boldsymbol{x}}) \ . \tag{2.3.17}$$

To see power conservation of the entire system, we first observe that the power equation for the coherent wave can be written, on referring to (2.1.7d,e) and using (2.3.6) and (2.2.15), as

$$\left(\frac{\partial}{\partial \hat{\boldsymbol{x}}} \cdot \hat{\boldsymbol{\alpha}} + \Gamma\right) U(\hat{\boldsymbol{x}}) = \Delta G(\hat{\boldsymbol{x}}) \ . \tag{2.3.18}$$

Here, $\hat{\boldsymbol{\alpha}} = (\alpha_1, \alpha_2, \alpha_3)$ is a vector operator, defined by the matrix elements,

$$\hat{\boldsymbol{\alpha}}(\hat{\boldsymbol{x}}|\hat{\boldsymbol{x}}_1;\hat{\boldsymbol{x}}_2) = (2\mathrm{i})^{-1}\delta(\hat{\boldsymbol{x}}_1 - \hat{\boldsymbol{x}})\left(\frac{\overleftarrow{\partial}}{\partial \hat{\boldsymbol{x}}} - \frac{\overrightarrow{\partial}}{\partial \hat{\boldsymbol{x}}}\right)\delta(\hat{\boldsymbol{x}} - \hat{\boldsymbol{x}}_2) \tag{2.3.19a}$$

$$= \delta(\hat{\boldsymbol{x}}|\hat{\boldsymbol{x}}_1;\hat{\boldsymbol{x}}_2)(2\mathrm{i})^{-1}\left(\frac{\partial}{\partial \hat{\boldsymbol{x}}_1} - \frac{\partial}{\partial \hat{\boldsymbol{x}}_2}\right) \ . \tag{2.3.19b}$$

Hence, substitution of (2.3.18) for $\Delta G(\hat{\boldsymbol{x}})$ in (2.3.17) leads to

$$\frac{\partial}{\partial \hat{\boldsymbol{x}}} \cdot \hat{\boldsymbol{\alpha}}(\hat{\boldsymbol{x}})UK = \Gamma(\hat{\boldsymbol{x}})(1 - UK) \ , \tag{2.3.20}$$

which can be written in terms of S, by multiplying both sides to the right with $(1 - UK)^{-1}$ and using (2.3.11a), as

$$\frac{\partial}{\partial \hat{\boldsymbol{x}}} \cdot \hat{\boldsymbol{\alpha}}(\hat{\boldsymbol{x}})US = \Gamma(\hat{\boldsymbol{x}}) \ . \tag{2.3.21}$$

Thus, with the power flux vector, $\widehat{\boldsymbol{w}}(\widehat{\boldsymbol{x}})$, of the entire system, written by using (2.3.19) as

$$\widehat{\boldsymbol{w}}(\widehat{\boldsymbol{x}}) = \widehat{\boldsymbol{\alpha}}(\widehat{\boldsymbol{x}})I = \widehat{\boldsymbol{\alpha}}(\widehat{\boldsymbol{x}})(U + USU) \ , \tag{2.3.22a}$$

we confirm that the power equation,

$$\frac{\partial}{\partial \widehat{\boldsymbol{x}}} \cdot \widehat{\boldsymbol{w}}(\widehat{\boldsymbol{x}}) = \Delta G(\widehat{\boldsymbol{x}}) \ , \tag{2.3.22b}$$

as required from the original wave equation, is reproduced in consequence of (2.3.18, 21).

Equation (2.3.15) is a local optical relation in the sense of ensuring power conservation at every point, $\widehat{\boldsymbol{x}}$, in space, while the $\widehat{\boldsymbol{x}}$-integrated relation is also meaningful in view of M and K having short-range matrix elements appreciable only over a distance of the medium-correlation distance, so that the integrated optical relation actually ensures power conservation when change of the wave intensity is negligibly small within the medium-correlation distance. Upon introduction of a new matrix, $\delta(1;2)$, defined by

$$\delta(1;2) = \int d\widehat{\boldsymbol{x}}\, \delta(\widehat{\boldsymbol{x}}|1;2) \ , \tag{2.3.23a}$$

the following relation,

$$\delta(1;2)[\Gamma(1;2) - \Delta GK(1;2)] = 0 \ , \tag{2.3.23b}$$

is obtained from (2.3.15, 16), and is shown, in fact, to be satisfied by M and K of (2.2.14, 19) for a system of independent particles, regardless of $M^* \neq M$, based on relation (2.2.10b).

2.3.2 Optical Expressions and Transport Equation

The incoherent part of I, say $\mathcal{I}$, is given by the second term of (2.3.8), and, from (2.3.10), is the solution of

$$\mathcal{I} \equiv USU = UK(U + \mathcal{I}) \ . \tag{2.3.24}$$

A specific expression of (2.3.24) is obtained in optical form by partially making the Fourier transformation. We first introduce relative coordinates, $\widehat{\boldsymbol{r}}$ and $\widehat{\boldsymbol{\rho}}$, defined by

$$\widehat{\boldsymbol{r}} = \widehat{\boldsymbol{x}}_2 - \widehat{\boldsymbol{x}}_1, \qquad \widehat{\boldsymbol{\rho}} = (\widehat{\boldsymbol{x}}_2 + \widehat{\boldsymbol{x}}_1)/2 \ , \tag{2.3.25a}$$

$$\widehat{\boldsymbol{x}}_1 = \widehat{\boldsymbol{\rho}} - \widehat{\boldsymbol{r}}/2, \qquad \widehat{\boldsymbol{x}}_2 = \widehat{\boldsymbol{\rho}} + \widehat{\boldsymbol{r}}/2 \ , \tag{2.3.25b}$$

$d\widehat{\boldsymbol{x}}_1\, d\widehat{\boldsymbol{x}}_2 = d\widehat{\boldsymbol{r}}\, d\widehat{\boldsymbol{\rho}}$, and also the corresponding Fourier variables, $\widehat{\boldsymbol{u}}$ and $\widehat{\boldsymbol{\lambda}}$, defined by

$$\widehat{\boldsymbol{u}} = (\widehat{\boldsymbol{\lambda}}_2 + \widehat{\boldsymbol{\lambda}}_1)/2, \qquad \widehat{\boldsymbol{\lambda}} = \widehat{\boldsymbol{\lambda}}_2 - \widehat{\boldsymbol{\lambda}}_1 \ , \tag{2.3.26a}$$

$$\widehat{\boldsymbol{\lambda}}_1 = \widehat{\boldsymbol{u}} - \widehat{\boldsymbol{\lambda}}/2, \qquad \widehat{\boldsymbol{\lambda}}_2 = \widehat{\boldsymbol{u}} + \widehat{\boldsymbol{\lambda}}/2 \ , \tag{2.3.26b}$$

so that

$$-\widehat{\boldsymbol{\lambda}}_1 \cdot \widehat{\boldsymbol{x}}_1 + \widehat{\boldsymbol{\lambda}}_2 \cdot \widehat{\boldsymbol{x}}_2 = \widehat{\boldsymbol{u}} \cdot \widehat{\boldsymbol{r}} + \widehat{\boldsymbol{\lambda}} \cdot \widehat{\boldsymbol{\rho}} \ . \tag{2.3.27}$$

Then the matrix elements of K can be written in the form

$$K(\widehat{\boldsymbol{x}}_1;\widehat{\boldsymbol{x}}_2|\widehat{\boldsymbol{x}}_1';\widehat{\boldsymbol{x}}_2') = K(\widehat{\boldsymbol{r}},\widehat{\boldsymbol{\rho}}|\widehat{\boldsymbol{r}}',\widehat{\boldsymbol{\rho}}') = K(\widehat{\boldsymbol{r}}|\widehat{\boldsymbol{\rho}}-\widehat{\boldsymbol{\rho}}'|\widehat{\boldsymbol{r}}') \ , \tag{2.3.28a}$$

in view of the translational invariance, though approximately in a bounded space; its Fourier transform is defined by

$$\begin{aligned} \widetilde{K}(\widehat{\boldsymbol{\lambda}}_1;\widehat{\boldsymbol{\lambda}}_2|\widehat{\boldsymbol{\lambda}}_1';\widehat{\boldsymbol{\lambda}}_2') = & \int d\widehat{\boldsymbol{x}}_1\, d\widehat{\boldsymbol{x}}_2 \int d\widehat{\boldsymbol{x}}_1'\, d\widehat{\boldsymbol{x}}_2' \exp\left[-\mathrm{i}(\widehat{\boldsymbol{\lambda}}_1 \cdot \widehat{\boldsymbol{x}}_1 - \widehat{\boldsymbol{\lambda}}_2 \cdot \widehat{\boldsymbol{x}}_2)\right] \\ & \times K(\widehat{\boldsymbol{x}}_1;\widehat{\boldsymbol{x}}_2|\widehat{\boldsymbol{x}}_1';\widehat{\boldsymbol{x}}_2') \exp\left[+\mathrm{i}(\widehat{\boldsymbol{\lambda}}_1' \cdot \widehat{\boldsymbol{x}}_1' - \widehat{\boldsymbol{\lambda}}_2' \cdot \widehat{\boldsymbol{x}}_2')\right] \ , \end{aligned} \tag{2.3.28b}$$

and therefore has the form

$$\widetilde{K}(\widehat{\boldsymbol{\lambda}}_1;\widehat{\boldsymbol{\lambda}}_2|\widehat{\boldsymbol{\lambda}}_1';\widehat{\boldsymbol{\lambda}}_2') = (2\pi)^3 \delta(\widehat{\boldsymbol{\lambda}} - \widehat{\boldsymbol{\lambda}}')\widetilde{K}(\widehat{\boldsymbol{u}}|\widehat{\boldsymbol{\lambda}}|\widehat{\boldsymbol{u}}') \ . \tag{2.3.29}$$

On the other hand, the corresponding Fourier transforms of wave quantities, e.g., $\widetilde{\mathcal{I}}(\widehat{\boldsymbol{u}},\widehat{\boldsymbol{\lambda}}|\widehat{\boldsymbol{u}}',\widehat{\boldsymbol{\lambda}}')$ of $\mathcal{I}(\widehat{\boldsymbol{r}},\widehat{\boldsymbol{\rho}}|\widehat{\boldsymbol{r}}',\widehat{\boldsymbol{\rho}}')$, cannot be written in the same form. We will hereafter use a composite expression, $\mathcal{I}(\widehat{\boldsymbol{u}},\widehat{\boldsymbol{\rho}}|\widehat{\boldsymbol{u}}',\widehat{\boldsymbol{\rho}}')$, by making the Fourier transformation only with respect to $\widehat{\boldsymbol{r}}$ and $\widehat{\boldsymbol{r}}'$.

As to the Fourier transform $\widetilde{U}$ of U from (2.3.6) and (2.2.15), we obtain

$$\begin{aligned} \widetilde{U}(\widehat{\boldsymbol{u}},\widehat{\boldsymbol{\lambda}}) &= \widetilde{G}^*(\widehat{\boldsymbol{u}} - \widehat{\boldsymbol{\lambda}}/2)\widetilde{G}(\widehat{\boldsymbol{u}} + \widehat{\boldsymbol{\lambda}}/2) \\ &\simeq \pi\delta(\widehat{\boldsymbol{u}}^2 - k^2)(k\gamma - \mathrm{i}\widehat{\boldsymbol{u}} \cdot \widehat{\boldsymbol{\lambda}})^{-1} \ , \end{aligned} \tag{2.3.30}$$

with

$$\widetilde{G}(\widehat{\boldsymbol{\lambda}}) = \{\widehat{\boldsymbol{\lambda}}^2 - [k^{(M)}]^2\}^{-1} \ , \tag{2.3.31a}$$

$$k^{(M)} = \left[k_0^2 + \widetilde{M}(\widehat{\boldsymbol{\lambda}})\right]^{1/2}, \qquad \mathrm{Im}\{k^{(M)}\} < 0 \ , \tag{2.3.31b}$$

$$\gamma = (2\mathrm{i}k)^{-1}(\widetilde{M}^* - \widetilde{M})(\widehat{\boldsymbol{\lambda}}) \ , \tag{2.3.31c}$$

and the approximation, $k = \mathrm{Re}\{k^{(M)}\} \simeq k_0$, excluding the case when the medium is intrinsically dispersive, Sect. 4.1.2. The expression (2.3.30) is a direct consequence of the identity,

$$\begin{aligned} &\{\widetilde{G}^{-1}(\widehat{\boldsymbol{\lambda}}_2) - [\widetilde{G}^*(\widehat{\boldsymbol{\lambda}}_1)]^{-1}\}\widetilde{U}(\widehat{\boldsymbol{u}},\widehat{\boldsymbol{\lambda}}) \\ &\quad = \widetilde{G}^*(\widehat{\boldsymbol{\lambda}}_1) - \widetilde{G}(\widehat{\boldsymbol{\lambda}}_2) \simeq 2\pi\mathrm{i}\delta(\widehat{\boldsymbol{u}}^2 - k^2) \ , \end{aligned} \tag{2.3.32}$$

and is valid under the condition, $|\hat{\boldsymbol{u}}| \sim k \gg |\hat{\boldsymbol{\lambda}}|$.

Hence, on writing the variable $\hat{\boldsymbol{u}}$ as $\hat{\boldsymbol{u}} = u\hat{\boldsymbol{\Omega}}$ with $u = |\hat{\boldsymbol{u}}|$, the unit vector, $\hat{\boldsymbol{\Omega}}$, $\hat{\boldsymbol{\Omega}}^2 = 1$, and the element of volume, $d\hat{\boldsymbol{u}} = u\,du\,d\hat{\boldsymbol{\Omega}}$, we obtain an important relation that, for any slowly changing function, $f(\hat{\boldsymbol{u}})$,

$$(2\pi)^{-3} \int d\hat{\boldsymbol{u}}\, \tilde{U}(\hat{\boldsymbol{u}}, \hat{\boldsymbol{\lambda}}) f(\hat{\boldsymbol{u}}) \simeq \int_{4\pi} d\hat{\boldsymbol{\Omega}}\, \tilde{U}(\hat{\boldsymbol{\Omega}}, \hat{\boldsymbol{\lambda}}) f(\hat{\boldsymbol{\Omega}}) \ , \qquad (2.3.33)$$

where

$$\tilde{U}(\hat{\boldsymbol{\Omega}}, \hat{\boldsymbol{\lambda}}) = (\gamma - \mathrm{i}\hat{\boldsymbol{\Omega}} \cdot \hat{\boldsymbol{\lambda}})^{-1} \ , \qquad (2.3.34)$$

$$f(\hat{\boldsymbol{\Omega}}) = (4\pi)^{-2} f(\hat{\boldsymbol{u}} = k\hat{\boldsymbol{\Omega}}), \qquad \gamma = \gamma(\hat{\boldsymbol{u}} = k\hat{\boldsymbol{\Omega}}) \ . \qquad (2.3.35)$$

Here, the $\hat{\boldsymbol{\lambda}}$ Fourier inversion of $\tilde{U}(\hat{\boldsymbol{\Omega}}, \hat{\boldsymbol{\lambda}})$, say $U(\hat{\boldsymbol{\Omega}}, \hat{\boldsymbol{\rho}})$, given by

$$U(\hat{\boldsymbol{\Omega}}, \hat{\boldsymbol{\rho}}) = (2\pi)^{-3} \int d\hat{\boldsymbol{\lambda}}\, \exp(-\mathrm{i}\hat{\boldsymbol{\lambda}} \cdot \hat{\boldsymbol{\rho}}) \tilde{U}(\hat{\boldsymbol{\Omega}}, \hat{\boldsymbol{\lambda}}) \ , \qquad (2.3.36)$$

becomes, using the integral representation,

$$\tilde{U}(\hat{\boldsymbol{\Omega}}, \hat{\boldsymbol{\lambda}}) = \int_0^\infty dt\, \exp\left[-t(\gamma - \mathrm{i}\hat{\boldsymbol{\Omega}} \cdot \hat{\boldsymbol{\lambda}})\right] \ , \qquad (2.3.37)$$

as

$$\begin{aligned} U(\hat{\boldsymbol{\Omega}}, \hat{\boldsymbol{\rho}}) &= \int_0^\infty dt\, \delta(\hat{\boldsymbol{\rho}} - t\hat{\boldsymbol{\Omega}}) \exp(-t\gamma) \\ &= |\hat{\boldsymbol{\rho}}|^{-2} \exp(-\gamma|\hat{\boldsymbol{\rho}}|) \delta^2(\hat{\boldsymbol{\Omega}} - \hat{\boldsymbol{\rho}}/|\hat{\boldsymbol{\rho}}|) \ , \end{aligned} \qquad (2.3.38)$$

where $\delta^2(\hat{\boldsymbol{\Omega}})$ is the (two-dimensional) δ function of solid angle $\hat{\boldsymbol{\Omega}}$. Equation (2.3.38) is the solution of

$$\left[\gamma + \hat{\boldsymbol{\Omega}} \cdot \frac{\partial}{\partial \hat{\boldsymbol{\rho}}}\right] U(\hat{\boldsymbol{\Omega}}, \hat{\boldsymbol{\rho}}) = \delta(\hat{\boldsymbol{\rho}}) \ . \qquad (2.3.39)$$

The Fourier transform of (2.3.10) for S becomes, using (2.3.29),

$$\begin{aligned} \tilde{S}(\hat{\boldsymbol{u}}, \hat{\boldsymbol{\lambda}}|\hat{\boldsymbol{u}}', \hat{\boldsymbol{\lambda}}') &= (2\pi)^3 \delta(\hat{\boldsymbol{\lambda}} - \hat{\boldsymbol{\lambda}}') \tilde{K}(\hat{\boldsymbol{u}}|\hat{\boldsymbol{\lambda}}|\hat{\boldsymbol{u}}') \\ &+ (2\pi)^{-3} \int d\hat{\boldsymbol{u}}''\, \tilde{K}(\hat{\boldsymbol{u}}|\hat{\boldsymbol{\lambda}}|\hat{\boldsymbol{u}}'') \tilde{U}(\hat{\boldsymbol{u}}'', \hat{\boldsymbol{\lambda}}) \tilde{S}(\hat{\boldsymbol{u}}'', \hat{\boldsymbol{\lambda}}|\hat{\boldsymbol{u}}', \hat{\boldsymbol{\lambda}}') \ . \end{aligned} \qquad (2.3.40)$$

Here, for the last integral, the optical transformation according to (2.3.33) is possible. Hence, upon introduction of optical expressions for K and S, defined according to (2.3.35), by

$$K(\hat{\boldsymbol{\Omega}}|\hat{\boldsymbol{\Omega}}') = (4\pi)^{-2}\widetilde{K}(\hat{\boldsymbol{u}}=k\hat{\boldsymbol{\Omega}}|\hat{\boldsymbol{\lambda}}|\hat{\boldsymbol{u}}'=k\hat{\boldsymbol{\Omega}}') \ , \tag{2.3.41}$$

$$\widetilde{S}(\hat{\boldsymbol{\Omega}},\hat{\boldsymbol{\lambda}}|\hat{\boldsymbol{\Omega}}',\hat{\boldsymbol{\lambda}}') = (4\pi)^{-2}\widetilde{S}(\hat{\boldsymbol{u}}=k\hat{\boldsymbol{\Omega}},\hat{\boldsymbol{\lambda}}|\hat{\boldsymbol{u}}'=k\hat{\boldsymbol{\Omega}}',\hat{\boldsymbol{\lambda}}') \ , \tag{2.3.42}$$

(2.3.40) is written as

$$\begin{aligned}\widetilde{S}(\hat{\boldsymbol{\Omega}},\hat{\boldsymbol{\lambda}}|\hat{\boldsymbol{\Omega}}',\hat{\boldsymbol{\lambda}}') &= (2\pi)^3\delta(\hat{\boldsymbol{\lambda}}-\hat{\boldsymbol{\lambda}}')\widetilde{K}(\hat{\boldsymbol{\Omega}}|\hat{\boldsymbol{\Omega}}')\\ &+\int_{4\pi} d\hat{\boldsymbol{\Omega}}''\widetilde{K}(\hat{\boldsymbol{\Omega}}|\hat{\boldsymbol{\Omega}}'')\widetilde{U}(\hat{\boldsymbol{\Omega}}'',\hat{\boldsymbol{\lambda}})\widetilde{S}(\hat{\boldsymbol{\Omega}}'',\hat{\boldsymbol{\lambda}}|\hat{\boldsymbol{\Omega}}',\hat{\boldsymbol{\lambda}}') \ ,\end{aligned} \tag{2.3.43}$$

which, in terms of

$$\widetilde{\mathcal{I}}(\hat{\boldsymbol{\Omega}},\hat{\boldsymbol{\lambda}}|\hat{\boldsymbol{\Omega}}',\hat{\boldsymbol{\lambda}}') = \widetilde{U}(\hat{\boldsymbol{\Omega}},\hat{\boldsymbol{\lambda}})\widetilde{S}(\hat{\boldsymbol{\Omega}},\hat{\boldsymbol{\lambda}}|\hat{\boldsymbol{\Omega}}',\hat{\boldsymbol{\lambda}}')\widetilde{U}(\hat{\boldsymbol{\Omega}}',\hat{\boldsymbol{\lambda}}') \ , \tag{2.3.44}$$

can be rewritten, using (2.3.34), as

$$\begin{aligned}(\gamma - \mathrm{i}\hat{\boldsymbol{\Omega}}\cdot\hat{\boldsymbol{\lambda}})\widetilde{\mathcal{I}}(\hat{\boldsymbol{\Omega}},\hat{\boldsymbol{\lambda}}|\hat{\boldsymbol{\Omega}}',\hat{\boldsymbol{\lambda}}') &= (2\pi)^3\delta(\hat{\boldsymbol{\lambda}}-\hat{\boldsymbol{\lambda}}')K(\hat{\boldsymbol{\Omega}}|\hat{\boldsymbol{\Omega}}')\widetilde{U}(\hat{\boldsymbol{\Omega}}',\hat{\boldsymbol{\lambda}}')\\ &+\int d\hat{\boldsymbol{\Omega}}''K(\hat{\boldsymbol{\Omega}}|\hat{\boldsymbol{\Omega}}'')\widetilde{\mathcal{I}}(\hat{\boldsymbol{\Omega}}'',\hat{\boldsymbol{\lambda}}|\hat{\boldsymbol{\Omega}}',\hat{\boldsymbol{\lambda}}') \ ;\end{aligned} \tag{2.3.45}$$

and by the $\hat{\boldsymbol{\lambda}}$- and $\hat{\boldsymbol{\lambda}}'$-Fourier inversions, leads to

$$\begin{aligned}\left(\gamma + \hat{\boldsymbol{\Omega}}\cdot\frac{\partial}{\partial\hat{\boldsymbol{\rho}}}\right)\mathcal{I}(\hat{\boldsymbol{\Omega}},\hat{\boldsymbol{\rho}}|\hat{\boldsymbol{\Omega}}',\hat{\boldsymbol{\rho}}') &= K(\hat{\boldsymbol{\Omega}}|\hat{\boldsymbol{\Omega}}')U(\hat{\boldsymbol{\Omega}}',\hat{\boldsymbol{\rho}}-\hat{\boldsymbol{\rho}}')\\ &+\int d\hat{\boldsymbol{\Omega}}''K(\hat{\boldsymbol{\Omega}}|\hat{\boldsymbol{\Omega}}'')\mathcal{I}(\hat{\boldsymbol{\Omega}}'',\hat{\boldsymbol{\rho}}|\hat{\boldsymbol{\Omega}}',\hat{\boldsymbol{\rho}}') \ ,\end{aligned} \tag{2.3.46}$$

i.e., the transport equation for the incoherent wave $\mathcal{I}(\hat{\boldsymbol{\Omega}},\hat{\boldsymbol{\rho}}|\hat{\boldsymbol{\Omega}}',\hat{\boldsymbol{\rho}}')$. Here, the dependence of $K(\hat{\boldsymbol{\Omega}}|\hat{\boldsymbol{\Omega}}')$ on $\hat{\boldsymbol{\lambda}}$ is negligible because the original K is a short-range function of the order of the medium-correlation distance, within which the change of $\mathcal{I}$ is negligibly small.

The intensity is given, in view of (2.3.44), by

$$\mathcal{I}(\hat{\boldsymbol{r}}=0,\hat{\boldsymbol{\rho}}|\hat{\boldsymbol{r}}'=0,\hat{\boldsymbol{\rho}}') = (4\pi)^{-2}\int d\hat{\boldsymbol{\Omega}}\int d\hat{\boldsymbol{\Omega}}'\mathcal{I}(\hat{\boldsymbol{\Omega}},\hat{\boldsymbol{\rho}}|\hat{\boldsymbol{\Omega}}',\hat{\boldsymbol{\rho}}') \ . \tag{2.3.47}$$

The power flux vector, $\hat{\boldsymbol{w}}^{(I)}(\hat{\boldsymbol{\rho}}|\hat{\boldsymbol{\rho}}')$, is similarly given from (2.3.22), by

$$\begin{aligned}\hat{\boldsymbol{w}}^{(I)}(\hat{\boldsymbol{\rho}}|\hat{\boldsymbol{\rho}}') &= \mathrm{i}\frac{\partial}{\partial\hat{\boldsymbol{r}}}\mathcal{I}(\hat{\boldsymbol{r}}(=0),\hat{\boldsymbol{\rho}}\,|\,\hat{\boldsymbol{r}}'=0,\hat{\boldsymbol{\rho}}')\\ &= (4\pi)^{-2}\int d\hat{\boldsymbol{\Omega}}\int d\hat{\boldsymbol{\Omega}}'k\hat{\boldsymbol{\Omega}}\mathcal{I}(\hat{\boldsymbol{\Omega}},\hat{\boldsymbol{\rho}}|\hat{\boldsymbol{\Omega}}',\hat{\boldsymbol{\rho}}') \ ,\end{aligned} \tag{2.3.48}$$

along with the equation of continuity,

$$\frac{\partial}{\partial\widehat{\rho}} \cdot \widehat{\boldsymbol{w}}^{(I)}(\widehat{\rho}|\widehat{\rho}') = (4\pi)^{-2} \int d\widehat{\boldsymbol{\Omega}}\, \gamma(\widehat{\boldsymbol{\Omega}}) U(\widehat{\boldsymbol{\Omega}}, \widehat{\rho} - \widehat{\rho}') \ , \tag{2.3.49}$$

which is ensured by the transport equation in consequence of the optical relation,

$$\int d\widehat{\boldsymbol{\Omega}}'\, K(\widehat{\boldsymbol{\Omega}}'|\widehat{\boldsymbol{\Omega}}) = \gamma(\widehat{\boldsymbol{\Omega}}) \ , \tag{2.3.50}$$

resulting from (2.3.23a,b). The left-hand side of (2.3.50) is derived, according to convention (2.3.14), from

$$\begin{aligned} &\int d\widehat{\boldsymbol{x}}_1'\, d\widehat{\boldsymbol{x}}_2' \int d\widehat{\boldsymbol{x}}_1\, d\widehat{\boldsymbol{x}}_2 \int d\widehat{\boldsymbol{x}}\, \Delta G(\widehat{\boldsymbol{x}}|\widehat{\boldsymbol{x}}_1;\widehat{\boldsymbol{x}}_2) K(\widehat{\boldsymbol{x}}_1;\widehat{\boldsymbol{x}}_2|\widehat{\boldsymbol{x}}_1';\widehat{\boldsymbol{x}}_2') \\ &\qquad\qquad \times \exp\left[\mathrm{i}(\widehat{\boldsymbol{\lambda}}_1' \cdot \widehat{\boldsymbol{x}}_1' - \widehat{\boldsymbol{\lambda}}_2' \cdot \widehat{\boldsymbol{x}}_2')\right] |_{\widehat{\boldsymbol{\lambda}}=0} \\ &= (2\pi)^{-3} \int d\widehat{\boldsymbol{u}}\, (2\mathrm{i})^{-1} \left[\widetilde{G}^*(\widehat{\boldsymbol{u}}) - \widetilde{G}(\widehat{\boldsymbol{u}})\right] \widetilde{K}(\widehat{\boldsymbol{u}}|\widehat{\boldsymbol{\lambda}}=0|\widehat{\boldsymbol{u}}') \\ &\simeq k \int d\widehat{\boldsymbol{\Omega}}\, K(\widehat{\boldsymbol{\Omega}}|\widehat{\boldsymbol{\Omega}}') \ , \end{aligned} \tag{2.3.51a}$$

with the aid of (2.3.32), and the right-hand side is from

$$\begin{aligned} &\int d\widehat{\boldsymbol{x}}_1'\, d\widehat{\boldsymbol{x}}_2' \int d\widehat{\boldsymbol{x}}\, \Gamma(\widehat{\boldsymbol{x}}|\widehat{\boldsymbol{x}}_1;\widehat{\boldsymbol{x}}_2) \\ &\times \exp\left[\mathrm{i}(\widehat{\boldsymbol{\lambda}}_1' \cdot \widehat{\boldsymbol{x}}_1' - \widehat{\boldsymbol{\lambda}}_2' \cdot \widehat{\boldsymbol{x}}_2')\right] |_{\widehat{\boldsymbol{\lambda}}=0} = k\gamma(\widehat{\boldsymbol{\Omega}}') \ , \end{aligned} \tag{2.3.51b}$$

with definition (2.3.31c) for γ.

For a medium of independent particles, specific expressions of $K(\widehat{\boldsymbol{\Omega}}|\widehat{\boldsymbol{\Omega}}')$ and $\gamma(\widehat{\boldsymbol{\Omega}})$ are obtained by substituting (2.2.19, 14b) in (2.3.41, 31c) as

$$K(\widehat{\boldsymbol{\Omega}}|\widehat{\boldsymbol{\Omega}}') = n(4\pi)^{-2} \left\langle |\widetilde{T}_a^M(\widehat{\boldsymbol{u}} = k\widehat{\boldsymbol{\Omega}}|\widehat{\boldsymbol{u}}' = k\widehat{\boldsymbol{\Omega}}')|^2 \right\rangle' \ , \tag{2.3.52a}$$

$$\gamma(\widehat{\boldsymbol{\Omega}}) = -k^{-1} n \,\mathrm{Im}\left\{\langle \widetilde{T}_a^M(k\widehat{\boldsymbol{\Omega}}|k\widehat{\boldsymbol{\Omega}})\rangle'\right\} \ . \tag{2.3.52b}$$

2.4 Random Layer with Free Boundaries

In the case of a random layer, distributed parallel to the horizontal plane, it is often assumed that the difference of the medium diffractive indices is, on average, negligible so that the waves are actually not scattered by the boundaries themselves. Even with this boundary condition, the solution of the BS equation is still general enough, in the sense that it provides a basic scattering matrix of the medium which permits the construction of solutions for the more general cases, including boundary scattering and/or two or more random layers with different properties, embedding a fixed scatterer, etc.. We hereafter employ the horizontal space coordinates, $\boldsymbol{\rho} = (x_1, x_2)$, and the vertical, $z = x_3$, so that $\hat{\boldsymbol{x}} = (\boldsymbol{\rho}, z)$, and denote the scalar product of two space vectors, $\hat{\boldsymbol{a}} = (\boldsymbol{a}, a_z)$ and $\hat{\boldsymbol{b}} = (\boldsymbol{b}, b_z)$, by $\hat{\boldsymbol{a}} \cdot \hat{\boldsymbol{b}} = \boldsymbol{a} \cdot \boldsymbol{b} + a_z b_z$ with $\boldsymbol{a} \cdot \boldsymbol{b} = a_1 b_1 + a_2 b_2$.

2.4.1 Boundary Condition

When no scattering from the boundaries occurs, $\mathcal{I}$ given by the Fourier inversion of (2.3.44), holds everywhere, both outside and inside of the layer. Here, the λ_z Fourier inversion of $\tilde{U}(\hat{\boldsymbol{\Omega}}, \hat{\boldsymbol{\lambda}})$, say $U(\hat{\boldsymbol{\Omega}}, z)$, is given from (2.3.34) by

$$\begin{aligned} U(\hat{\boldsymbol{\Omega}}, z) &= (2\pi)^{-1} \int d\lambda_z \exp(-\mathrm{i}\lambda_z z) \tilde{U}(\hat{\boldsymbol{\Omega}}, \hat{\boldsymbol{\lambda}}) \\ &= \begin{cases} |\Omega_z|^{-1} \exp\left[-\Omega_z^{-1}(\gamma - \mathrm{i}\boldsymbol{\Omega} \cdot \boldsymbol{\lambda}) z\right], & \Omega_z z > 0, \\ 0, \quad \Omega_z z < 0, \end{cases} \end{aligned} \tag{2.4.1}$$

so that, when the medium is distributed over the range $z < 0$, we obtain for the boundary condition at $z = 0$,

$$\mathcal{I}(\hat{\boldsymbol{\Omega}}, z{=}0 | \hat{\boldsymbol{\Omega}}', z') = 0, \qquad \Omega_z < 0 \ . \tag{2.4.2}$$

2.4.2 Addition of Scattering Matrices

Suppose that a random layer is composed of two independent layers $K^{(A)}$ and $K^{(B)}$, as illustrated in Fig. 2.1, so that the second order Green's function in this case, $I^{(A+B)}$, is the solution of the BS equation [2.2]:

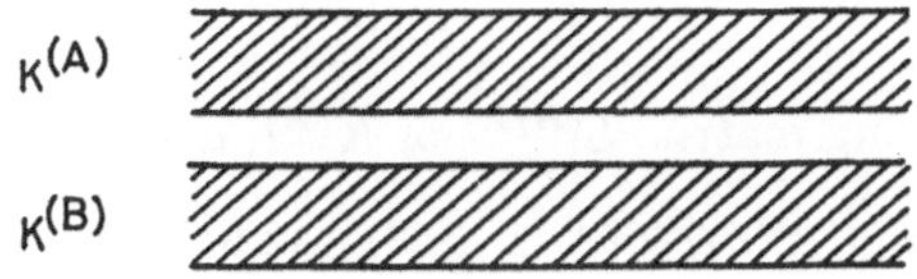

Fig. 2.1. Layer composed of two independent layers of $K^{(A)}$ and $K^{(B)}$ for (2.4.3)

$$I^{(A+B)} = U\left[1 + (K^{(A)} + K^{(B)}) I^{(A+B)}\right] \ . \tag{2.4.3}$$

Here the second order Green's functions for the respective media $K^{(A)}$ and $K^{(B)}$, given by $I^{(A)}$ and $I^{(B)}$, are also the solutions of

$$I^{(A)} = U[1 + K^{(A)} I^{(A)}] \ , \tag{2.4.4a}$$

$$I^{(B)} = U[1 + K^{(B)} I^{(B)}] \ , \tag{2.4.4b}$$

where U is the coherent propagator of (2.3.6) and is the same in all the equations. The solutions to these equations are

$$I^{(A)} = U + U S^{(A)} U \ , \tag{2.4.5a}$$

$$I^{(B)} = U + U S^{(B)} U \ , \tag{2.4.5b}$$

where $S^{(A)}$ and $S^{(B)}$ are the scattering matrices of $K^{(A)}$ and $K^{(B)}$, respectively, and are defined by

$$S^{(A)} U = K^{(A)} I^{(A)}, \qquad S^{(B)} U = K^{(B)} I^{(B)} \ , \tag{2.4.6}$$

and given by

$$S^{(A)} = K^{(A)}[1 + U S^{(A)}] = [1 - K^{(A)} U]^{-1} K^{(A)} \ , \tag{2.4.7a}$$

$$S^{(B)} = [1 - K^{(B)} U]^{-1} K^{(B)} \ , \tag{2.4.7b}$$

in the same fashion as the scattering matrix, T_a^b, for a coherent wave is given by (2.2.9). Now, writing $I^{(A+B)}$ in the form

$$I^{(A+B)} = U + U S^{(A+B)} U \ , \tag{2.4.8}$$

the resultant scattering matrix, $S^{(A+B)}$, can be obtained in terms of $S^{(A)}$ and $S^{(B)}$, as follows:

Equation (2.4.3) can be rewritten in terms of the solution $I^{(A)}$ of (2.4.4a), as

$$I^{(A+B)} = I^{(A)}[1 + K^{(B)} I^{(A+B)}] \ , \tag{2.4.9}$$

and hence the solution as

$$I^{(A+B)} = I^{(A)} + I^{(A)} S^{(B/A)} I^{(A)} \ , \tag{2.4.10}$$

in terms of a scattering matrix, $S^{(B/A)}$ of $K^{(B)}$, defined by

$$S^{(B/A)} I^{(A)} = K^{(B)} I^{(A+B)} \ , \tag{2.4.11}$$

and given by

$$S^{(B/A)} = K^{(B)}[1 + I^{(A)} S^{(B/A)}] \ , \tag{2.4.12}$$

which can be written in terms of $S^{(B)}$, upon substitution of (2.4.4a) for $I^{(A)}$, and followed by the use of (2.4.7b), as

$$S^{(B/A)} = S^{(B)}[1 + U S^{(A)} U S^{(B/A)}] \ . \tag{2.4.13}$$

Hence the solution is

$$S^{(B/A)} = [1 - S^{(B)}US^{(A)}U]^{-1}S^{(B)} \ , \qquad (2.4.14)$$

where $S^{(B/A)}$ is expressed in terms of $S^{(A)}$ and $S^{(B)}$. Now by substitution of (2.4.5a) in (2.4.10), we finally find that

$$S^{(A+B)} = S^{(A)} + (1 + S^{(A)}U)S^{(B/A)}(US^{(A)} + 1) \ . \qquad (2.4.15)$$

Here, $S^{(A+B)}$ can be expanded in a series with terms of the form, $(K^{()}U)^n K^{()}$, $n = 1, 2, 3, \ldots$, where the $K^{(\ )}$'s may not be the same; so that $I^{(A+B)}$ as given by (2.4.10) can be written as a series with terms of the form, $U(K^{(\ )}U)^n$, showing that the optical transformation of $I^{(A+B)}$, according to (2.3.33–35), is possible for each term of the series. Thus the addition formula of incoherent scattering matrices (2.4.15) holds true, not only as a coordinate matrix equation, but also as an optical ($\widehat{\Omega}$–$\widehat{\rho}$ matrix) equation, and may be utilized to construct the scattering matrix, S, by the successive addition of layers of small width, or doubling the width at every time of the addition. Similar equations were obtained also in [2.3–5], based on the transport equation.

Two relations exist between $S^{(B/A)}$ and $S^{(A/B)}$; one directly from (2.4.14),

$$S^{(B/A)} = S^{(B)} + S^{(B)}US^{(A/B)}US^{(B)} \ , \qquad (2.4.16)$$

and the other from (2.4.11) and the transposed version of (2.4.6),

$$S^{(A/B)}US^{(B)} = S^{(A)}US^{(B/A)} = K^{(A)}I^{(A+B)}K^{(B)} \ , \qquad (2.4.17)$$

which leads to another expression of $I^{(A+B)}$ as

$$I^{(A+B)} = (1 + US^{(A)})(1 - US^{(B)}US^{(A)})^{-1}I^{(B)} \ . \qquad (2.4.18)$$

The proof is straightforward by using (2.4.14) for $S^{(B/A)}$ in (2.4.17), and dropping the common factors, $K^{(A)}$ and $K^{(B)}$, from both sides of (2.4.17) by the use of relations (2.4.6, 7a). The expression (2.4.18) may be convenient to use in finding an averall structure of the transmitted wave through the layer of $K^{(A)} + K^{(B)}$.

Here, it may be convenient to introduce two "attenuation" factors, $F^{(A)}$ and $\overline{F}^{(A)}$ of $I^{(A)}$, defined by

$$I^{(A)} = F^{(A)}U = U\overline{F}^{(A)} \ , \qquad (2.4.19)$$

i.e., the attenuation relative to U where, from (2.4.5),

$$F^{(A)} = 1 + US^{(A)}, \qquad \overline{F}^{(A)} = 1 + S^{(A)}U \ . \qquad (2.4.20)$$

Hence (2.4.18) can be written as

$$I^{(A+B)} = F^{(A+B)}U = U\overline{F}^{(A+B)} , \tag{2.4.21}$$

with

$$F^{(A+B)} = F^{(A)}(1 - US^{(B)}US^{(A)})^{-1}F^{(B)} , \tag{2.4.22}$$

and $\overline{F}^{(A+B)}$ given by the transposition of $F^{(A+B)}$; while $S^{(A+B)}$ from (2.4.15) is written as

$$S^{(A+B)} = S^{(A)} + \overline{F}^{(A)}S^{(B/A)}F^{(A)} . \tag{2.4.23}$$

When the layer is composed of three layers with an additional layer of $K^{(C)}$, other addition formulas are available by observing that $I^{(A)}$ is changed to $I^{(A+C)}$ by the replacement of $U \to I^{(C)}$ and $S^{(A)} \to S^{(A/C)}$ in (2.4.5a), in view of (2.4.10); and, therefore, with the same replacement, $I^{(A+B)}$ is changed to $I^{(A+B+C)}$ so that, from (2.4.15),

$$S^{(A+B/C)} = S^{(A/C)} + [1 + S^{(A/C)}I^{(C)}]S^{(B/A+C)}[I^{(C)}S^{(A/C)} + 1] , \tag{2.4.24}$$

with $S^{(B/A+C)}$ obtained from (2.4.14) by the same replacement procedure, as

$$S^{(B/A+C)} = [1 - S^{(B/C)}I^{(C)}S^{(A/C)}I^{(C)}]^{-1}S^{(B/C)} . \tag{2.4.25}$$

The relation resulting from (2.4.17) is

$$\begin{aligned} S^{(A/B+C)}I^{(C)}S^{(B/C)} &= S^{(A/C)}I^{(C)}S^{(B/A+C)} \\ &= K^{(A)}I^{(A+B+C)}K^{(B)} , \end{aligned} \tag{2.4.26}$$

which, with the aid of relations (2.4.11, 12), leads to an expression,

$$\begin{aligned} I^{(A+B+C)} = &\left[1 + I^{(C)}S^{(A/C)}\right] \\ &\times \left[1 - I^{(C)}S^{(B/C)}I^{(C)}S^{(A/C)}\right]^{-1} I^{(B+C)} , \end{aligned} \tag{2.4.27}$$

similar in form to (2.4.18) for $I^{(A+B)}$. Alternatively, writing

$$I^{(A+C)} = F^{(A/C)}I^{(C)} = I^{(C)}\overline{F}^{(A/C)} , \tag{2.4.28}$$

in terms of a new factor, $F^{(A/C)}$ [meaning the attenuation of $I^{(A+C)}$ relative to $I^{(C)}$, so that its scheme of definition differs from that for $S^{(B/A)}$ by (2.4.11)] defined by

$$F^{(A/C)} = 1 + I^{(C)}S^{(A/C)}, \qquad F^{(A+C)} = F^{(A/C)}F^{(C)} , \tag{2.4.29}$$

then, (2.4.27, 24) can be presented in a symmetrical form by

$$F^{(A+B/C)} = F^{(A/C)}\left[1 - I^{(C)}S^{(B/C)}I^{(C)}S^{(A/C)}\right]^{-1}F^{(B/C)} , \tag{2.4.30}$$

$$S^{(A+B/C)} = S^{(A/C)} + \overline{F}^{(A/C)}S^{(B/A+C)}F^{(A/C)} , \tag{2.4.31}$$

in agreement with those which would be directly obtained from (2.4.22, 23), respectively, with the same replacement.

2.5 Eigenfunction Expansions and Diffusion Approximation

To solve the integral equation (2.3.43) for $\widetilde{S}(\widehat{\boldsymbol{\Omega}},\widehat{\boldsymbol{\lambda}}|\widehat{\boldsymbol{\Omega}}',\widehat{\boldsymbol{\lambda}}')$, it is often convenient to expand it by a set of eigenfunctions, $f_A(\widehat{\boldsymbol{\Omega}},\widehat{\boldsymbol{\lambda}})$ and $\overline{f}_A(\widehat{\boldsymbol{\Omega}},\widehat{\boldsymbol{\lambda}})$ of the cross section $K(\widehat{\boldsymbol{\Omega}}|\widehat{\boldsymbol{\Omega}}')$, defined by the eigenvalue equations [2.6],

$$\int d\widehat{\boldsymbol{\Omega}}'\, K(\widehat{\boldsymbol{\Omega}}|\widehat{\boldsymbol{\Omega}}')\widetilde{U}(\widehat{\boldsymbol{\Omega}}',\widehat{\boldsymbol{\lambda}})f_A(\widehat{\boldsymbol{\Omega}}',\widehat{\boldsymbol{\lambda}}) = A(\widehat{\boldsymbol{\lambda}})f_A(\widehat{\boldsymbol{\Omega}},\widehat{\boldsymbol{\lambda}}) \;, \tag{2.5.1a}$$

$$\int d\widehat{\boldsymbol{\Omega}}'\, \overline{f}_A(\widehat{\boldsymbol{\Omega}}',\widehat{\boldsymbol{\lambda}})\widetilde{U}(\widehat{\boldsymbol{\Omega}}',\widehat{\boldsymbol{\lambda}})K(\widehat{\boldsymbol{\Omega}}'|\widehat{\boldsymbol{\Omega}}) = A(\widehat{\boldsymbol{\lambda}})\overline{f}_A(\widehat{\boldsymbol{\Omega}},\widehat{\boldsymbol{\lambda}}) \;, \tag{2.5.1b}$$

with the normalization,

$$\int d\widehat{\boldsymbol{\Omega}}\, \overline{f}_A(\widehat{\boldsymbol{\Omega}},\widehat{\boldsymbol{\lambda}})\widetilde{U}(\widehat{\boldsymbol{\Omega}},\widehat{\boldsymbol{\lambda}})f_B(\widehat{\boldsymbol{\Omega}},\widehat{\boldsymbol{\lambda}}) = \delta_{AB} \;. \tag{2.5.2}$$

Here, $\widetilde{U}(\widehat{\boldsymbol{\Omega}},\widehat{\boldsymbol{\lambda}})$ is regarded as a weighting function when making the $\widehat{\boldsymbol{\Omega}}$ integration, and $A(\widehat{\boldsymbol{\lambda}})$ is the eigenvalue with branch points at $\lambda = \pm i\gamma$ on the $\lambda = |\widehat{\boldsymbol{\lambda}}|$ plane resulting from (continuous) poles of the factor, $\widetilde{U}(\widehat{\boldsymbol{\Omega}},\widehat{\boldsymbol{\lambda}})$, given by (2.3.34), and tends to zero as $\lambda \to \infty$ (Fig. 2.2). In terms of the eigenvalues and the eigenfunctions, the cross section can be exhibited by the series,

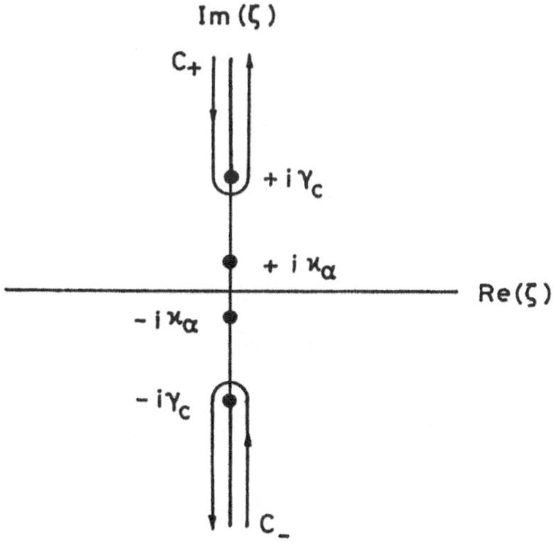

Fig. 2.2. Poles $\pm i\varkappa_\alpha$ and integration contours $C_\pm$ for each term in (2.5.60)

$$K(\widehat{\Omega}|\widehat{\Omega}') = \sum_A A(\widehat{\boldsymbol{\lambda}}) f_A(\widehat{\Omega}, \widehat{\boldsymbol{\lambda}}) \overline{f}_A(\widehat{\Omega}', \widehat{\boldsymbol{\lambda}}) \ . \tag{2.5.3}$$

A similar expansion is also possible for the scattering matrix, $\widetilde{S}$, in the form

$$\widetilde{S}(\widehat{\Omega}, \widehat{\boldsymbol{\lambda}}|\widehat{\Omega}', \widehat{\boldsymbol{\lambda}}') = \sum_A f_A(\widehat{\Omega}, \widehat{\boldsymbol{\lambda}}) \widetilde{S}_A(\widehat{\boldsymbol{\lambda}}|\widehat{\boldsymbol{\lambda}}') \overline{f}_A(\widehat{\Omega}', \widehat{\boldsymbol{\lambda}}') \ , \tag{2.5.4}$$

in view of (2.3.43), and, with (2.5.3) and the normalization (2.5.2), leads to an equation for each $\widetilde{S}_A$,

$$[1 - A(\widehat{\boldsymbol{\lambda}})] \widetilde{S}_A(\widehat{\boldsymbol{\lambda}}|\widehat{\boldsymbol{\lambda}}') = A(\widehat{\boldsymbol{\lambda}})(2\pi)^3 \delta(\widehat{\boldsymbol{\lambda}} - \widehat{\boldsymbol{\lambda}}') \ , \tag{2.5.5}$$

or, by the Fourier inversion,

$$[1 - A(\mathrm{i}\partial/\partial\widehat{\rho})] S_A(\widehat{\rho}|\widehat{\rho}') = A(\mathrm{i}\partial/\partial\widehat{\rho}) \delta(\widehat{\rho} - \widehat{\rho}') \ . \tag{2.5.6}$$

And, from (2.5.4)

$$S(\widehat{\Omega}, \widehat{\rho}|\widehat{\Omega}', \widehat{\rho}') = \sum_A f_A(\widehat{\Omega}, \mathrm{i}\overrightarrow{\partial}/\partial\widehat{\rho}) S_A(\widehat{\rho}|\widehat{\rho}') \overline{f}_A(\widehat{\Omega}', -\mathrm{i}\overleftarrow{\partial}/\partial\widehat{\rho}') \ , \tag{2.5.7}$$

where the arrows $\rightarrow$ and $\leftarrow$ mean the operation on the right-hand and left-hand sides, respectively.

The resulting series for the incoherent wave, $\mathcal{I}$ given by (2.3.44), is written in terms of another set of eigenfunctions, $\phi_A(\widehat{\Omega}, \widehat{\boldsymbol{\lambda}})$ and $\overline{\phi}_A(\widehat{\Omega}, \widehat{\boldsymbol{\lambda}})$, defined by

$$\phi_A(\widehat{\Omega}, \widehat{\boldsymbol{\lambda}}) = \widetilde{U}(\widehat{\Omega}, \widehat{\boldsymbol{\lambda}}) f_A(\widehat{\Omega}, \widehat{\boldsymbol{\lambda}}) \ , \tag{2.5.8a}$$

$$\overline{\phi}_A(\widehat{\Omega}, \widehat{\boldsymbol{\lambda}}) = \overline{f}_A(\widehat{\Omega}, \widehat{\boldsymbol{\lambda}}) \widetilde{U}(\widehat{\Omega}, \widehat{\boldsymbol{\lambda}}) \ , \tag{2.5.8b}$$

as:

$$\mathcal{I}(\widehat{\Omega}, \widehat{\rho}|\widehat{\Omega}', \widehat{\rho}') = \sum_A \mathcal{I}_A(\widehat{\Omega}, \widehat{\rho}|\widehat{\Omega}', \widehat{\rho}') \ , \tag{2.5.9}$$

$$\mathcal{I}_A(\widehat{\Omega}, \widehat{\rho}|\widehat{\Omega}', \widehat{\rho}') = \phi_A(\widehat{\Omega}, \mathrm{i}\overrightarrow{\partial}/\partial\widehat{\rho}) S_A(\widehat{\rho}|\widehat{\rho}') \overline{\phi}_A(\widehat{\Omega}', -\mathrm{i}\overleftarrow{\partial}/\partial\widehat{\rho}') \ . \tag{2.5.10}$$

Here, ϕ_A and $\overline{\phi}_A$ are subject to the normalization,

$$\int d\widehat{\Omega}\, \overline{\phi}_A \widetilde{U}^{-1} \phi_B(\widehat{\Omega}, \widehat{\boldsymbol{\lambda}}) = \int d\widehat{\Omega}\, \overline{\phi}_A f_B(\widehat{\Omega}, \widehat{\boldsymbol{\lambda}}) = \delta_{AB} \ , \tag{2.5.11}$$

and hence

$$\widetilde{U}(\widehat{\boldsymbol{\Omega}},\widehat{\boldsymbol{\lambda}})\delta^2(\widehat{\boldsymbol{\Omega}}-\widehat{\boldsymbol{\Omega}}')=\sum_A \phi_A(\widehat{\boldsymbol{\Omega}},\widehat{\boldsymbol{\lambda}})\overline{\phi}_A(\widehat{\boldsymbol{\Omega}}',\widehat{\boldsymbol{\lambda}}) \ . \tag{2.5.12}$$

In the diffusion region far from the source, where change of the wave intensity is sufficiently small within the wave coherence distance, γ^{-1}, so that $|\widehat{\boldsymbol{\lambda}}/\gamma| \ll 1$, the convergence of the series (2.5.9) becomes good enough to be approximated by the first term and with the eigenfunction and the eigenvalue, given by the first non-vanishing terms in their series expansions with respect to $\widehat{\boldsymbol{\lambda}}/\gamma$ (Sect. 2.5.2), in view of the other terms of the eigenfunction series which are short-range functions negligible in the diffusion region (Sect. 2.5.3).

The eigenvalue, $A(\widehat{\boldsymbol{\lambda}})$, is closely connected to the power flux of the eigenfunction term $\mathcal{I}_A$. We first observe that the $\widehat{\boldsymbol{\Omega}}$ integration of (2.5.1a) yields, in view of (2.3.50),

$$\int d\widehat{\boldsymbol{\Omega}}\,\gamma\widetilde{U}f_A(\widehat{\boldsymbol{\Omega}},\widehat{\boldsymbol{\lambda}})=A(\widehat{\boldsymbol{\lambda}})\int d\widehat{\boldsymbol{\Omega}}\,f_A(\widehat{\boldsymbol{\Omega}},\widehat{\boldsymbol{\lambda}}) \ , \tag{2.5.13}$$

which, upon substitution of

$$\gamma\widetilde{U}=\mathrm{i}\widehat{\boldsymbol{\Omega}}\cdot\widehat{\boldsymbol{\lambda}}\widetilde{U}+1 \tag{2.5.14}$$

from (2.3.34), in the left-hand side , can be rewritten as

$$1-A(\widehat{\boldsymbol{\lambda}})=-\mathrm{i}\int d\widehat{\boldsymbol{\Omega}}\,\widehat{\boldsymbol{\lambda}}\cdot\widehat{\boldsymbol{\Omega}}\phi_A(\widehat{\boldsymbol{\Omega}},\widehat{\boldsymbol{\lambda}}) \Big/ \int d\widehat{\boldsymbol{\Omega}}\,f_A(\widehat{\boldsymbol{\Omega}},\widehat{\boldsymbol{\lambda}}) \tag{2.5.15a}$$

$$\equiv -\mathrm{i}\widehat{\boldsymbol{\lambda}}\cdot\widehat{\boldsymbol{p}}_A(\widehat{\boldsymbol{\lambda}}) \ , \tag{2.5.15b}$$

where the last equation defines the vector, $\widehat{\boldsymbol{p}}_A(\widehat{\boldsymbol{\lambda}})$. Hence the govering equation (2.5.6) for S_A can be written in terms of $\widehat{\boldsymbol{p}}_A$, as

$$\frac{\partial}{\partial\widehat{\boldsymbol{\rho}}}\cdot\widehat{\boldsymbol{p}}_A(\mathrm{i}\partial/\partial\widehat{\boldsymbol{\rho}})S_A(\widehat{\boldsymbol{\rho}}|\widehat{\boldsymbol{\rho}}')=A(\mathrm{i}\partial/\partial\widehat{\boldsymbol{\rho}})\delta(\widehat{\boldsymbol{\rho}}-\widehat{\boldsymbol{\rho}}') \ , \tag{2.5.16}$$

meaning the power equation for each $\mathcal{I}_A$; i.e., with the power flux vector, $\widehat{\boldsymbol{w}}_A$, defined by

$$\widehat{\boldsymbol{w}}_A(\widehat{\boldsymbol{\rho}}|\widehat{\boldsymbol{\Omega}}',\widehat{\boldsymbol{\rho}}')\equiv\int d\widehat{\boldsymbol{\Omega}}\,\widehat{\boldsymbol{\Omega}}\mathcal{I}_A(\widehat{\boldsymbol{\Omega}},\widehat{\boldsymbol{\rho}}|\widehat{\boldsymbol{\Omega}}',\widehat{\boldsymbol{\rho}}')$$

$$=\widehat{\boldsymbol{p}}_A(\mathrm{i}\overrightarrow{\partial}/\partial\widehat{\boldsymbol{\rho}})\int d\widehat{\boldsymbol{\Omega}}\,f_A(\widehat{\boldsymbol{\Omega}},\mathrm{i}\overrightarrow{\partial}/\partial\widehat{\boldsymbol{\rho}})S_A(\widehat{\boldsymbol{\rho}}|\widehat{\boldsymbol{\rho}}')\overline{\phi}_A(\widehat{\boldsymbol{\Omega}}',-\mathrm{i}\overleftarrow{\partial}/\partial\widehat{\boldsymbol{\rho}}') \ , \tag{2.5.17a}$$

(2.5.16) can be written as

$$\frac{\partial}{\partial\widehat{\boldsymbol{\rho}}}\cdot\widehat{\boldsymbol{w}}_A(\widehat{\boldsymbol{\rho}}|\widehat{\boldsymbol{\Omega}}',\widehat{\boldsymbol{\rho}}')$$

$$=A(\mathrm{i}\partial/\partial\widehat{\boldsymbol{\rho}})\int d\widehat{\boldsymbol{\Omega}}\,f_A(\widehat{\boldsymbol{\Omega}},\mathrm{i}\overrightarrow{\partial}/\partial\widehat{\boldsymbol{\rho}})\delta(\widehat{\boldsymbol{\rho}}-\widehat{\boldsymbol{\rho}}')\phi_A(\widehat{\boldsymbol{\Omega}}',-\mathrm{i}\overleftarrow{\partial}/\partial\widehat{\boldsymbol{\rho}}') \ . \tag{2.5.17b}$$

Here, except for the diffusion term, the source term on the right-hand side is zero when integrated over the entire space, as we will see later by (2.5.32), meaning that the power is not carried away as a whole by all the $\mathcal{I}_A$'s, excluding the diffusion's.

To investigate the local aspect of the power equation, on the other hand, we observe that (2.5.5) can be rewritten, using (2.5.15a), as

$$-\mathrm{i}\int d\widehat{\boldsymbol{\Omega}}\,\widehat{\boldsymbol{\lambda}}\cdot\widehat{\boldsymbol{\Omega}}\phi_A(\widehat{\boldsymbol{\Omega}},\widehat{\boldsymbol{\lambda}})\widetilde{S}_A(\widehat{\boldsymbol{\lambda}}|\widehat{\boldsymbol{\lambda}}') = A(\widehat{\boldsymbol{\lambda}})\int d\widehat{\boldsymbol{\Omega}}\,f_A(\widehat{\boldsymbol{\Omega}},\widehat{\boldsymbol{\lambda}})(2\pi)^3\delta(\widehat{\boldsymbol{\lambda}}-\widehat{\boldsymbol{\lambda}}') \ , \tag{2.5.18}$$

which, on multiplying both sides with $\overline{\phi}_A(\widehat{\boldsymbol{\Omega}}',\widehat{\boldsymbol{\lambda}}')$ and summing up over all the A's, leads to

$$-\mathrm{i}\int d\widehat{\boldsymbol{\Omega}}\,\widehat{\boldsymbol{\lambda}}\cdot\widehat{\boldsymbol{\Omega}}\widetilde{\mathcal{I}}(\widehat{\boldsymbol{\Omega}},\widehat{\boldsymbol{\lambda}}|\widehat{\boldsymbol{\Omega}}',\widehat{\boldsymbol{\lambda}}') = \int d\widehat{\boldsymbol{\Omega}}\,K(\widehat{\boldsymbol{\Omega}}|\widehat{\boldsymbol{\Omega}}')\widetilde{U}(\widehat{\boldsymbol{\Omega}}',\widehat{\boldsymbol{\lambda}}')(2\pi)^3\delta(\widehat{\boldsymbol{\lambda}}-\widehat{\boldsymbol{\lambda}}') \ , \tag{2.5.19}$$

in view of (2.5.9, 3), and hence, by the Fourier inversion, to

$$\frac{\partial}{\partial\widehat{\boldsymbol{\rho}}}\cdot\int d\widehat{\boldsymbol{\Omega}}\,\widehat{\boldsymbol{\Omega}}\mathcal{I}(\widehat{\boldsymbol{\Omega}},\widehat{\boldsymbol{\rho}}|\widehat{\boldsymbol{\Omega}}',\widehat{\boldsymbol{\rho}}') = \gamma(\widehat{\boldsymbol{\Omega}}')U(\widehat{\boldsymbol{\Omega}}',\widehat{\boldsymbol{\rho}}-\widehat{\boldsymbol{\rho}}') \ , \tag{2.5.20}$$

reproducing what is given by the $\widehat{\boldsymbol{\Omega}}$ integration of the transport equation (2.3.46).

2.5.1 Symmetries of the Eigenfunctions and Eigenvalues

Since the deterministic Green's function satisfies the reciprocity, $g^T = g$, the same is true also for the statistical Green's functions, being symmetrical coordinate matrices subject to $G^T = G$ and $I^T = I$. Hence, it follows from the governing equations (2.2.15) and (2.3.7), that $M^T = M$ and $K^T = K$, which leads to the relations,

$$\gamma(-\widehat{\boldsymbol{\Omega}}) = \gamma(\widehat{\boldsymbol{\Omega}}), \qquad K(-\widehat{\boldsymbol{\Omega}}'|-\widehat{\boldsymbol{\Omega}}) = K(\widehat{\boldsymbol{\Omega}}|\widehat{\boldsymbol{\Omega}}') \ . \tag{2.5.21}$$

Now, with these symmetries and the relations,

$$\widetilde{U}(\widehat{\boldsymbol{\Omega}},\widehat{\boldsymbol{\lambda}}) = \widetilde{U}(-\widehat{\boldsymbol{\Omega}},-\widehat{\boldsymbol{\lambda}}) = \widetilde{U}^*(\widehat{\boldsymbol{\Omega}},-\widehat{\boldsymbol{\lambda}}) \ , \tag{2.5.22}$$

it is straightforward to find that $f_A(-\widehat{\boldsymbol{\Omega}},-\widehat{\boldsymbol{\lambda}})$ is an eigenfunction of (2.5.1b) with the eigenvalue, $A(-\widehat{\boldsymbol{\lambda}})$, i.e.,

$$\int d\hat{\Omega}' f_A(-\hat{\Omega}', -\hat{\lambda})\tilde{U}(\hat{\Omega}', \hat{\lambda})K(\hat{\Omega}'|\hat{\Omega}) = A(-\hat{\lambda})f_A(-\hat{\Omega}, -\hat{\lambda}) \ . \quad (2.5.23a)$$

Hence, with (2.5.1a), the following relation holds:

$$[A(-\hat{\lambda}) - B(\hat{\lambda})] \int d\hat{\Omega}\, f_A(-\hat{\Omega}, -\hat{\lambda})\tilde{U}(\hat{\Omega}, \hat{\lambda})f_B(\hat{\Omega}, \hat{\lambda}) = 0 \ ,$$

which shows that if

$$\int d\hat{\Omega}\, f_A(-\hat{\Omega}, -\hat{\lambda})\tilde{U}(\hat{\Omega}, \hat{\lambda})f_B(\hat{\Omega}, \hat{\lambda}) \neq 0 \ , \quad (2.5.23b)$$

then $A(-\hat{\lambda}) = B(\hat{\lambda})$; which is reduced to $a = b$ when $\hat{\lambda} = 0$ (Sect. 2.5.2), implying, therefore, that

$$A(-\hat{\lambda}) = A(\hat{\lambda}) \ . \quad (2.5.23c)$$

Hence, when $f_A(\hat{\Omega}, \hat{\lambda})$ is not degenerate (having one eigenfunction for each of the different eigenvalues), we can set

$$\overline{f}_A(\hat{\Omega}, \hat{\lambda}) = N^{(A)}(\hat{\lambda})f_A(-\hat{\Omega}, -\hat{\lambda}) \ , \quad (2.5.24a)$$

with the factor, $N^{(A)}(\hat{\lambda}) = N^{(A)}(-\hat{\lambda})$, determined by

$$N^{(A)}(\hat{\lambda}) \int d\hat{\Omega}\, f_A(-\hat{\Omega}, -\hat{\lambda})\tilde{U}(\hat{\Omega}, \hat{\lambda})f_A(\hat{\Omega}, \hat{\lambda}) = 1 \ . \quad (2.5.24b)$$

Even when $f_A(\hat{\Omega}, \hat{\lambda})$ is degenerate at $\hat{\lambda} \neq 0$, having several eigenfunctions, $f_{A,j}$, $j = 1, 2, \ldots$, for the same eigenvalue, $A(\hat{\lambda})$, we can set

$$\overline{f}_{A,i}(\hat{\Omega}, \hat{\lambda}) = \sum_j N_{ij}^{(A)}(\hat{\lambda})f_{A,j}(-\hat{\Omega}, -\hat{\lambda}) \ ,$$

and determine the factors, $N_{ij}^{(A)}$, by

$$\int d\hat{\Omega}\, \overline{f}_{A,i}(\hat{\Omega}, \hat{\lambda})\tilde{U}(\hat{\Omega}, \hat{\lambda})f_{A,j}(\hat{\Omega}, \hat{\lambda}) = \delta_{ij} \ .$$

Equation (2.5.23a) also holds true in view of (2.5.22) for the replacement of $f_A(-\hat{\Omega}, -\hat{\lambda}) \to f_A^*(-\hat{\Omega}, \hat{\lambda})$ and $A(-\hat{\lambda}) \to A^*(\hat{\lambda})$, showing that $A^*(\hat{\lambda})$ is a member of the $A(\hat{\lambda})$'s. And, from (2.5.23b), $A^*(\hat{\lambda}) = A(\hat{\lambda})$ if

$$\int d\hat{\Omega}\, f_A^*(-\hat{\Omega}, \hat{\lambda})\tilde{U}(\hat{\Omega}, \hat{\lambda})f_A(\hat{\Omega}, \hat{\lambda}) \neq 0 \ ,$$

wherein the integrand is a real function of $\hat{\Omega}$; hence the condition is ensured when $f_A(\hat{\Omega}, \hat{\lambda})$ is an even function of $\hat{\Omega}$.

2.5.2 Diffusion Eigenfunctions

From (2.5.1) the eigenvalue equations for ϕ_A and $\overline{\phi}_A$ are

$$\int d\widehat{\Omega}' \, K(\widehat{\Omega}|\widehat{\Omega}')\phi_A(\widehat{\Omega}',\widehat{\lambda}) = A(\widehat{\lambda})\widetilde{U}^{-1}(\widehat{\Omega},\widehat{\lambda})\phi_A(\widehat{\Omega},\widehat{\lambda}) \ , \tag{2.5.25a}$$

$$\int d\widehat{\Omega}' \, \overline{\phi}_A(\widehat{\Omega}',\widehat{\lambda})K(\widehat{\Omega}'|\widehat{\Omega}) = A(\widehat{\lambda})\overline{\phi}_A(\widehat{\Omega},\widehat{\lambda})\widetilde{U}^{-1}(\widehat{\Omega},\widehat{\lambda}) \ . \tag{2.5.25b}$$

Here

$$\widetilde{U}^{-1}(\widehat{\Omega},\widehat{\lambda}) = \gamma - \mathrm{i}\widehat{\Omega}\cdot\widehat{\lambda} \ , \tag{2.5.25c}$$

and the normalization (2.5.2) is written by

$$\int d\widehat{\Omega}\, \overline{\phi}_A \widetilde{U}^{-1}\phi_B = \int d\widehat{\Omega}\, \overline{\phi}_A f_B = \int d\widehat{\Omega}\, \overline{f}_A \phi_B = \delta_{AB} \ . \tag{2.5.26}$$

Hence, when $\widehat{\lambda} = 0$, ϕ_A and $\overline{\phi}_A$ are reduced to $\phi_a(\widehat{\Omega})$ and $\overline{\phi}_a(\widehat{\Omega})$ with the eigenvalue a, respectively, defined by

$$\int d\widehat{\Omega}' \, K(\widehat{\Omega}|\widehat{\Omega}')\phi_a(\widehat{\Omega}') = a\gamma\phi_a(\widehat{\Omega}) \ , \tag{2.5.27a}$$

$$\int d\widehat{\Omega}' \, \overline{\phi}_a(\widehat{\Omega}')K(\widehat{\Omega}'|\widehat{\Omega}) = a\overline{\phi}_a\gamma(\widehat{\Omega}) \ , \tag{2.5.27b}$$

and subjected to the normalization,

$$(\overline{\phi}_a\gamma\phi_b) \equiv \int d\widehat{\Omega}\, \overline{\phi}_a\gamma\phi_b(\widehat{\Omega}) = \delta_{ab} \ . \tag{2.5.28}$$

Here, a uniform distribution, $\overline{\phi}_1$, is an eigenfunction of (2.5.27b) and has the eigenvalue, $a = 1$, in view of the optical relation (2.3.50). Hence, according to (2.5.28), we can set

$$\overline{\phi}_1(\widehat{\Omega}) = (4\pi)^{-1}, \qquad \phi_1(\widehat{\Omega}) = \overline{\gamma}^{-1} \ , \tag{2.5.29a}$$

where $\overline{\gamma}$ is the angle-averaged value of $\gamma(\widehat{\Omega})$, and, from (2.5.8),

$$\overline{f}_1(\widehat{\Omega}) = (4\pi)^{-1}\gamma(\widehat{\Omega}), \qquad f_1(\widehat{\Omega}) = \overline{\gamma}^{-1}\gamma(\widehat{\Omega}) \ , \tag{2.5.29b}$$

which shows that $f_1 = 1$ when the cross section is subject to the rotational invariance. The $\widehat{\Omega}$ integration of (2.5.27a) on both sides leads to the important relation,

$$(1-a)\int d\widehat{\Omega}\, \gamma\phi_a(\widehat{\Omega}) = 0 \ , \tag{2.5.30}$$

which shows that

$$a = 1 \quad \text{if} \quad \int d\hat{\Omega}\, \phi_a(\hat{\Omega}) = \int d\hat{\Omega}\, f_a(\hat{\Omega}) \neq 0 \ , \tag{2.5.31}$$

whereas,

$$\int d\hat{\Omega}\, f_a(\hat{\Omega}) = 0 \quad \text{if} \quad a \neq 1 \ ; \tag{2.5.32}$$

the same relation also holds for $\overline{f}_a(\hat{\Omega})$, in view of (2.5.24a). Here $|a| \leq 1$ since, by (2.5.27a),

$$\int d\hat{\Omega}\, K(\hat{\Omega}|\hat{\Omega}')|\phi_a(\hat{\Omega}')| \geq |a|\gamma|\phi_a(\hat{\Omega})|, \qquad \gamma > 0 \ ,$$

and hence, by the $\hat{\Omega}$ integration on both sides,

$$(1 - |a|) \int d\hat{\Omega}\, |\phi_a(\hat{\Omega})| \geq 0 \ ,$$

which shows that $|a| \leq 1$.

To obtain ϕ_A in a series of the ϕ_a's, we set

$$\phi_A = \sum_b \phi_b C_{bA}, \qquad C_{aA} = 1 \ , \tag{2.5.33}$$

which, substituting into (2.5.25a) and using (2.5.27, 28), leads to

$$(A - b)C_{bA} = \mathrm{i}A \sum_c \langle \hat{\Omega} \cdot \hat{\boldsymbol{\lambda}} \rangle_{bc} C_{cA} \ , \tag{2.5.34}$$

in terms of the notation,

$$\langle \hat{\Omega} \cdot \hat{\boldsymbol{\lambda}} \rangle_{bc} = \int d\hat{\Omega}\, \overline{\phi}_b \hat{\Omega} \cdot \hat{\boldsymbol{\lambda}} \phi_c \ . \tag{2.5.35}$$

Equation (2.5.34) gives, to the first order of $\hat{\boldsymbol{\lambda}}$,

$$C_{bA} = \mathrm{i}(a - b)^{-1} a \langle \hat{\Omega} \cdot \hat{\boldsymbol{\lambda}} \rangle_{ba}, \qquad b \neq a \ . \tag{2.5.36}$$

Hence, particularly when $a = 1$ (diffusion term),

$$\phi_A = \phi_1 + \mathrm{i} \sum_{b(\neq 1)} \phi_b (1 - b)^{-1} \langle \hat{\Omega} \cdot \hat{\boldsymbol{\lambda}} \rangle_{b1} \ , \tag{2.5.37}$$

and the eigenvalue, $A(\hat{\boldsymbol{\lambda}})$, is given from (2.5.34) by

$$1-A(\widehat{\lambda}) = \sum_{b(\neq 1)} (1-b)^{-1}\langle \widehat{\Omega}\cdot\widehat{\lambda}\rangle_{1b}\langle \widehat{\Omega}\cdot\widehat{\lambda}\rangle_{b1} \qquad (2.5.38)$$

$$= -\mathrm{i}(4\pi)^{-1}\int d\widehat{\Omega}\,(\widehat{\lambda}\cdot\widehat{\Omega})\phi_A(\widehat{\Omega},\widehat{\lambda}) \ , \qquad (2.5.39)$$

(where $\langle\widehat{\Omega}\cdot\widehat{\lambda}\rangle_{11}=0$), which provides the non-vanishing $\widehat{\lambda}$ term of $A(\widehat{\lambda})$ to the lowest order. The same result can also be directly obtained from (2.5.15), in consequence of relation (2.5.29b).

Thus, by definition (2.5.15),

$$\widehat{p}_A(\widehat{\lambda}) = \mathrm{i}\sum_{b(\neq 1)}(1-b)^{-1}\langle\widehat{\Omega}\rangle_{1b}\langle\widehat{\Omega}\rangle_{b1}\cdot\widehat{\lambda} \ , \qquad (2.5.40a)$$

and, from (2.5.16), the governing equation for the diffusion term can be written as

$$\frac{\partial}{\partial\widehat{\rho}}\cdot\widehat{p}_A(\mathrm{i}\partial/\partial\widehat{\rho})S_A(\widehat{\rho}|\widehat{\rho}') = \delta(\widehat{\rho}-\widehat{\rho}') \ , \qquad (2.5.40b)$$

with the approximation, $A\simeq 1$, on the right-hand side .

Also, for $\overline{\phi}_A$, we similarly obtain the expansion,

$$\overline{\phi}_A = \overline{\phi}_a + \sum_{b\neq a}\overline{C}_{Ab}\overline{\phi}_b, \qquad \overline{C}_{Aa}=1 \ ,$$

$$\int d\widehat{\Omega}\,\overline{\phi}_A\gamma\phi_A = 1+\sum_{b\neq a}\overline{C}_{Ab}C_{bA} \ .$$

Hence, from (2.5.36), the present method is valid under the condition,

$$|\overline{C}_{Ab}C_{bA}| \sim \left|(a-b)^{-1}a\langle\widehat{\Omega}\cdot\widehat{\lambda}\rangle_{ba}\right|^2 \ll 1 \ , \qquad (2.5.41)$$

or $(1-b/a)^{-2}|\widehat{\lambda}/\gamma|^2 \ll 1$, which is certainly ensured, unless $b/a\sim 1$, to the diffusion range where $|\widehat{\lambda}/\gamma|^2 \ll 1$.

When assuming the rotational invariance of the cross section of the form, $K(\widehat{\Omega}|\widehat{\Omega}') = K(\widehat{\Omega}\cdot\widehat{\Omega}')$, the cross section can be expanded in terms of the Legendre functions, $P_n(x)$, $n=0,1,2,\ldots$, by

$$K(\cos\theta) = (4\pi)^{-1}\gamma\left[1+\sum_{n=1}^{\infty}(2n+1)P_n(\cos\theta)a_n\right] \ , \qquad (2.5.42)$$

where $\gamma=\overline{\gamma}$ is presently constant, and

$$a_n = \gamma^{-1}\int d\widehat{\Omega}\,P_n(\cos\theta)K(\cos\theta), \qquad d\widehat{\Omega} = 2\pi\sin\theta\,d\theta \ ,$$

$$P_n(1)=1, \qquad a_0 = 1 \ . \qquad (2.5.43)$$

Here, by (2.5.42), we can directly show that the spherical harmonics,

$$Y_n^m(\widehat{\boldsymbol{\Omega}}) \equiv Y_n^m(\theta,\varphi) = \left.\begin{matrix}\cos m\varphi \\ \sin m\varphi\end{matrix}\right\} P_n^m(\cos\theta), \qquad n \geq m \geq 0,$$

$$\int d\widehat{\boldsymbol{\Omega}} \left[Y_n^m(\widehat{\boldsymbol{\Omega}})\right]^2 = 4\pi(2n+1)^{-1}\frac{(n+m)!}{(n-m)!}\begin{cases} 1, & m=0 \\ 1/2, & m \neq 0, \end{cases} \tag{2.5.44}$$

are the eigenfunctions, $f_a(\widehat{\boldsymbol{\Omega}})$, of (2.5.1a) when $\widehat{\boldsymbol{\lambda}} = 0$, and have the eigenvalues, a_n's, i.e.,

$$\int d\widehat{\boldsymbol{\Omega}}\, K(\widehat{\boldsymbol{\Omega}} \cdot \widehat{\boldsymbol{\Omega}}')\gamma^{-1} Y_n^m(\widehat{\boldsymbol{\Omega}}') = a_n Y_n^m(\widehat{\boldsymbol{\Omega}}) \ . \tag{2.5.45}$$

Hence, in (2.5.38), we can set

$$\phi_b(\widehat{\boldsymbol{\Omega}}) = \Omega_j, \qquad \overline{\phi}_b(\widehat{\boldsymbol{\Omega}}) = (3/4\pi\gamma)\Omega_j, \qquad j = 1,2,3 \ , \tag{2.5.46}$$

$\widehat{\boldsymbol{\Omega}} = (\Omega_1, \Omega_2, \Omega_3)$, which is consistent with the normalization (2.5.28); the other eigenfunctions do not make any contribution in view of the matrix elements given by (2.5.35). Hence,

$$\langle\widehat{\boldsymbol{\Omega}} \cdot \widehat{\boldsymbol{\lambda}}\rangle_{1b} = \frac{1}{3}\lambda_j, \qquad \langle\widehat{\boldsymbol{\Omega}} \cdot \widehat{\boldsymbol{\lambda}}\rangle_{b1} = \gamma^{-2}\lambda_j \ , \tag{2.5.47}$$

and the eigenvalue, b, is found directly from (2.5.43) [or by substitution of (2.5.47) into (2.5.27a)] to be

$$b = a_1 \equiv \gamma^{-1}\int d\widehat{\boldsymbol{\Omega}}\, K(\widehat{\boldsymbol{\Omega}} \cdot \widehat{\boldsymbol{\Omega}}')\widehat{\boldsymbol{\Omega}} \cdot \widehat{\boldsymbol{\Omega}}' \ . \tag{2.5.48}$$

Thus, in terms of a quantity, D, defined by

$$D = (3\gamma)^{-1}(1 - a_1)^{-1} \ , \tag{2.5.49}$$

(2.5.38, 40a) can be written as

$$1 - A(\widehat{\boldsymbol{\lambda}}) = \gamma^{-1}D\widehat{\boldsymbol{\lambda}}^2, \qquad \widehat{\boldsymbol{p}}_A(\widehat{\boldsymbol{\lambda}}) = \mathrm{i}\gamma^{-1}D\widehat{\boldsymbol{\lambda}} \ , \tag{2.5.50}$$

and, from (2.5.37),

$$\gamma\phi_A(\widehat{\boldsymbol{\Omega}}) = 4\pi\overline{\phi}_A(\widehat{\boldsymbol{\Omega}}) = 1 + \mathrm{i}3D\widehat{\boldsymbol{\Omega}} \cdot \widehat{\boldsymbol{\lambda}} \ , \tag{2.5.51}$$

which is consistent with ϕ_1 and $\overline{\phi}_1$ of (2.5.29a).

Thus the diffusion equation (2.5.40b) is reduced to

$$\gamma^{-1}\left[\gamma^{(ab)} - D\left(\frac{\partial}{\partial \widehat{\rho}}\right)^2\right] S_A(\widehat{\rho}|\widehat{\rho}') = \delta(\widehat{\rho} - \widehat{\rho}') , \tag{2.5.52}$$

with an additional term, $\gamma^{(ab)}$, to represent an intrinsic dissipation by the medium. Here the associated power flux is given from (2.5.17a), by

$$\widehat{w}_A(\widehat{\rho}|\widehat{\Omega}', \widehat{\rho}') = -\gamma^{-1} D \frac{\partial}{\partial \widehat{\rho}} S_A(\widehat{\rho}|\widehat{\rho}') \overline{\phi}_A(\widehat{\Omega}', -\mathrm{i}\overleftarrow{\partial}/\partial \widehat{\rho}') . \tag{2.5.53}$$

When the scattering is made mostly in a forward direstion, as in the case of light wave propagation in a turbulent medium, then (2.5.43) shows that $a_n \sim 1$ over the first several orders of n, while D of (2.5.49) tends to infinity as $a_1 \to 1$. This reflects the fact that the present perturbative method is valid only under the condition of (2.5.41), which is not satisfied when $a_n \sim a_0 = 1 \quad n = 1, 2, 3, \ldots$.

2.5.3 Mode Expansions in a Homogeneous Random Medium

Solving (2.3.43) in matrix form

$$\widetilde{S} = K + K\widetilde{U}\widetilde{S} = (1 - K\widetilde{U})^{-1} K \tag{2.5.54a}$$

$$= K + (1 - K\widetilde{U})^{-1} K\widetilde{U} K , \tag{2.5.54b}$$

whose matrix elements in a homogeneous random medium can be written as

$$\widetilde{S}(\widehat{\Omega}|\widehat{\lambda}|\widehat{\Omega}') = K(\widehat{\Omega}|\widehat{\Omega}') + \widetilde{S}^{(1)}(\widehat{\Omega}|\widehat{\lambda}|\widehat{\Omega}') , \tag{2.5.55}$$

$$\widetilde{S}^{(1)}(\widehat{\Omega}|\widehat{\lambda}|\widehat{\Omega}') = (1 - K\widetilde{U})^{-1} K\widetilde{U} K(\widehat{\Omega}|\widehat{\lambda}|\widehat{\Omega}')$$

$$= \sum_A \widetilde{S}_A^{(1)}(\widehat{\Omega}|\widehat{\lambda}|\widehat{\Omega}') . \tag{2.5.56}$$

Here, using (2.5.3, 2),

$$\widetilde{S}_A^{(1)}(\widehat{\Omega}|\widehat{\lambda}|\widehat{\Omega}') = f_A(\widehat{\Omega}, \widehat{\lambda})(1 - A)^{-1} A^2(\widehat{\lambda}) \overline{f}_A(\widehat{\Omega}', \widehat{\lambda}) , \tag{2.5.57}$$

where, as $|\widehat{\lambda}| \to \infty$,

$$A(\widehat{\lambda}) \sim O\{|\widehat{\lambda}|^{-1}\}, \qquad f_A \overline{f}_A \sim O\{|\widehat{\lambda}|\},$$

$$\widetilde{S}_A^{(1)} \sim O\{|\widehat{\lambda}|^{-1}\} , \tag{2.5.58}$$

in view of (2.3.34).

Here, we write $\widehat{\lambda} = (\lambda, \zeta)$, $\zeta = \lambda_z$, $\widehat{\lambda}^2 = \lambda^2 + \zeta^2$, $A(\widehat{\lambda}) = A(\zeta)$, and $\widetilde{S}_A^{(1)}(\widehat{\Omega}|\widehat{\lambda}|\widehat{\Omega}') = \widetilde{S}_A^{(1)}(\widehat{\Omega}|\zeta|\widehat{\Omega}')$, and assume that for each eigenvalue, A,

$1 - A(\zeta) = 0$ has a pair of roots at $\zeta = \pm i\varkappa_\alpha(\boldsymbol{\lambda})$, $\operatorname{Im}\{\varkappa_\alpha\} < 0$, or at $\widehat{\boldsymbol{\lambda}} = \widehat{\boldsymbol{\lambda}}_{\pm\alpha} = (\boldsymbol{\lambda}, \pm i\varkappa_\alpha)$, $\alpha = 1, 2, \ldots$, so that $\widetilde{S}_A^{(1)}(\widehat{\boldsymbol{\Omega}}|\zeta|\widehat{\boldsymbol{\Omega}}')$ is regular everywhere in the ζ plane, except a pair of poles due to the roots and two branch points at $\zeta = \pm i\gamma_C(\boldsymbol{\lambda})$, $\gamma_C(\boldsymbol{\lambda}) = (\gamma^2 + \boldsymbol{\lambda}^2)^{1/2}$ (Fig. 2.2, page 27). Here the branch points result from the poles of the factor, $\widetilde{U}(\widehat{\boldsymbol{\Omega}}, \widehat{\boldsymbol{\lambda}})$, involved in the eigenvalue equations (2.5.1). In the case of isotropic scatterers, the eigenvalue is

$$A(\widehat{\boldsymbol{\lambda}}) = \frac{i\gamma}{2|\widehat{\boldsymbol{\lambda}}|} \ln\left[\frac{\gamma - i|\widehat{\boldsymbol{\lambda}}|}{\gamma + i|\widehat{\boldsymbol{\lambda}}|}\right] = \frac{\gamma}{|\widehat{\boldsymbol{\lambda}}|} \arctan\left[\frac{|\widehat{\boldsymbol{\lambda}}|}{\gamma}\right] . \tag{2.5.59}$$

Here, in view of the asymptotic forms (2.5.58), use of Cauchy's integral formula leads to a singular expansion of $\widetilde{S}_A^{(1)}(\widehat{\boldsymbol{\Omega}}|\zeta|\widehat{\boldsymbol{\Omega}}')$ as

$$\begin{aligned}\widetilde{S}_A^{(1)}(\widehat{\boldsymbol{\Omega}}|\zeta|\widehat{\boldsymbol{\Omega}}') &= \frac{1}{2\pi i}\oint d\zeta' \frac{\widetilde{S}_A^{(1)}(\widehat{\boldsymbol{\Omega}}|\zeta'|\widehat{\boldsymbol{\Omega}}')}{\zeta' - \zeta} \\ &= \frac{f_A(\widehat{\boldsymbol{\Omega}}, \widehat{\boldsymbol{\lambda}}_{+\alpha})\overline{f}_A(\widehat{\boldsymbol{\Omega}}', \widehat{\boldsymbol{\lambda}}_{+\alpha})}{(\zeta - i\varkappa_\alpha)[-\partial A/\partial\zeta]_{+\alpha}} + \frac{f_A(\widehat{\boldsymbol{\Omega}}, \widehat{\boldsymbol{\lambda}}_{-\alpha})\overline{f}_A(\widehat{\boldsymbol{\Omega}}', \widehat{\boldsymbol{\lambda}}_{-\alpha})}{(\zeta + i\varkappa_\alpha)[-\partial A/\partial\zeta]_{-\alpha}} \\ &\quad + \frac{1}{2\pi i}\left[\int_{C_+} d\zeta' + \int_{C_-} d\zeta'\right] \frac{\widetilde{S}_A^{(1)}(\widehat{\boldsymbol{\Omega}}|\zeta'|\widehat{\boldsymbol{\Omega}}')}{\zeta - \zeta'} .\end{aligned} \tag{2.5.60}$$

Here $|_{\pm\alpha}$ means the setting, $\zeta = \pm i\varkappa_\alpha$, and the last two integrations are made along the contour paths, $C_\pm$, drawn from $\pm i\gamma_C$ to $\pm i\infty$, respectively, as shown in Fig. 2.2. By the Fourier inversion of (2.5.60), the first two terms yield a mode wave, $S_\alpha(\widehat{\boldsymbol{\Omega}}|\widehat{\boldsymbol{\rho}} - \widehat{\boldsymbol{\rho}}'|\widehat{\boldsymbol{\Omega}}')$, to be defined in the following.

For the diffusion mode term, $S_{\alpha=1}$ with $\varkappa_\alpha = \varkappa_1 = \lambda \equiv |\boldsymbol{\lambda}|$, a specific expression can be obtained by using (2.5.50, 51). Hence, on using

$$-\partial A/\partial\zeta \,|_{\alpha=\pm1} = \pm i2\lambda\gamma^{-1}D , \tag{2.5.61}$$

it is straightforward to find the mode wave by the Fourier inversion of (2.5.60), i.e.,

$$\begin{aligned}S_1(\widehat{\boldsymbol{\Omega}}|\widehat{\boldsymbol{\rho}} - \widehat{\boldsymbol{\rho}}'|\widehat{\boldsymbol{\Omega}}') &= (2\pi)^{-2}\int d\boldsymbol{\lambda}\, f_A(\widehat{\boldsymbol{\Omega}}, \widehat{\boldsymbol{\lambda}}_{\mp1})\overline{f}_A(\widehat{\boldsymbol{\Omega}}', \widehat{\boldsymbol{\lambda}}_{\mp1}) \\ &\quad \times\gamma D^{-1}(2\lambda)^{-1}\exp\left[-i\boldsymbol{\lambda}\cdot(\widehat{\boldsymbol{\rho}} - \widehat{\boldsymbol{\rho}}') - \lambda|z - z'|\right] ,\end{aligned} \tag{2.5.62a}$$

where the signs, $\mp$, are chosen depending on $z - z' \gtrless 0$, respectively. Thus

$$S_1(\widehat{\boldsymbol{\Omega}}|\widehat{\boldsymbol{\rho}} - \widehat{\boldsymbol{\rho}}'|\widehat{\boldsymbol{\Omega}}') = f_A(\widehat{\boldsymbol{\Omega}}, i\overrightarrow{\partial}/\partial\widehat{\boldsymbol{\rho}})S_1(\widehat{\boldsymbol{\rho}} - \widehat{\boldsymbol{\rho}}')\overline{f}_A(\widehat{\boldsymbol{\Omega}}', -i\overleftarrow{\partial}/\partial\widehat{\boldsymbol{\rho}}') , \tag{2.5.62b}$$

where

$$S_1(\hat{\rho}-\hat{\rho}') = \gamma D^{-1}|4\pi(\hat{\rho}-\hat{\rho}')|^{-1} \tag{2.5.63}$$

and $f_A(\hat{\Omega}, \mathrm{i}\partial/\partial\hat{\rho})$ should be understood as a power series of $\mathrm{i}\partial/\partial\hat{\rho}$. It may be noticed here that the diffusion mode wave (2.5.62) is exact, in view of using the exact poles given by (2.5.50), independent of the higher order terms, and that the entire $S_A^{(1)}$, as defined by the Fourier inversion of $\widetilde{S}_A^{(1)}$, is obtained by adding the contribution from (one of) the contour integrals in (2.5.60), say $S_A^{(1,C)}(\hat{\Omega}|\hat{\rho}-\hat{\rho}'|\hat{\Omega}')$, which are short-range functions in view of $\gamma_C \geq \gamma$. Whereas as long as (2.5.50) is used for $A(\hat{\boldsymbol{\lambda}})$, (2.5.6) provides only an approximate equation for S_A, so that the solution should be improved with the higher order terms of $A(\hat{\boldsymbol{\lambda}})$.

Summarizing, from (2.5.55, 60),

$$S(\hat{\Omega}|\hat{\rho}-\hat{\rho}'|\hat{\Omega}') = K(\hat{\Omega}|\hat{\Omega}')\delta(\hat{\rho}-\hat{\rho}') + \sum_A S_A^{(1,C)}(\hat{\Omega}|\hat{\rho}-\hat{\rho}'|\hat{\Omega}')$$
$$+ \sum_\alpha S_\alpha(\hat{\Omega}|\hat{\rho}-\hat{\rho}'|\hat{\Omega}') \ , \tag{2.5.64}$$

where all the terms in the first line are short-range functions. While, by definition (2.3.24),

$$\mathcal{I}(\hat{\Omega}|\hat{\rho}-\hat{\rho}'|\hat{\Omega}')$$
$$= \int d\hat{\rho}''\,d\hat{\rho}'''U(\hat{\Omega},\hat{\rho}-\hat{\rho}'')S(\hat{\Omega}|\hat{\rho}''-\hat{\rho}'''|\hat{\Omega}')U(\hat{\Omega}',\hat{\rho}'''-\hat{\rho}') \ , \tag{2.5.65}$$

wherein each mode term, $\mathcal{I}_\alpha(\hat{\Omega}|\hat{\rho}-\hat{\rho}'|\hat{\Omega}')$, is given in view of (2.5.62, 8), by

$$\mathcal{I}_\alpha(\hat{\Omega}|\hat{\rho}-\hat{\rho}'|\hat{\Omega}') = \phi_A(\hat{\Omega}, \mathrm{i}\overrightarrow{\partial}/\partial\hat{\rho})S_\alpha(\hat{\rho}-\hat{\rho}')\overline{\phi}_A(\hat{\Omega}', -\mathrm{i}\overleftarrow{\partial}/\partial\hat{\rho}') \ . \tag{2.5.66}$$

To investigate the meaning of the mode term in more detail, we rewrite (2.5.54b) as

$$\widetilde{S} = [1 + K\widetilde{U} + \cdots + (K\widetilde{U})^{N-1}]K + \widetilde{S}^{(N)} \ , \tag{2.5.67}$$

with

$$\widetilde{S}^{(N)} = (1-K\widetilde{U})^{-1}(K\widetilde{U})^N K = \sum_A \widetilde{S}_A^{(N)} \ , \tag{2.5.68}$$

$$\widetilde{S}_A^{(N)}(\hat{\Omega}|\hat{\boldsymbol{\lambda}}|\hat{\Omega}') = f_A(\hat{\Omega},\hat{\boldsymbol{\lambda}})(1-A)^{-1}A^{N+1}\overline{f}_A(\hat{\Omega}',\hat{\boldsymbol{\lambda}}) \ . \tag{2.5.69}$$

Here, since $\widetilde{S}_A^{(N)} \sim O\{|\hat{\boldsymbol{\lambda}}|^{-N}\}$ for $|\hat{\boldsymbol{\lambda}}| \sim \infty$, the singular expansion of $\widetilde{S}_A^{(N)}(\hat{\Omega}|\zeta|\hat{\Omega}')$ is possible in the same way that $\widetilde{S}_A^{(1)}(\hat{\Omega}|\zeta|\hat{\Omega}')$ is given by

(2.5.60), simply with the replacement of $\widetilde{S}_A^{(1)}(\widehat{\boldsymbol{\Omega}}|\zeta'|\widehat{\boldsymbol{\Omega}}')$ in the last integrand with $\widetilde{S}_A^{(N)}(\widehat{\boldsymbol{\Omega}}|\zeta'|\widehat{\boldsymbol{\Omega}}')$. Note that the mode terms, given by the first two terms, are exactly the same as those of $\widetilde{S}_A^{(1)}$, and the difference is only in the contour integrals, $S_A^{(N,C)}$, having an additional factor, $A^{N-1}(\zeta')$, in the integrand as given by (2.5.69).

Generally, the residual evaluation of (2.5.60) leads to the integral representation for the mode waves, including the diffusion, as

$$S_\alpha(\widehat{\boldsymbol{\Omega}}|\widehat{\rho}-\widehat{\rho}'|\widehat{\boldsymbol{\Omega}}') = (2\pi)^{-2}\int d\boldsymbol{\lambda}\, f_A(\widehat{\boldsymbol{\Omega}},\widehat{\boldsymbol{\lambda}}_{\mp\alpha})\overline{f}_A(\widehat{\boldsymbol{\Omega}}',\widehat{\boldsymbol{\lambda}}_{\mp\alpha})$$
$$\times[\mp \mathrm{i}\partial A/\partial\zeta]^{-1}_{\mp\alpha}\exp[-\mathrm{i}\boldsymbol{\lambda}\cdot(\boldsymbol{\rho}-\boldsymbol{\rho}')-\varkappa_\alpha|z-z'|] \ , \qquad (2.5.70)$$

where the residue value is $[-\partial A/\partial\zeta]^{-1}_{\mp\alpha}$, depending on $z-z' \gtrless 0$, respectively. Here, as is proved below,

$$-\mathrm{i}\partial A/\partial\zeta\,\big|_{\mp\alpha} = (\overline{\phi}_{\mp\alpha}\Omega_z\phi_{\mp\alpha}) = -(\overline{\phi}_{\pm\alpha}\Omega_z\phi_{\pm\alpha}) \ , \qquad (2.5.71)$$

where use has been made of the notation,

$$\phi_{\pm\alpha}(\widehat{\boldsymbol{\Omega}}) = \phi_A(\widehat{\boldsymbol{\Omega}},\widehat{\boldsymbol{\lambda}}_{\pm\alpha}), \qquad \overline{\phi}_{\pm\alpha}(\widehat{\boldsymbol{\Omega}}) = \overline{\phi}_A(\widehat{\boldsymbol{\Omega}},\widehat{\boldsymbol{\lambda}}_{\pm\alpha}) \ . \qquad (2.5.72)$$

Hence, the mode wave terms of (2.5.66) are given specifically by

$$\mathcal{I}_\alpha(\widehat{\boldsymbol{\Omega}}|\widehat{\rho}-\widehat{\rho}'|\widehat{\boldsymbol{\Omega}}') = (2\pi)^{-2}\int d\boldsymbol{\lambda}\,(\overline{\phi}_{-\alpha}\Omega_z\phi_{-\alpha})^{-1}$$
$$\times\phi_{\mp\alpha}(\widehat{\boldsymbol{\Omega}})\overline{\phi}_{\mp\alpha}(\widehat{\boldsymbol{\Omega}}')\exp[-\mathrm{i}\boldsymbol{\lambda}\cdot(\boldsymbol{\rho}-\boldsymbol{\rho}')-\varkappa_\alpha|z-z'|] \ , \qquad (2.5.73)$$

$z-z' \gtrless 0$. The relation (2.5.71) is proved by writing (2.5.25) in matrix form as

$$(K - A\widetilde{U}^{-1})\phi_A = 0, \qquad (2.5.74a)$$
$$\overline{\phi}_A(K - A\widetilde{U}^{-1}) = 0, \quad \widetilde{U}^{-1} = \gamma - \mathrm{i}\widehat{\boldsymbol{\Omega}}\cdot\widehat{\boldsymbol{\lambda}} \ , \qquad (2.5.74b)$$

so that

$$\frac{\partial}{\partial\zeta}[\overline{\phi}_A(K - A\widetilde{U}^{-1})\phi_A]\,\big|_\alpha = 0$$

is reduced to

$$[\overline{\phi}_\alpha(-\partial A/\partial\zeta\,\big|_\alpha\,\widetilde{U}_\alpha^{-1} + \mathrm{i}\Omega_z)\phi_\alpha] = 0 \ ,$$

yielding (2.5.71), as a consequence of

$$(\overline{\phi}_\alpha\widetilde{U}_\alpha^{-1}\phi_\alpha) = (\overline{\phi}_\alpha f_\alpha) = (\overline{\phi}_A f_A)\,\big|_\alpha = 1 \ , \qquad (2.5.75)$$

and with $\partial A/\partial\zeta$ being an odd function of ζ.

2.5.4 Orthogonality and Power Carried by the Mode Waves

The ϕ_α's are orthogonal, i.e.,

$$(\overline{\phi}_\alpha \Omega_z \phi_\beta) = 0, \qquad \alpha \neq \beta \ . \tag{2.5.76}$$

The proof is given by using the equations,

$$(K - \widetilde{U}_\beta^{-1})\phi_\beta = \overline{\phi}_\alpha (K - \widetilde{U}_\alpha^{-1}) = 0 \tag{2.5.77}$$

from (2.5.74), which shows that, with $\widetilde{U}_\alpha = \widetilde{U} \mid_\alpha$,

$$(\overline{\phi}_\alpha K \phi_\beta) = (\overline{\phi}_\alpha \widetilde{U}_\beta^{-1} \phi_\beta) = (\overline{\phi}_\alpha \widetilde{U}_\alpha^{-1} \phi_\beta) \ , \tag{2.5.78a}$$

and hence,

$$(\varkappa_\alpha - \varkappa_\beta)(\overline{\phi}_\alpha \Omega_z \phi_\beta) = 0 \ , \tag{2.5.78b}$$

in view of $\widetilde{U}^{-1}$ as given in (2.5.74b). Equaitons (2.5.77) indicate that the ϕ_α's are solutions of the homogeneous transport equation (2.3.46).

Here, by (2.5.37),

$$\overline{\phi}_{\alpha=1} \mid_{\boldsymbol{\lambda}=0} = \overline{\phi}_{a=1} \ , \tag{2.5.79}$$

hence, the orthogonality condition (2.5.76) shows that,

$$(\overline{\phi}_1 \Omega_z \phi_{\beta\neq 1}) \mid_{\boldsymbol{\lambda}=0} = 0, \qquad \varkappa_\beta \neq 0 \ , \tag{2.5.80}$$

in consistent with the result from (2.5.15) when $\boldsymbol{\lambda} = 0$,

$$(p_{\alpha\neq 1})_z \mid_{\boldsymbol{\lambda}=0} = 0, \qquad A \mid_\alpha = 1 \ . \tag{2.5.81}$$

Hence, from (2.5.17a), the $\boldsymbol{\rho}$ integrated power of each mode wave of $\alpha \neq 1$ is found to be zero,

$$\begin{aligned} \int d\boldsymbol{\rho} (W_\alpha)_z &= (W_A)_z \mid_{\alpha,\boldsymbol{\lambda}=0} \\ &= \int d\boldsymbol{\rho} \int d\widehat{\Omega}\, \Omega_z \mathcal{I}_\alpha(\widehat{\Omega}|\widehat{\boldsymbol{\rho}} - \widehat{\boldsymbol{\rho}}'|\widehat{\Omega}') = 0, \qquad \alpha \neq 1 \ , \end{aligned} \tag{2.5.82}$$

and confirms that all the mode waves, other than the diffusion, do not propagate the power to the infinity.

Equation (2.5.82) can be generalized, by using (2.5.73) directly, to the case of including the diffusion mode wave, as

$$\begin{aligned} &\int d\boldsymbol{\rho} \int d\widehat{\Omega}\, \Omega_z \mathcal{I}_\alpha(\widehat{\Omega}|\widehat{\boldsymbol{\rho}} - \widehat{\boldsymbol{\rho}}'|\widehat{\Omega}') \\ &= 4\pi \left[(\overline{\phi}_1 \Omega_z \phi_{\mp\alpha})(\overline{\phi}_{-\alpha} \Omega_z \phi_{-\alpha})^{-1} \overline{\phi}_{\mp\alpha}(\widehat{\Omega}') \right]_{\boldsymbol{\lambda}=0} = \pm \frac{1}{2} \delta_{1\alpha} \end{aligned} \tag{2.5.83a}$$

for $z - z' \gtrless 0$, respectively, in view of (2.5.29a), and

$$(\overline{\phi}_1 \Omega_z \phi_{-1}) \mid_{\alpha\sim 0} = \frac{1}{2}(\overline{\phi}_{-1} \Omega_z \phi_{-1}) \mid_{\alpha\sim 0} \sim O[|\hat{\boldsymbol{\lambda}}|] \qquad (2.5.83b)$$

from (2.5.37). Thus the total integrated power due to the diffusion term, $\mathcal{I}_{\alpha=1}$, out of two boundary planes enclosing a source inside, is exactly what is given by the volume integration of the entire power equation (2.5.20) over the same enclosed space, under the condition that the boundaries are separated enough from the source at z', such that $|z - z'| \gg \gamma^{-1}$. Here, the disagreement in the case of the integration over $|z - z'| \lesssim \gamma^{-1}$ should be corrected by the contribution from the other $\mathcal{I}_{\alpha\neq 1}$'s and $\mathcal{I}_A^{(C)}$ terms, which are short-range functions.

3. Random Rough Boundaries

To investigate boundary-value problems of a wave in a random medium (Sect. 4), we prepare now basic equations of both untransmissible (one-sided) and transmissible (two-sided) rough boundaries, and show that they can be obtained in the same form as those in a random medium.

3.1 Rough Surface (One-Sided Boundary)

A rough surface, S, is assumed to be planar on the average, and the coordinate vector in three-dimensional space is denoted by $\widehat{\boldsymbol{x}} = (x_1, x_2, x_3) = (\boldsymbol{\rho}, z)$, with the horizontal being $\boldsymbol{\rho} = (x_1, x_2)$, and the vertical, $z = x_3$, taken in the direction normal to the average surface, as illustrated in Fig. 3.1. The differential operator, $\partial/\partial\widehat{\boldsymbol{x}}$, will often be denoted by $(\partial_\rho, \partial_z)$, with $\partial_\rho = \partial/\partial\boldsymbol{\rho}$ and $\partial_z = \partial/\partial z$. The same scalar wave, $\psi(\widehat{\boldsymbol{x}})$, as in Chap. 2 is assumed so that the wave equation is given by (2.1.7), or, by suppressing the $\boldsymbol{\rho}$ coordinates,

$$(-\partial_z^2 - h^2)\psi(z) = j(z) \ , \tag{3.1.1}$$

with a $\boldsymbol{\rho}$ operator h defined by

$$h = (k_0^2 + \partial_\rho^2)^{1/2} \ . \tag{3.1.2}$$

Here, $\mathrm{Im}\,\{h\} < 0$, in terms of the eigenvalues given by the Fourier transform, to be given by (3.1.8).

In free space, the solution, say $\psi_0(z)$, is therefore given by

$$\psi_0(z) = \int dz'\, G^{(0)}(z - z')j(z') \ , \tag{3.1.3}$$

$$G^{(0)}(z) = (2\mathrm{i}h)^{-1} \exp(-\mathrm{i}h|z|) \ , \tag{3.1.4}$$

in exactly the same form as the corresponding solution in one-dimensional space. Here, the full coordinate expression, $\psi_0(\widehat{\boldsymbol{x}}) \equiv \psi_0(\boldsymbol{\rho}, z)$ of $\psi_0(z)$, is obtained therefrom as

$$\psi_0(\widehat{\boldsymbol{x}}) = \int d\widehat{\boldsymbol{x}}'\, G^{(0)}(\widehat{\boldsymbol{x}} - \widehat{\boldsymbol{x}}')j(\widehat{\boldsymbol{x}}') \ , \tag{3.1.5}$$

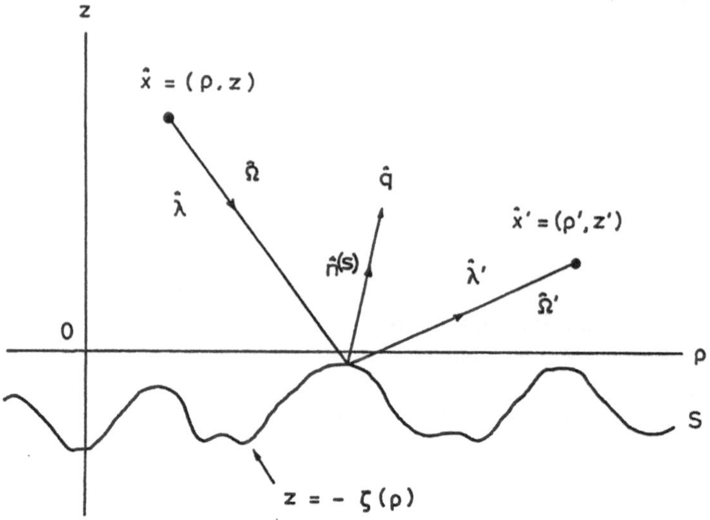

Fig. 3.1. Surface profile and notations for (3.1.11, 81)

in terms of the three-dimensional Green's function given as matrix elements of the operator, $G^{(0)}(z)$, by

$$\begin{aligned} G^{(0)}(\boldsymbol{\rho}-\boldsymbol{\rho}', z-z') &= G^{(0)}(z-z')\delta(\boldsymbol{\rho}-\boldsymbol{\rho}') \\ &= (2\pi)^{-2}\int d\boldsymbol{\lambda}\,\left[2\mathrm{i}\widetilde{h}(\boldsymbol{\lambda})\right]^{-1}\exp\left\{-\mathrm{i}\left[\widetilde{h}(\boldsymbol{\lambda})|z-z'|+\boldsymbol{\lambda}\cdot(\boldsymbol{\rho}-\boldsymbol{\rho}')\right]\right\} \quad (3.1.6) \\ &= \left|4\pi(\widehat{\boldsymbol{x}}-\widehat{\boldsymbol{x}}')\right|^{-1}\exp(-\mathrm{i}k_0|\widehat{\boldsymbol{x}}-\widehat{\boldsymbol{x}}'|)\ , \quad (3.1.7) \end{aligned}$$

where

$$\widetilde{h}(\boldsymbol{\lambda}) = \begin{cases} (k_0^2-\boldsymbol{\lambda}^2)^{1/2}, & |\boldsymbol{\lambda}| \le k_0 \\ -\mathrm{i}(\boldsymbol{\lambda}^2-k_0^2)^{1/2}, & |\boldsymbol{\lambda}| > k_0\ . \end{cases} \quad (3.1.8)$$

Generally, any $\boldsymbol{\rho}$ operator, say, $A(\boldsymbol{\rho}, \frac{\partial}{\partial \boldsymbol{\rho}})$, can be represented by a $\boldsymbol{\rho}$-coordinate matrix with the matrix elements, $A(\boldsymbol{\rho}|\boldsymbol{\rho}')$, defined by

$$\begin{aligned} A(\boldsymbol{\rho}|\boldsymbol{\rho}') &= A\delta(\boldsymbol{\rho}-\boldsymbol{\rho}') \\ &= (2\pi)^{-2}\int_{-\infty}^{\infty} d\boldsymbol{\lambda}\, A\exp\left[-\mathrm{i}\boldsymbol{\lambda}\cdot(\boldsymbol{\rho}-\boldsymbol{\rho}')\right]\ , \end{aligned} \quad (3.1.9)$$

(where A affects only $\boldsymbol{\rho}$ and not $\boldsymbol{\rho}'$), so that it is subject to the multiplication rule,

$$AB(\boldsymbol{\rho}|\boldsymbol{\rho}') = \int d\boldsymbol{\rho}''A(\boldsymbol{\rho}|\boldsymbol{\rho}'')B(\boldsymbol{\rho}''|\boldsymbol{\rho}')\ . \quad (3.1.10)$$

The entire rough surface, S, is assumed to be described by $z = -\zeta(\boldsymbol{\rho}) < 0$ (Fig. 3.1), and also to be perfectly non-dissipative for the time being. Here, as will be seen in Sect. 3.1.5, it is not necessary for the boundary condition to be given on the real boundary, but it can be transferred onto a reference boundary plane chosen, say, at $z = 0$ in the form [3.1],

$$-\partial_z \psi \big|_{z=0} = B\psi \big|_{z=0}, \qquad B = B_0 + b \ , \tag{3.1.11a}$$

with a $\boldsymbol{\rho}$ operator, B, or, written as a $\boldsymbol{\rho}$ matrix equation according to (3.1.10),

$$-\partial_z \psi(\boldsymbol{\rho}, z) \big|_{z=0} = \int d\boldsymbol{\rho}' \, B(\boldsymbol{\rho}|\boldsymbol{\rho}')\psi(\boldsymbol{\rho}') \ , \tag{3.1.11b}$$

where $\psi(\boldsymbol{\rho})$ denotes $\psi(\widehat{\boldsymbol{x}})$ bounded on the reference plane, $z = 0$. Here, B_0 is the term when the boundary is a smooth plane at $z = 0$, whereas b is the change necessary to meet the real boundary S; and, throughout Sect. 3.1, the boundary equation on S is assumed to be

$$-\widehat{\boldsymbol{n}}^{(S)} \cdot \frac{\partial}{\partial \widehat{\boldsymbol{x}}} \psi \big|_S = B_0 \psi \big|_S \ , \tag{3.1.12}$$

with a positive constant, B_0. Here, $\widehat{\boldsymbol{n}}^{(S)}$ is the unit vector directed normal to S, and, generally speaking, B_0 is also ζ-dependent, and is an operator of two-dimensional coordinates $\boldsymbol{s}$ on S, so that the right-hand side can be written in the form,

$$\int d\boldsymbol{s}' \, B_0(\boldsymbol{s}|\boldsymbol{s}')\psi(\boldsymbol{s}') \ , \qquad B_0(\boldsymbol{s}|\boldsymbol{s}') \sim B_0(\boldsymbol{s} - \boldsymbol{s}') \ , \tag{3.1.13}$$

in the same fashion as in (3.1.11b). The matrix, B, is hereafter referred to as the surface impedance.

In Sect. 3.1.5, an explicit expression of B is obtained in an operator form for arbitrarily given $\zeta(\boldsymbol{\rho})$, and shows that, to the first order in ζ,

$$b = \partial_\rho \cdot [\zeta(\boldsymbol{\rho})\partial_\rho] + (k_0^2 + B_0^2)\zeta(\boldsymbol{\rho}) \ . \tag{3.1.14}$$

Here, ∂_ρ affects all the factors to be placed on its right-hand side. In [3.2], the corresponding b was given by

$$b = \left[\partial\zeta(\boldsymbol{\rho})/\partial\boldsymbol{\rho}\right] \cdot \partial_\rho - \zeta(\boldsymbol{\rho})\partial_z^2 - B_0\zeta(\boldsymbol{\rho})\partial_z \ ,$$

which agrees with (3.1.14) by the replacement of $-\partial_z^2 \to k_0^2 + \partial_\rho^2$ and $-\partial_z \to B_0$, showing the equivalence of the two expessions.

Here, we introduce a particular solution, $\psi^{(1)}(z)$, of the wave equation (3.1.1) subjected to (3.1.11), by writing it in the form,

$$\psi^{(1)}(z) = \left[\exp(ihz) - \exp(-ihz)\right](2ih)^{-1} + \exp(-ihz)g_b \ . \tag{3.1.15}$$

Here, the coefficient, g_b, is also a ρ operator, depending on both ρ and $\partial/\partial\rho$, and is therefore not commutable with h; the order of the factors in the equation is very important. Note that $\psi^{(1)} \to g_b$ as $z \to 0$. By the substitution in the boundary equation (3.1.11a), g_b is found to be the solution of

$$-1 + \mathrm{i}hg_b = Bg_b \ ,$$

or,

$$(\mathrm{i}h - B)g_b = 1, \qquad B = B_0 + b \ , \tag{3.1.16a}$$

which has the same form as the governing equation for the Green's function in a medium, $B = B_0 + b$, with the random part, b. Hereafter, g_b will be referred to as the surface Green's function because it gives the boundary-value of a three-dimensional Green's function of the wave equation (3.1.1), subjected to boundary condition (3.1.11). $g_b(z|z')$ is given in view of (3.1.4) by

$$g_b(z|z') = \Big\{\exp(-\mathrm{i}h|z-z'|) - \exp\big[-\mathrm{i}h(z+z')\big]\Big\}(2\mathrm{i}h)^{-1}$$
$$+\exp(-\mathrm{i}hz)g_b\exp(-\mathrm{i}hz') \ , \tag{3.1.16b}$$

$$-\partial_z g_b(z|z')\,\big|_{z=0} = Bg_b(z=0|z') \ . \tag{3.1.16c}$$

The power flux vector, $\widehat{\boldsymbol{w}} = (\boldsymbol{w}, w_z)$, is defined by

$$\widehat{\boldsymbol{w}}(\widehat{\boldsymbol{x}}) = \frac{1}{2\mathrm{i}}\psi^*(\widehat{\boldsymbol{x}})\Big(\frac{\overleftarrow{\partial}}{\partial\widehat{\boldsymbol{x}}} - \frac{\overrightarrow{\partial}}{\partial\widehat{\boldsymbol{x}}}\Big)\psi(\widehat{\boldsymbol{x}}) \ , \tag{3.1.17}$$

with the equation of continuity,

$$\frac{\partial}{\partial\widehat{\boldsymbol{x}}}\cdot\widehat{\boldsymbol{w}}(\widehat{\boldsymbol{x}}) = \frac{1}{2\mathrm{i}}\left[\psi^* j(\widehat{\boldsymbol{x}}) - \psi j^*(\widehat{\boldsymbol{x}})\right] \ . \tag{3.1.18}$$

Here the vertical component at the boundary, $w_z(\boldsymbol{\rho})$, can be expressed in terms of the matrix, B, as

$$w_z(\boldsymbol{\rho}) = (2\mathrm{i})^{-1}\psi^*(\widehat{\boldsymbol{x}})(\overleftarrow{\partial}_z - \overrightarrow{\partial}_z)\psi(\widehat{\boldsymbol{x}})\,\big|_{z=0} \tag{3.1.19a}$$
$$= (2\mathrm{i})^{-1}\left[\psi^*(\boldsymbol{\rho})(B\psi)(\boldsymbol{\rho}) - \psi(\boldsymbol{\rho})(B^*\psi^*)(\boldsymbol{\rho})\right] \ . \tag{3.1.19b}$$

Hence, the total integrated power out of the boundary is

$$\int d\boldsymbol{\rho}\, w_z(\boldsymbol{\rho}) = (2\mathrm{i})^{-1}\int d\boldsymbol{\rho}\int d\boldsymbol{\rho}'$$
$$\times\left[B(\boldsymbol{\rho}|\boldsymbol{\rho}') - B^*(\boldsymbol{\rho}'|\boldsymbol{\rho})\right]\psi^*(\boldsymbol{\rho})\psi(\boldsymbol{\rho}') \ , \tag{3.1.20a}$$

which should be zero for an arbitrary solution, ψ, showing, therefore, that

$$B^{\dagger}(\rho|\rho') \equiv B^{*}(\rho'|\rho) = B(\rho|\rho') \ , \qquad (3.1.20b)$$

or, $B^{\dagger} = B$, i.e., the matrix, B, is Hermitian whenever the boundary is non-dissipative. In fact, b, given by (3.1.14) to the first order in ζ, is exactly Hermitian. Thus the situation becomes the same as in the case of a non-dissipative random medium, $q = q^{\dagger}$ (Chap.2).

Futhermore, B is a symmetrical matrix, as will be seen, so that symmetries to which B and g_b are subjected in accord with (3.1.16a) are summarized as:

$$B^T = B = B^{*}, \qquad g_b^T = g_b \ , \qquad (3.1.21)$$

wherein $B = B^{*}$ holds only when the system is non-dissipative. The proof for $B^T = B$ is given by applying the Green's theorem to the space enclosed by the plane, $z = 0$, and the boundary, S, with arbitrary solutions, ψ' and ψ'', subject to (3.1.12). Therein the surface integral on S becomes zero, so that the integral on the plane, $z = 0$, should also be zero, i.e.,

$$\begin{aligned}&\int d\rho\, \psi'(\overleftarrow{\partial}_z - \overrightarrow{\partial}_z)\psi'' \big|_{z=0}\\ &= \int d\rho[\psi'(B\psi'') - (\psi' B^T)\psi''](\rho) = 0 \ ,\end{aligned}$$

showing that $B^T = B$.

In the case of $b = 0$, g_b becomes the "free surface" Green's function, g_0, with its z-continued version, $g_0(z)$, given by

$$g_0(z) = g_0 \exp(-ihz), \qquad g_0 = (ih - B_0)^{-1} \ , \qquad (3.1.22)$$

which should not be confused with the free space Green's function, $G^{(0)}(z)$, defined by (3.1.4). Here, the full coordinate expression of $g_0(z)$, is given in the same fashion as in (3.1.7), by

$$\begin{aligned}g_0(\rho - \rho', z) = (2\pi)^{-2} \int d\boldsymbol{\lambda} \big[i\widetilde{h}(\boldsymbol{\lambda}) - \widetilde{B}_0(\boldsymbol{\lambda})\big]^{-1}\\ \times \exp\Big\{-i\big[\widetilde{h}(\boldsymbol{\lambda})z + \boldsymbol{\lambda}\cdot(\rho - \rho')\big]\Big\} \ , \qquad (3.1.23)\end{aligned}$$

whose integral has been investigated in detail in connection with the ground wave propagation over a flat earth [3.3]. The result is written in the form,

$$g_0(\rho, z) = (2\pi r)^{-1} \exp(-ik_0 r) A(\rho, z) \ . \qquad (3.1.24)$$

Here, $r = (\rho^2 + z^2)^{1/2}$ and the factor, $A(\rho, z)$, (meaning an attenuation coefficient relative to the double times free space value) is given, when $|k_0\rho| \gg 1$ and B_0 is constant, by

$$A(\rho, z) = 1 - \pi^{1/2}\xi w(\xi + \eta) \ , \tag{3.1.25}$$

where

$$\xi = -B_0(r/2k_0)^{1/2}e^{-i\pi/4} \ , \tag{3.1.26a}$$

$$\eta = z(k_0/2r)^{1/2}e^{i\pi/4} \ , \tag{3.1.26b}$$

$$w(x) = \frac{2}{\sqrt{\pi}}e^{x^2}\int_x^\infty du\, e^{-u^2} \ . \tag{3.1.27}$$

Here when $|\xi| \ll 1$, $A \simeq 1$; whereas, when $|\xi| \gg 1$ and $|\arg(\xi)| < 3\pi/4$, the factor, A, tends to the asymptotic expression,

$$A(\rho, z) \sim \begin{cases} (2\xi^2)^{-1}, & |\eta| \ll |2\xi|^{-1} \quad (3.1.28a) \\ \frac{z/r}{z/r + iB_0/k_0}, & |\eta| \gg |2\xi|^{-1} \ . \quad (3.1.28b) \end{cases}$$

Here, with the aid of the relation,

$$w(x) = 2\exp(x^2) - w(-x) \ , \tag{3.1.29}$$

(3.1.25) can be rewritten in another form,

$$A(\rho, z) = 1 + \pi^{1/2}\xi w(-\xi - \eta) + A_s(\rho, z) \ , \tag{3.1.30}$$

with

$$A_s(\rho, z) = -2\pi^{1/2}\xi \exp\left\{-B_0 z - i\left[B_0^2(r/2k_0) - k_0 z^2(2r)^{-1}\right]\right\} \ , \tag{3.1.31}$$

Here, when $B_0 > 0$, the term, $A_s(\rho, z)$, decreases exponentially with height, z, from the surface, and its contribution to $g_0(\rho, z)$ decreases with the horizontal distance by a factor, $|\rho|^{-1/2}$, having, therefore, all the characteristics of the conventional surface wave. In fact, in the region of $|\xi| \gg 1$, the first two terms on the right-hand side of (3.1.30) are combined to be given by the asymptotic expression similar to (3.1.28a), letting the surface wave term, $A_s(\rho, z)$, be dominant within the range, $B_0 z \lesssim 1$, $B_0 > 0$.

The surface Green's function for the coherent wave, $G(\rho - \rho') = \langle g_b(\rho|\rho') \rangle$, is also provided by (3.1.24–31) with the replacement of B_0 by $B_0 + \widetilde{M}$ (Sect. 3.2). Here, $\widetilde{M}$ is an effective value of the operator, b, and is very close to a real positive value when $B_0 = 0$ and the correlation distance of the surface change is small enough compared with the wavelength, showing that the coherent wave supports the conventional surface wave, which is dominant only in the region $|\xi| \gg 1$ and $\mathrm{Re}\{\widetilde{M}\}z \lesssim 1$.

3.1.1 Reflection Coefficient

A reflection coefficient matrix, R, is defined here by setting

$$\psi^{(1)}(z) = \left[\exp(\mathrm{i}hz) + \exp(-\mathrm{i}hz)R\right](2\mathrm{i}h)^{-1} \ , \tag{3.1.32a}$$

for $\psi^{(1)}(z)$ of (3.1.15), so that $g_b(z|z')$ of (3.1.16b) can be written as

$$\begin{aligned} g_b(z|z') = &\exp(-\mathrm{i}h|z-z'|)(2\mathrm{i}h)^{-1} \\ &+ \exp(-\mathrm{i}hz)R(2\mathrm{i}h)^{-1}\exp(-\mathrm{i}hz') \ . \end{aligned} \tag{3.1.32b}$$

Hence,

$$R = -1 + 2\mathrm{i}g_b h \ , \tag{3.1.33a}$$

$$g_b = (1+R)(2\mathrm{i}h)^{-1} \ , \tag{3.1.33b}$$

which, upon substitution of g_b from (3.1.16a), gives

$$R = (\mathrm{i}h - B)^{-1}(\mathrm{i}h + B) \ , \tag{3.1.34a}$$

$$B = \mathrm{i}h(R-1)(R+1)^{-1}, \qquad B = B_0 + b \ . \tag{3.1.34b}$$

Here, since h and B are non-commutable operators, the order of the factors in (3.1.32–34) are again important, although (3.1.34) has the same form as when the surface is smooth.

From (3.1.34a),

$$R^{-1} = (\mathrm{i}h + B)^{-1}(\mathrm{i}h - B) \ , \tag{3.1.35a}$$

$$R^* = (\mathrm{i}h^* + B)^{-1}(\mathrm{i}h^* - B), \qquad B^* = B \ , \tag{3.1.35b}$$

showing that, when the non-optical range (wherein $h^* \neq h$) is negligible,

$$R^{-1} = R^*, \qquad h^* = h \ , \tag{3.1.35c}$$

which is also consistent with (3.1.34b) for B [a specific aspect is given by (3.1.72)]. Note that $R^T \neq R$; instead, a scattering matrix, T, defined by

$$T = 2\mathrm{i}hR = T^T \ , \tag{3.1.36a}$$

satisfies the reciprocity in view of (3.1.33a, 21), and enables $g_b(z|z')$ of (3.1.32b) to be written as

$$g_b(z|z') = G^{(0)}(z-z') + G^{(0)}(z)TG^{(0)}(-z') \ , \tag{3.1.36b}$$

in terms of the free space Green's function, $G^{(0)}(z)$, of (3.1.4).

3.1.2 Scattering Matrix of a Small Boss on a Smooth Boundary

In the boundary equation (3.1.11a), $b = B - B_0$ is nonzero only in the region $\zeta(\rho) \neq 0$, and, therefore, when $\zeta(\rho)$ is nonzero in only a small area of ρ, the term b involved in (3.1.16a) for g_b can be treated in the same way as the corresponding term of a small scatterer in three-dimensional space (Sect. 2.1).

Hence, in terms of the free surface Green's function, g_0, the solution, g_b, of (3.1.16) can be written in the form,

$$g_b = g_0 + g_0 T_b g_0 \ . \tag{3.1.37}$$

Here, T_b is a scattering matrix of b, defined by

$$b g_b = T_b g_0 \ , \tag{3.1.38a}$$

and given by [cf. (2.2.9)]

$$T_b = b(1 + g_0 T_b) = (1 - b g_0)^{-1} b \tag{3.1.38b}$$

$$= b + b g_0 b + b g_0 b g_0 b + \dots \ . \tag{3.1.38c}$$

Equation (3.1.37) is explicitly written by

$$g_b(\rho|\rho') = g_0(\rho - \rho') + \int d\rho'' d\rho''' g_0(\rho - \rho'') \times T_b(\rho'' - \rho_b|\rho''' - \rho_b) g_0(\rho''' - \rho') \ . \tag{3.1.39}$$

Here, ρ_b is the center coordinates of the scatterer, b, and $g_0(\rho) \equiv g_0(\rho, z{=}0)$ is given by (3.1.24–28) when $k_0|\rho| \gg 1$, which shows that the change of $g_0(\rho)$ is mostly due to the phase factor, $\exp(-\mathrm{i}k_0|\rho|)$. Hence, in the region where the points ρ and ρ' are both separated far enough from the scatterer so that $|k_0(\rho - \rho_b)|, |k_0(\rho_b - \rho')| \gg 1$, but their ξ-values, by (3.1.26), are small enough so that $A(\rho - \rho_b) \simeq A(\rho_b - \rho') \simeq 1$, the expression (3.1.39) can be written in the asymptotic form,

$$g_b(\rho|\rho') \sim g_0(\rho - \rho') + |\rho - \rho_b|^{-1} a(\Omega|\Omega') g_0(\rho_b - \rho') \ . \tag{3.1.40}$$

Here, in terms of the Fourier transform, $\tilde{T}_b$, defined by

$$\tilde{T}_b(\boldsymbol{\lambda}|\boldsymbol{\lambda}') = \int d\rho\, d\rho' \exp[\mathrm{i}(\boldsymbol{\lambda} \cdot \rho - \boldsymbol{\lambda}' \cdot \rho')] T_b(\rho|\rho') \ , \tag{3.1.41}$$

the factor, $a(\Omega|\Omega')$, is given by

$$a(\Omega|\Omega') = (2\pi)^{-1} \tilde{T}_b(k_0\Omega|k_0\Omega') \ , \tag{3.1.42}$$

where $\boldsymbol{\Omega}$ and $\boldsymbol{\Omega}'$ are unit vectors in the directions of $\boldsymbol{\rho} - \boldsymbol{\rho}_b$ and $\boldsymbol{\rho}_b - \boldsymbol{\rho}'$, respectively. Equation (3.1.41) shows that $a(\boldsymbol{\Omega}|\boldsymbol{\Omega}')$ gives a scattering amplitude of the boss, b, in the same sense as that of a scatterer in free space, and may be experimentally estimated for a boss of complicated shape. In case of semi-spherical bosses, a specific investigation was made in [3.4].

The above treatment can be extended to the more general case where many small bosses are independently distributed over the plane, $z = 0$, so that

$$b = \sum_a b_a \ , \tag{3.1.43}$$

where b_a denotes the contribution from the a-th boss. The situation is again the same as when discrete scatterers are randomly distributed in space, and, in Sect. 3.2.1, the result will be briefly summarized by using the previous method of Sect. 2.2.

3.1.3 Optical Conditions

The vertical component of power flux for the solution, $\psi^{(1)}(z)$, of (3.1.15), say, $w_z^{(1)}(\boldsymbol{\rho})$ on the plane $z = 0$, is given, according to (3.1.19a), by

$$\begin{aligned} w_z^{(1)}(\boldsymbol{\rho}) = {} & 2^{-1}(h^* + h) g_b^*(\boldsymbol{\rho}|\boldsymbol{\rho}_1) g_b(\boldsymbol{\rho}|\boldsymbol{\rho}_2) \\ & - (2\mathrm{i})^{-1}\left[g_b^*(\boldsymbol{\rho}|\boldsymbol{\rho}_1) - g_b(\boldsymbol{\rho}|\boldsymbol{\rho}_2)\right] \ , \end{aligned} \tag{3.1.44a}$$

in terms of the surface Green's functions g_b^* and g_b. Here, h^* and h are independent operators operating only on g_b^* and g_b, respectively; $\boldsymbol{\rho}_1$ and $\boldsymbol{\rho}_2$ are arbitrary coordinates which are not necessarily equal to each other when regarded as the matrix coordinates. With a $\boldsymbol{\rho}$ matrix, $\delta(\boldsymbol{\rho}|1;2)$, similar to $\delta(\widehat{\boldsymbol{x}}|1;2)$ of (2.3.14), (3.1.44a) can be written as

$$\begin{aligned} w_z^{(1)}(\boldsymbol{\rho}) = \delta(\boldsymbol{\rho}|1;2)\Big\{ & 2^{-1}\left[h^*(1) + h(2)\right] g_b^*(1) g_b(2) \\ & - (2\mathrm{i})^{-1}\left[g_b^*(1) - g_b(2)\right]\Big\} \ . \end{aligned} \tag{3.1.44b}$$

On the other hand, the condition (3.1.20b) that B be Hermitian is expressed by

$$B = \mathrm{i}h - g_b^{-1} = -\mathrm{i}h^\dagger - (g_b^\dagger)^{-1} \ , \tag{3.1.45}$$

which, by multiplication of both sides, to the left with $(2\mathrm{i})^{-1} g_b^\dagger$, and to the right with g_b, leads to

$$2^{-1} g_b^\dagger (h^\dagger + h) g_b - (2\mathrm{i})^{-1}(g_b^\dagger - g_b) = 0 \ , \tag{3.1.46a}$$

which shows, from (3.1.44b), that

$$w_z^{(1)} \equiv \int d\rho\, w_z^{(1)}(\rho) = 0 \ . \tag{3.1.46b}$$

being a reproduction of (3.1.20a,b).

For convenience, we prepare the Fourier transform of (3.1.46a) according to definition (3.1.41), with the aid of the multiplication rule,

$$\widetilde{A}\widetilde{B}(\boldsymbol{\lambda}|\boldsymbol{\lambda}') = (2\pi)^{-2} \int d\boldsymbol{\lambda}''\, \widetilde{A}(\boldsymbol{\lambda}|\boldsymbol{\lambda}'')\widetilde{B}(\boldsymbol{\lambda}''|\boldsymbol{\lambda}') \ . \tag{3.1.47}$$

Hence,

$$\begin{aligned}(2\pi)^{-2} \int_{|\boldsymbol{\lambda}|\le k_0} & d\boldsymbol{\lambda}\, \widetilde{h}(\boldsymbol{\lambda})\widetilde{g}_b^*(\boldsymbol{\lambda}|\boldsymbol{\lambda}_1)\widetilde{g}_b(\boldsymbol{\lambda}|\boldsymbol{\lambda}_2) \\ &= (2\mathrm{i})^{-1}\left[\widetilde{g}_b^*(\boldsymbol{\lambda}_2|\boldsymbol{\lambda}_1) - \widetilde{g}_b(\boldsymbol{\lambda}_1|\boldsymbol{\lambda}_2)\right] \ ,\end{aligned} \tag{3.1.48}$$

where the integration is made only over the optical range, $|\boldsymbol{\lambda}| \le k_0$, in consequence of

$$2^{-1}\left[\widetilde{h}^*(\boldsymbol{\lambda}) + \widetilde{h}(\boldsymbol{\lambda})\right] = \begin{cases} (k_0^2 - \boldsymbol{\lambda}^2)^{1/2}, & |\boldsymbol{\lambda}| \le k_0 \\ 0, & |\boldsymbol{\lambda}| > k_0 \end{cases} \ , \tag{3.1.49}$$

from (3.1.8). The relation can be rewritten in terms of the reflection coefficient, R, by using (3.1.33b), with the result that

$$\begin{aligned}(2\pi)^{-2} \int_{|\boldsymbol{\lambda}|\le k_0} & d\boldsymbol{\lambda}\, \widetilde{h}(\boldsymbol{\lambda})\widetilde{R}^*(\boldsymbol{\lambda}|\boldsymbol{\lambda}_1)\widetilde{R}(\boldsymbol{\lambda}|\boldsymbol{\lambda}_2) \\ &= 2^{-1}(\widetilde{h}^* + \widetilde{h})(\boldsymbol{\lambda}_1)(2\pi)^2\delta(\boldsymbol{\lambda}_1 - \boldsymbol{\lambda}_2) \\ &\quad + 2^{-1}(\widetilde{h}^* - \widetilde{h})(\boldsymbol{\lambda}_1)\widetilde{R}(\boldsymbol{\lambda}_1|\boldsymbol{\lambda}_2) \\ &\quad - 2^{-1}(\widetilde{h}^* - \widetilde{h})(\boldsymbol{\lambda}_2)\widetilde{R}^*(\boldsymbol{\lambda}_2|\boldsymbol{\lambda}_1) \ .\end{aligned} \tag{3.1.50}$$

Here, the last two terms become zero when $\boldsymbol{\lambda}_1$ and $\boldsymbol{\lambda}_2$ are both in the optical range.

3.1.4 Surface Wave and Power Equation

The boundary condition has been given, not on the real boundary, but on the reference plane at $z = 0$ by (3.1.11a,b), in terms of the surface impedance, B, to be evaluated in detail in the following section. Therefore, it is not immediately clear how, in the neighborhood of the reference plane, the wave power is transmitted and its continuity is guaranteed.

On the plane, $z = 0$, the ρ integrated vertical power flux is exactly zero from (3.1.20, 21), but this does not mean that $w_z(\rho)$ is zero at every point of $\boldsymbol{\rho}$. In fact, in the case of b given by (3.1.14) to the first order in ζ, for example, (3.1.19b), leads to

$$w_z(\boldsymbol{\rho}) = -\frac{\partial}{\partial \boldsymbol{\rho}} \cdot [\zeta(\boldsymbol{\rho})(\psi^* \boldsymbol{\alpha} \psi)(\boldsymbol{\rho})] \ , \tag{3.1.51}$$

in terms of the operator, $\boldsymbol{\alpha}$, defined by (4.1.2b) and (2.3.19), showing that $w_z(\boldsymbol{\rho})$ is not zero everywhere, and is expressed by the space divergence of a two-dimensional vector in the $\boldsymbol{\rho}$ space.

The above divergence form of $w_z(\boldsymbol{\rho})$ is generally true as long as B is a Hermitian operator, and ensures the zero of the $\boldsymbol{\rho}$ integrated value of $w_z(\boldsymbol{\rho})$; i.e., as will be shown later, separately, it is always possible, in virtue of the Hermitian condition (3.1.20b), to write $w_z(\boldsymbol{\rho})$ in the form,

$$\begin{aligned} w_z(\boldsymbol{\rho}) &= -(2\mathrm{i})^{-1} \int d\boldsymbol{\rho}' \left[\psi^*(\boldsymbol{\rho}')B(\boldsymbol{\rho}'|\boldsymbol{\rho})\psi(\boldsymbol{\rho}) - \psi^*(\boldsymbol{\rho})B(\boldsymbol{\rho}|\boldsymbol{\rho}')\psi(\boldsymbol{\rho}')\right] \\ &= -\frac{\partial}{\partial \boldsymbol{\rho}} \cdot \int d\boldsymbol{\rho}_1\, d\boldsymbol{\rho}_2\, \boldsymbol{\beta}^{(B)}(\boldsymbol{\rho}|\boldsymbol{\rho}_1;\boldsymbol{\rho}_2)\psi^*(\boldsymbol{\rho}_1)\psi(\boldsymbol{\rho}_2) \ , \end{aligned} \tag{3.1.52a}$$

or, in matrix form,

$$w_z(\boldsymbol{\rho}) = -\frac{\partial}{\partial \boldsymbol{\rho}} \cdot \boldsymbol{\beta}^{(B)}(\boldsymbol{\rho}|1;2)\psi^*(1)\psi(2) \ . \tag{3.1.52b}$$

On the other hand, although power equation (3.1.18) is established in the region $z > 0$, it is generally not so on the reference plane, $z = 0$; the equation is subject to a change so that it holds everywhere, including the reference plane. This is simply accomplished by assuming $w_z(\widehat{\boldsymbol{x}}) = 0$ in the region $z < 0$ beyong the surface, so that, in (3.1.18), $\partial_z w_z(\widehat{\boldsymbol{x}})$ results in the additional term, $\delta(z)w_z(\boldsymbol{\rho})$, which can be cancelled out on using $w_z(\boldsymbol{\rho})$ of (3.1.52b) by subtracting the same term from the power equation. The continuity of the power flux is guaranteed everywhere with a new power equation, given by

$$\begin{aligned} \frac{\partial}{\partial \boldsymbol{\rho}} \cdot \left[\boldsymbol{w}(\widehat{\boldsymbol{x}}) + \delta(z)\boldsymbol{\beta}^{(B)}\psi^*\psi(\boldsymbol{\rho})\right] + \frac{\partial}{\partial z} w_z(\widehat{\boldsymbol{x}}) \\ = (2\mathrm{i})^{-1}\left[\psi^* j(\widehat{\boldsymbol{x}}) - \psi j^*(\widehat{\boldsymbol{x}})\right] \ , \end{aligned} \tag{3.1.53}$$

which shows that coutinuity is guaranteed only with an additional flux given by the $\boldsymbol{\beta}^{(B)}$ term, or by a surface wave.

To prove (3.1.52a,b), we first introduce a Fourier transform, $\widetilde{b}(\boldsymbol{u}, \boldsymbol{\rho}')$ of $b(\boldsymbol{\rho}|\boldsymbol{\rho}')$, defined by

$$\begin{aligned} \widetilde{b}(\boldsymbol{u}, \boldsymbol{\rho}') &= \int d\boldsymbol{\rho}\, \exp\left[\mathrm{i}\boldsymbol{u} \cdot (\boldsymbol{\rho} - \boldsymbol{\rho}')\right] b(\boldsymbol{\rho}|\boldsymbol{\rho}') \\ &= \sum_{n=0}^{\infty} (n!)^{-1} (\mathrm{i}\boldsymbol{u})^n b_n(\boldsymbol{\rho}') \ . \end{aligned}$$

Here, the last expression is the power series expansion in iu, upon writing the variable, u, as if it were scalar, for short, and

$$b_n(\rho') = \int d\rho\,(\rho-\rho')^n b(\rho|\rho'), \qquad n = 0, 1, 2 \ldots\ .$$

Hence, by the Fourier inversion, we obtain a series expression of $b(\rho|\rho')$ as

$$\begin{aligned} b(\rho|\rho') &= (2\pi)^{-2}\int du\, \tilde{b}(u,\rho')\exp\left[-iu\cdot(\rho-\rho')\right] \\ &= \sum_{n=0}^{\infty}(n!)^{-1}(-\partial_\rho)^n\delta(\rho-\rho')b_n(\rho')\ , \end{aligned}$$

which says that when a function, say $\psi(\rho)$, is multiplied to the left by the ρ matrix, b, the same result can be obtained by the operation of an operator, $\overrightarrow{b}$, defined by

$$\overrightarrow{b} = \sum_{n=0}^{\infty}(n!)^{-1}(-\overrightarrow{\partial}_\rho)^n b_n(\rho)\ ;$$

and, when multiplied to the right by the matrix, b, the same result can be obtained by the operation of an operator, $\overleftarrow{b}$, defined by

$$\overleftarrow{b} = \sum_{n=0}^{\infty}(n!)^{-1}(+\overleftarrow{\partial}_\rho)^n b_n(\rho)\ .$$

Thus, applying the result to (3.1.52a), the right-hand side is expressed, except for the factor $-(2i)^{-1}$, by

$$\psi^*(\rho)(\overleftarrow{b}-\overrightarrow{b})\psi(\rho) = \sum_{n=0}^{\infty}(n!)^{-1}\psi^*(\rho)\left[(+\overleftarrow{\partial}_\rho)^n - (-\overrightarrow{\partial}_\rho)^n\right]b_n(\rho)\psi(\rho)\ ,$$

leading to the expression (3.1.52b) in divergence form as

$$\partial_\rho\cdot\left\{\psi^* b_1\psi + 2^{-1}\psi^*(\overleftarrow{\partial}_\rho - \overrightarrow{\partial}_\rho)b_2\psi + \cdots\right\}\ .$$

Here, in the case of b given by (3.1.14) to the first order of ζ, $b_1 = \partial\zeta/\partial\rho$, $2^{-1}b_2 = \zeta$, and $b_n = 0, n \geq 3$; hence, $w_z(\rho)$ of (3.1.51) is reproduced.

3.1.5 Derivation of Surface Impedance

First, all parts of the rough surface, S, are assumed to be given by $z = -\zeta(\rho) < 0$, as illustrated in Fig. 3.1, so that the space unit vector, $\hat{n}^{(S)} = (n^{(S)}, n_z^{(S)})$, $n_z > 0$, inward normal to S, is given by

$$n_z^{(S)} = \left[1 + (\partial\zeta/\partial\rho)^2\right]^{-1/2}, \qquad n^{(S)} = n_z^{(S)}\frac{\partial\zeta}{\partial\rho}\ . \tag{3.1.54}$$

The boundary equation on S is to be given by (3.1.12) in which B_0 will be assumed to be a real constant for the sake of simplicity, although this restriction should not essentially change the final form (Sect. 3.3.2). The purpose of this section is to transfer the boundary equation on S onto the plane, $z = 0$, in the form,

$$-\partial_z \psi(z) = B(\rho, \partial_\rho)\psi(z), \qquad z = 0 \ , \tag{3.1.55}$$

(where $\partial_z = \partial/\partial z$ and $\partial_\rho = \partial/\partial \rho$), and to find the unknown B explicitly in terms of given $\zeta(\rho)$ and B_0.

We first introduce an operator, $\mathcal{T}(\rho)$, defined by

$$\mathcal{T}(\rho) = \exp[-\zeta(\rho)\partial_z] \ , \tag{3.1.56}$$

to exhibit $\psi(z)$ on S by

$$\psi(z)\mid_S = \psi(-\zeta) = \mathcal{T}\psi(z)\mid_{z=0} \ , \tag{3.1.57}$$

as is shown by expanding it in a power series of $-\zeta$. The same is also possible for the boundary equation (3.1.12), i.e.,

$$-\mathcal{T}\widehat{\boldsymbol{n}}^{(S)} \cdot \frac{\partial}{\partial \widehat{\boldsymbol{x}}}\psi(z)\Big|_{z=0} = B_0 \mathcal{T}\psi(z)\mid_{z=0} \ , \tag{3.1.58}$$

which is correct, including the general case of (3.1.13) where B_0 is an operator dependent on the surface coordinates, $\boldsymbol{s}$. Note that $\mathcal{T}$ is not commutable with ∂_ρ.

On the other hand, since $\psi(z)$ is a solution of the wave equation (3.1.1), wherein $j(z) = 0$ in the region $0 \geq z \geq -\zeta(\rho)$,

$$(\partial_z)^{2n}\psi = (-h^2)^n\psi = (\mathrm{i}h)^{2n}\psi \ , \tag{3.1.59a}$$

$n = 0, 1, 2, \dots$, whereas, which the aid of (3.1.55),

$$(\partial_z)^{2n+1}\psi = \partial_z(-h^2)^n\psi = (\mathrm{i}h)^{2n}(-B)\psi \ , \tag{3.1.59b}$$

at $z = 0$. Hence, the right-hand side of (3.1.57) can be written, using (3.1.56), as

$$\begin{aligned} \mathcal{T}\psi &= [\cosh(\zeta\partial_z) - \sinh(\zeta\partial_z)]\psi\mid_{z=0} \\ &= \mathcal{N}[\cos(\zeta h) + \sin(\zeta h)h^{-1}B]\psi\mid_{z=0} \ . \end{aligned} \tag{3.1.60a}$$

Here, the symbol $\mathcal{N}$ means that when the referred function is expanded in the power series of h (and/or ∂_ρ, if any), the coefficients containing $\zeta(\rho)$ are understood to be plased to the left of all the h's (and/or $\partial'_\rho s$), and B is placed to the right of all the factors. For example,

$$\mathcal{N}[B\zeta^n h^m] = \mathcal{N}[h^m \zeta^n B] = \zeta^n h^m B \ . \tag{3.1.60b}$$

In the same way,

$$\mathcal{T}\partial_z \psi \big|_{z=0} = \mathcal{N}[-\cos(\zeta h)B + \sin(\zeta h)h]\psi \big|_{z=0} \ . \tag{3.1.61}$$

For the left-hand side of (3.1.58), using (3.1.54), we obtain the expression,

$$-\mathcal{T}\widehat{\boldsymbol{n}}^{(S)} \cdot \frac{\partial}{\partial \widehat{\boldsymbol{x}}}\psi \Big|_{z=0} = -n_z^{(S)}[\mathcal{T}\partial_z + (\partial\zeta/\partial\boldsymbol{\rho}) \cdot \mathcal{T}\partial_{\boldsymbol{\rho}}]\psi \big|_{z=0} \ . \tag{3.1.62}$$

Here, for the first term on the right-hand side, (3.1.61) can be used directly, whereas, for the second term, (3.1.60a) can be used to write it as

$$\mathcal{T}\partial_{\boldsymbol{\rho}}\psi \big|_{z=0} = \mathcal{N}\Big\{\partial_{\boldsymbol{\rho}}\big[\cos(\zeta h) + \sin(\zeta h)h^{-1}B\big]\Big\}\psi \ . \tag{3.1.63}$$

Thus, the boundary equation (3.1.58) is written finally as

$$\begin{aligned} \mathcal{N}\Big[\!\Big[n_z^{(S)}\Big\{&\cos(\zeta h)B - \sin(\zeta h)h \\ -(\partial\zeta/\partial\boldsymbol{\rho}) \cdot \partial_{\boldsymbol{\rho}}&\big[\cos(\zeta h) + \sin(\zeta h)h^{-1}B\big]\Big\}\Big]\!\Big]\psi \\ &= \mathcal{N}\Big\{B_0\big[\cos(\zeta h) + \sin(\zeta h)h^{-1}B\big]\Big\}\psi \ . \end{aligned} \tag{3.1.64}$$

Here, since ψ is quite arbitrary, the above equation provides us with an equation to determine the unknown operator, B, yielding

$$\begin{aligned} B = \Big\{&\mathcal{N}\big[n_z^{(S)}\cos(\zeta h) - (B_0 + \boldsymbol{n}^{(S)} \cdot \partial_{\boldsymbol{\rho}})\sin(\zeta h)h^{-1}\big]\Big\}^{-1} \\ &\times\mathcal{N}\big[(B_0 + \boldsymbol{n}^{(S)} \cdot \partial_{\boldsymbol{\rho}})\cos(\zeta h) + n_z^{(S)}\sin(\zeta h)h\big] \ ; \end{aligned} \tag{3.1.65}$$

which is an even function of h, guaranteeing the condition $B^* = B$, and, although derived under the condition of $\zeta(\boldsymbol{\rho}) > 0$, it may be analytically continued into the range, $\zeta(\boldsymbol{\rho}) < 0$, in some cases. In the special case of ζ being constant, the result agrees with that which is directly derived, as it should be.

Example: Case of Small Surface Displacement
To the first order of ζ, (3.1.65) becomes

$$\begin{aligned} B \sim\ &\mathcal{N}\big[n_z^{(S)} - (B_0 + \boldsymbol{n}^{(S)} \cdot \partial_{\boldsymbol{\rho}})\zeta\big]^{-1} \\ &\times \mathcal{N}\big[B_0 + \boldsymbol{n}^{(S)} \cdot \partial_{\boldsymbol{\rho}} + n_z^{(S)}\zeta h^2\big] \ . \end{aligned} \tag{3.1.66}$$

Here, $n_z^{(S)} \sim 1$, $\boldsymbol{n}^{(S)} \sim \partial\zeta/\partial\boldsymbol{\rho}$, whence, according to the definition of the symbol $\mathcal{N}$,

$$\begin{aligned} B &\sim (1 - B_0\zeta)^{-1}(B_0 + \partial\zeta/\partial\boldsymbol{\rho} \cdot \partial_{\boldsymbol{\rho}} + \zeta h^2) \\ &\sim B_0 + \zeta(B_0^2 + h^2) + \partial\zeta/\partial\boldsymbol{\rho} \cdot \partial_{\boldsymbol{\rho}} \ , \end{aligned} \tag{3.1.67}$$

which is reduced to (3.1.14), written in an explicitly Hermitian form.

3.1.6 Integral Equation for the Reflection Coefficient

The reflection coefficient, R, is given in terms of B, by (3.1.34a), or, as the solution of

$$(\mathrm{i}h - B)R = \mathrm{i}h + B \ , \tag{3.1.68}$$

which, upon substitution of (3.1.65) for B, leads to

$$\begin{aligned}&\mathcal{N}\{\exp(\mathrm{i}\zeta h)[\mathrm{i}n_z^{(S)}h - B_0 - \boldsymbol{n}^{(S)}\cdot\partial_\rho]\}R\\&\quad = \mathcal{N}\{\exp(-\mathrm{i}\zeta h)[\mathrm{i}n_z^{(S)}h + B_0 + \boldsymbol{n}^{(S)}\cdot\partial_\rho]\} \ ;\end{aligned} \tag{3.1.69}$$

and an explicit operator expression of R is directly obtained therefrom. It is not an even function of h any longer, in contrast to B of (3.1.65).

Equation (3.1.69) can be transformed to an ordinary integral equation for matrix elements of R (free from any operator, and hence, also from the symbol $\mathcal{N}$), by multiplication of both sides to the right with $\exp(-\mathrm{i}\boldsymbol{\lambda}\cdot\boldsymbol{\rho})$, and followed by the use of

$$\begin{aligned}R\exp(-\mathrm{i}\boldsymbol{\lambda}\cdot\boldsymbol{\rho}) &= \int d\boldsymbol{\rho}'\, R(\boldsymbol{\rho}|\boldsymbol{\rho}')\exp(-\mathrm{i}\boldsymbol{\lambda}\cdot\boldsymbol{\rho}')\\&= (2\pi)^{-2}\int d\boldsymbol{\lambda}'\,\exp(-\mathrm{i}\boldsymbol{\lambda}'\cdot\boldsymbol{\rho})\widetilde{R}(\boldsymbol{\lambda}'|\boldsymbol{\lambda}) \ ,\end{aligned} \tag{3.1.70}$$

where the Fourier transform, $\widetilde{R}(\boldsymbol{\lambda}'|\boldsymbol{\lambda})$, is defined by (3.1.41). Hence, according to the definition of the symbol, $\mathcal{N}$, (3.1.69) is transformed to

$$\begin{aligned}&(2\pi)^{-2}\int d\boldsymbol{\lambda}'\,\left[\mathrm{i}n_z^{(S)}(\boldsymbol{\rho})\widetilde{h}(\boldsymbol{\lambda}') - B_0 + \mathrm{i}\boldsymbol{n}^{(S)}(\boldsymbol{\rho})\cdot\boldsymbol{\lambda}'\right]\\&\qquad\times\exp\left\{\mathrm{i}\left[\zeta(\boldsymbol{\rho})\widetilde{h}(\boldsymbol{\lambda}') - \boldsymbol{\lambda}'\cdot\boldsymbol{\rho}\right]\right\}\widetilde{R}(\boldsymbol{\lambda}'|\boldsymbol{\lambda})\\&\quad = \left[\mathrm{i}n_z^{(S)}(\boldsymbol{\rho})\widetilde{h}(\boldsymbol{\lambda}) + B_0 - \mathrm{i}\boldsymbol{n}^{(S)}(\boldsymbol{\rho})\cdot\boldsymbol{\lambda}\right]\\&\qquad\times\exp\left\{-\mathrm{i}\left[\zeta(\boldsymbol{\rho})\widetilde{h}(\boldsymbol{\lambda}) + \boldsymbol{\lambda}\cdot\boldsymbol{\rho}\right]\right\} \ ,\end{aligned} \tag{3.1.71}$$

which, with (3.1.54), provides us with a specific integral equation of $\widetilde{R}(\boldsymbol{\lambda}'|\boldsymbol{\lambda})$ for given $\zeta(\boldsymbol{\rho}) > 0$.

Here, when the variables $\boldsymbol{\lambda}$ and $\boldsymbol{\lambda}'$ can be limited to the optical range so that $\widetilde{h}(\boldsymbol{\lambda})$ and $\widetilde{h}(\boldsymbol{\lambda}')$ are both real in the integral equation, the solution is subject to the relation,

$$\widetilde{R}^{-1}(\boldsymbol{\lambda}|\boldsymbol{\lambda}') = \left[\widetilde{R}(-\boldsymbol{\lambda}| - \boldsymbol{\lambda}')\right]^* = \widetilde{R}^*(\boldsymbol{\lambda}|\boldsymbol{\lambda}') \ , \tag{3.1.72}$$

which gives a specific aspect of (3.1.35c).

3.1.7 Tangent Plane Method – Case of a Large-Scale Rough Surface

When a rough surface changes over a large-scale compared with the wavelength, the tangent plane method [3.2, 5] has been employed with the simple result that the scattering cross section depends only on the probability density function of the tangent of the rough surface. But, a crucial point is that the results fail to satisfy power conservation in the scattering, among other things, although not much attention has been paid on that point. Without changing the basic idea of the tangent plane method, however, the cross section can be obtained to be strictly consistent with the optical relation to ensure power conservation [3.1].

Throughout this section, $\zeta(\boldsymbol{\rho})$ is assumed to be given by the tangent plane at a point, $\boldsymbol{\rho}_0$, i.e.,

$$\zeta(\boldsymbol{\rho}) = \zeta_0 + \zeta_0' \cdot (\boldsymbol{\rho} - \boldsymbol{\rho}_0), \qquad \zeta_0' = \left.\frac{\partial \zeta}{\partial \boldsymbol{\rho}}\right|_{\boldsymbol{\rho}=\boldsymbol{\rho}_0}, \tag{3.1.73}$$

and the integral equation (3.1.71) for $\widetilde{R}$ is first solved exactly. To this end, we multiply both sides of the equation with $\exp(\mathrm{i}\boldsymbol{\lambda}'' \cdot \boldsymbol{\rho})$, and make the $\boldsymbol{\rho}$ integration; hence, the left-hand side becomes

$$(2\pi)^{-2} \int d\boldsymbol{\lambda}' \int d\boldsymbol{\rho} \exp\left\{\mathrm{i}[(\boldsymbol{\lambda}'' - \boldsymbol{\lambda}') \cdot \boldsymbol{\rho} + \widetilde{h}(\boldsymbol{\lambda}')\zeta(\boldsymbol{\rho})]\right\}$$
$$\times \left[\mathrm{i}n_z^{(S)}(\boldsymbol{\rho})\widetilde{h}(\boldsymbol{\lambda}') - B_0 + \mathrm{i}\boldsymbol{n}^{(S)}(\boldsymbol{\rho}) \cdot \boldsymbol{\lambda}'\right] \widetilde{R}(\boldsymbol{\lambda}'|\boldsymbol{\lambda}), \tag{3.1.74}$$

which, upon substitution of (3.1.73) and integrating with respect to $\boldsymbol{\rho}$, becomes

$$\int d\boldsymbol{\lambda}' \, \delta[\boldsymbol{\lambda}'' - \boldsymbol{\lambda}' + \zeta_0' \widetilde{h}(\boldsymbol{\lambda}')] \exp\left\{\mathrm{i}[(\boldsymbol{\lambda}'' - \boldsymbol{\lambda}') \cdot \boldsymbol{\rho}_0 + \widetilde{h}(\boldsymbol{\lambda}')\zeta_0]\right\}$$
$$\times \left[\mathrm{i}n_z^{(S)}(\boldsymbol{\rho}_0)\widetilde{h}(\boldsymbol{\lambda}') - B_0 + \mathrm{i}\boldsymbol{n}^{(S)}(\boldsymbol{\rho}_0) \cdot \boldsymbol{\lambda}'\right] \widetilde{R}(\boldsymbol{\lambda}'|\boldsymbol{\lambda}) \,. \tag{3.1.75a}$$

In the same way, the right-hand side of (3.1.71) is transformed to

$$(2\pi)^2 \delta[\boldsymbol{\lambda}'' - \boldsymbol{\lambda} - \zeta_0' \widetilde{h}(\boldsymbol{\lambda})] \exp\left\{\mathrm{i}[(\boldsymbol{\lambda}'' - \boldsymbol{\lambda}) \cdot \boldsymbol{\rho}_0 - \widetilde{h}(\boldsymbol{\lambda})\zeta_0]\right\}$$
$$\times \left[\mathrm{i}n_z^{(S)}(\boldsymbol{\rho}_0)\widetilde{h}(\boldsymbol{\lambda}) + B_0 - \mathrm{i}\boldsymbol{n}^{(S)}(\boldsymbol{\rho}_0) \cdot \boldsymbol{\lambda}\right] . \tag{3.1.75b}$$

Here, to perform the $\boldsymbol{\lambda}'$ integration in (3.1.75a), we utilize the lemma,

$$\int d\boldsymbol{\lambda} \, \delta[\boldsymbol{a}(\boldsymbol{\lambda})] = |\partial \boldsymbol{a}/\partial \boldsymbol{\lambda}|^{-1}, \qquad \boldsymbol{a}(\boldsymbol{\lambda}) = 0 \,, \tag{3.1.76}$$

where $|\partial\boldsymbol{a}/\partial\boldsymbol{\lambda}| = |\,(\partial a_1/\partial\boldsymbol{\lambda}_1)(\partial a_2/\partial\boldsymbol{\lambda}_2) - (\partial a_1/\partial\boldsymbol{\lambda}_2)(\partial a_2/\partial\boldsymbol{\lambda}_1)\,|$ is the Jacobian; hence,

$$\int d\boldsymbol{\lambda}'\,\delta[\boldsymbol{\lambda}'' - \boldsymbol{\lambda}' + \zeta_0'\widetilde{h}(\boldsymbol{\lambda}')] = [1 + \zeta_0' \cdot \boldsymbol{\lambda}'\widetilde{h}(\boldsymbol{\lambda}')^{-1}]^{-1} \ ,$$
$$\boldsymbol{\lambda}'' - \boldsymbol{\lambda}' + \zeta_0'\widetilde{h}(\boldsymbol{\lambda}') = 0 \ . \tag{3.1.77}$$

Thus, by setting the two eq. (3.1.75a,b) equal, we obtain the solution,

$$\widetilde{R}(\boldsymbol{\lambda}'|\boldsymbol{\lambda}) = (2\pi)^2\delta\Big\{\boldsymbol{\lambda}' - \boldsymbol{\lambda} - \zeta_0'[\widetilde{h}(\boldsymbol{\lambda}') + \widetilde{h}(\boldsymbol{\lambda})]\Big\}$$
$$\times\,[1 + \zeta_0' \cdot \boldsymbol{\lambda}'\widetilde{h}(\boldsymbol{\lambda}')^{-1}]\left[\frac{\mathrm{i}n_z^{(S)}\widetilde{h}(\boldsymbol{\lambda}) + B_0 - \mathrm{i}\boldsymbol{n}^{(S)} \cdot \boldsymbol{\lambda}}{\mathrm{i}n_z^{(S)}\widetilde{h}(\boldsymbol{\lambda}') - B_0 + \mathrm{i}\boldsymbol{n}^{(S)} \cdot \boldsymbol{\lambda}'}\right]$$
$$\times \exp\Big\{-\mathrm{i}(\zeta_0 - \zeta_0' \cdot \boldsymbol{\rho}_0)[\widetilde{h}(\boldsymbol{\lambda}') + \widetilde{h}(\boldsymbol{\lambda})]\Big\} \ . \tag{3.1.78}$$

Equation (3.1.78) can be written in a simple form as

$$\widetilde{R}(\boldsymbol{\lambda}'|\boldsymbol{\lambda}) = (2\pi)^2\delta(\zeta_0' - \boldsymbol{t})|\partial\boldsymbol{t}/\partial\boldsymbol{\lambda}'|$$
$$\times\, R_n \exp[-\mathrm{i}(\zeta_0 - \zeta_0' \cdot \boldsymbol{\rho}_0)q_z] \ . \tag{3.1.79}$$

Here, the factor, R_n, is the reflection coefficient of the tangent plane, and is given by

$$R_n = (\mathrm{i}2^{-1}q + B_0)(\mathrm{i}2^{-1}q - B_0)^{-1} \ , \tag{3.1.80}$$

where $q = |\widehat{\boldsymbol{q}}|$ is given from a three-dimensional vector, $\widehat{\boldsymbol{q}} = (\boldsymbol{q}, q_z) = \widehat{\boldsymbol{\lambda}}' - \widehat{\boldsymbol{\lambda}}$, with the components (Fig. 3.1)

$$\boldsymbol{q} = \boldsymbol{\lambda}' - \boldsymbol{\lambda}, \qquad q_z = \widetilde{h}(\boldsymbol{\lambda}') + \widetilde{h}(\boldsymbol{\lambda}) \ , \tag{3.1.81}$$
$$\widehat{\boldsymbol{\lambda}}' = [\boldsymbol{\lambda}', +\widetilde{h}(\boldsymbol{\lambda}')], \qquad \widehat{\boldsymbol{\lambda}} = [\boldsymbol{\lambda}, -\widetilde{h}(\boldsymbol{\lambda})] \ , \tag{3.1.82}$$

by

$$q^2 = \widehat{\boldsymbol{q}}^2 = 2[k_0^2 - \boldsymbol{\lambda}' \cdot \boldsymbol{\lambda} + \widetilde{h}(\boldsymbol{\lambda}')\widetilde{h}(\boldsymbol{\lambda})] \ ; \tag{3.1.83}$$

and

$$\boldsymbol{t} = \boldsymbol{q}/q_z \ , \tag{3.1.84}$$

whose Jacobians are given by

$$\widetilde{h}(\boldsymbol{\lambda}')|\partial\boldsymbol{t}/\partial\boldsymbol{\lambda}'| = \widetilde{h}(\boldsymbol{\lambda})|\partial\boldsymbol{t}/\partial\boldsymbol{\lambda}| = 2^{-1}q^2q_z^{-3} \ . \tag{3.1.85}$$

(The q should not be confused with the random medium, q, introduced in other sections.) Also, since the vector, $\widehat{\boldsymbol{q}}$, is in the direction of $\widehat{\boldsymbol{n}}^{(S)}$ (Fig. 3.1),

$$\widehat{\boldsymbol{n}}^{(S)} \cdot \widehat{\boldsymbol{\lambda}}' = -\widehat{\boldsymbol{n}}^{(S)} \cdot \widehat{\boldsymbol{\lambda}} = 2^{-1}q \ . \tag{3.1.86}$$

Here, in virtue of relation (3.1.85), $\widetilde{R}$ of (3.1.79) can be shown to strictly satisfy the optical relation (3.1.50); i.e., when B_0 is real and $|\boldsymbol{\lambda}_1|, |\boldsymbol{\lambda}_2| \leq k_0$ so that $|R_n| = 1$ from (3.1.80),

$$\begin{aligned}(2\pi)^{-2}\int_{|\boldsymbol{\lambda}'|\leq k_0} & d\boldsymbol{\lambda}'\,\widetilde{h}(\boldsymbol{\lambda}')\,\widetilde{R}^*(\boldsymbol{\lambda}'|\boldsymbol{\lambda}_1)\widetilde{R}(\boldsymbol{\lambda}'|\boldsymbol{\lambda}_2)\\ &=(2\pi)^2\delta(\boldsymbol{\lambda}_1-\boldsymbol{\lambda}_2)\widetilde{h}(\boldsymbol{\lambda}_2)\ .\end{aligned} \tag{3.1.87}$$

The proof is given by noting that, since $|R_n| = 1$, the left-hand side of (3.1.87) can be written, using (3.1.79), as

$$(2\pi)^2\int d\boldsymbol{\lambda}'\widetilde{h}(\boldsymbol{\lambda}')|\partial t_1/\partial\boldsymbol{\lambda}'|\,|\partial t_2/\partial\boldsymbol{\lambda}'|\delta(\zeta_0'-t_1)\delta(\zeta_0'-t_2)\ ,$$

where t_1 and t_2 are to be given by t of (3.1.84), with $\boldsymbol{\lambda} \to \boldsymbol{\lambda}_1$ and $\boldsymbol{\lambda}_2$, respectively. Hence, upon changing the variable of integration from $\boldsymbol{\lambda}'$ to t_1 for given $\boldsymbol{\lambda}_1$, the above expression becomes, in view of $dt_1 = |\partial t_1/\partial\boldsymbol{\lambda}'|d\boldsymbol{\lambda}'$,

$$(2\pi)^2\widetilde{h}(\boldsymbol{\lambda}')|\partial t_2/\partial\boldsymbol{\lambda}'|\delta(t_1-t_2)\ ,$$

(where $t_1 = \zeta_0'$), which is reduced to the right-hand side of (3.1.87), in consequence of

$$\begin{aligned}\widetilde{h}(\boldsymbol{\lambda}')|\partial t_2/\partial\boldsymbol{\lambda}'| &= \widetilde{h}(\boldsymbol{\lambda}_2)|\partial t_2/\partial\boldsymbol{\lambda}_2|\ ,\\ |\partial t_2/\partial\boldsymbol{\lambda}_2|\delta(t_1-t_2) &= \delta(\boldsymbol{\lambda}_1-\boldsymbol{\lambda}_2)\ .\end{aligned}$$

a) Scattering Cross Section

Using (3.1.36b) for $g_b(z|z')$, the mutual coherence function, $I(z_1;z_2|z_1';z_2')$, is defined by

$$\begin{aligned}I(z_1;z_2|z_1';z_2') &= \langle g_b^*(z_1|z_1')g_b(z_2|z_2')\rangle\\ &= G^{(0)*}(z_1-z_1')G^{(0)}(z_2-z_2')\\ +G^{(0)*}(z_1)G^{(0)}(z_2)&\langle T^*(1)T(2)\rangle\, G^{(0)*}(-z_1')G^{(0)}(-z_2')\ ,\end{aligned} \tag{3.1.88a}$$

on neglect of the interference terms between the direct and reflected waves, and with the subscripts 1 and 2 to distinguish the coordinates and quantities of the complex-conjugate, and the original wave functions, respectively. The corresponding full coordinate expression, $I(\widehat{\boldsymbol{x}}_1;\widehat{\boldsymbol{x}}_2|\widehat{\boldsymbol{x}}_1';\widehat{\boldsymbol{x}}_2')$, is hence written by

$$\begin{aligned}I(\widehat{\boldsymbol{x}}_1;\widehat{\boldsymbol{x}}_2|\widehat{\boldsymbol{x}}_1';\widehat{\boldsymbol{x}}_2') &= G^{(0)*}(\widehat{\boldsymbol{x}}_1-\widehat{\boldsymbol{x}}_1')G^{(0)}(\widehat{\boldsymbol{x}}_2-\widehat{\boldsymbol{x}}_2')\\ +\int d\boldsymbol{\rho}_1''d\boldsymbol{\rho}_2''&\int d\boldsymbol{\rho}_1'''d\boldsymbol{\rho}_2'''G^{(0)*}(\widehat{\boldsymbol{x}}_1-\boldsymbol{\rho}_1'')G^{(0)}(\widehat{\boldsymbol{x}}_2-\boldsymbol{\rho}_2'')\\ \times\langle T^*(\boldsymbol{\rho}_1''|\boldsymbol{\rho}_1''')T(\boldsymbol{\rho}_2''|\boldsymbol{\rho}_2''')\rangle &G^{(0)*}(\boldsymbol{\rho}_1'''-\widehat{\boldsymbol{x}}_1')G^{(0)}(\boldsymbol{\rho}_2'''-\widehat{\boldsymbol{x}}_2')\ ,\end{aligned} \tag{3.1.88b}$$

in terms of the matrix elements, $T(\rho|\rho')$ of T.

Here, to evaluate the scattered wave intensity at a point, $\hat{x}_1 = \hat{x}_2 = \hat{x}'$, far from a point source placed at $\hat{x}'_1 = \hat{x}'_2 = \hat{x}$ (on interchanging the primed and unprimed coordinates to be consistent with those used in the previous equations), say $I^{(S)}(\hat{x}'|\hat{x})$, given by the second term in (3.1.88b), we utilize the relative coordinates, $\boldsymbol{r}$ and $\boldsymbol{\rho}$, of (2.3.25–27) to write

$$G^{(0)*}(\hat{x} - \rho_1)G^{(0)}(\hat{x} - \rho_2) \sim \left|G^{(0)}(\hat{x} - \rho)\right|^2 \exp(ik_0 \boldsymbol{\Omega} \cdot \boldsymbol{r}) \ , \qquad (3.1.89)$$

where $\boldsymbol{r} = \rho_2 - \rho_1$, $|\boldsymbol{r}| \ll |\hat{x} - \rho|$, and $\boldsymbol{\Omega}$ is the horizontal component of the unit vector, $\hat{\boldsymbol{\Omega}} = (\boldsymbol{\Omega}, \Omega_z)$, in the direction of $\hat{x} - \rho$. The integral in (3.1.88b) is now obtained in the asymptotic form as

$$I^{(S)}(\hat{x}'|\hat{x}) \sim \int d\rho' \int d\rho \, |\hat{x}' - \rho'|^{-2} \times \sigma^{(T)}(\hat{\boldsymbol{\Omega}}', \rho'|\hat{\boldsymbol{\Omega}}, \rho)\left|G^{(0)}(\rho - \hat{x})\right|^2 \ . \qquad (3.1.90)$$

Here, $\hat{\boldsymbol{\Omega}}' = (\boldsymbol{\Omega}', +\Omega'_z)$ and $\hat{\boldsymbol{\Omega}} = (\boldsymbol{\Omega}, -\Omega_z)$ are the unit vectors in the directions of $\hat{x}' - \rho'$ and $\rho - \hat{x}$, respectively, and, with the matrix, T, defined by (3.1.36a),

$$\sigma^{(T)}(\hat{\boldsymbol{\Omega}}', \rho'|\hat{\boldsymbol{\Omega}}, \rho) = (4\pi)^{-2} \int d\boldsymbol{r}' d\boldsymbol{r} \times \exp\left[ik_0(\boldsymbol{\Omega}' \cdot \boldsymbol{r}' - \boldsymbol{\Omega} \cdot \boldsymbol{r})\right] \langle T^* T(\boldsymbol{r}', \rho'|\boldsymbol{r}, \rho)\rangle \ , \qquad (3.1.91)$$

where the Fourier transform of the last factor is

$$\left\langle \widetilde{T}^* \widetilde{T}(\boldsymbol{u}', \boldsymbol{\lambda}'|\boldsymbol{u}, \boldsymbol{\lambda})\right\rangle = \left\langle \widetilde{T}^*(\boldsymbol{\lambda}'_1|\boldsymbol{\lambda}_1)\widetilde{T}(\boldsymbol{\lambda}'_2|\boldsymbol{\lambda}_2)\right\rangle \ , \qquad (3.1.92)$$

which is given, except for the factor $4\widetilde{h}^*(\boldsymbol{\lambda}'_1)\widetilde{h}(\boldsymbol{\lambda}'_2)$ [(3.1.36a)], by $\langle \widetilde{R}^*(\boldsymbol{\lambda}'_1|\boldsymbol{\lambda}_1) \times \widetilde{R}(\boldsymbol{\lambda}'_2|\boldsymbol{\lambda}_2)\rangle$ and latter by (3.1.97). The factor $\sigma^{(T)}(\hat{\boldsymbol{\Omega}}', \rho'|\hat{\boldsymbol{\Omega}}, \rho)$ is a short-range function of $\rho' - \rho$, being appreciable only within a range of the order of the surface correlation distance, so that (3.1.90) can be reduced, in terms of the scattering cross section per unit area, $\sigma^{(T)}(\hat{\boldsymbol{\Omega}}'|\hat{\boldsymbol{\Omega}})$, defined by

$$\sigma^{(T)}(\hat{\boldsymbol{\Omega}}'|\hat{\boldsymbol{\Omega}}) = \int d\rho' \, \sigma^{(T)}(\hat{\boldsymbol{\Omega}}', \rho'|\hat{\boldsymbol{\Omega}}, \rho) \ , \qquad (3.1.93)$$

to:

$$I^{(S)}(\hat{x}'|\hat{x}) \sim \int d\rho \, |\hat{x}' - \rho|^{-2} \sigma^{(T)}(\hat{\boldsymbol{\Omega}}'|\hat{\boldsymbol{\Omega}})\left|G^{(0)}(\rho - \hat{x})\right|^2 \ . \qquad (3.1.94)$$

Here, to guarantee power conservation in the scattering, the cross section is required to satisfy the optical relation,

$$\int d\widehat{\Omega}'\,\sigma^{(T)}(\widehat{\Omega}'|\widehat{\Omega}) = \Omega_z > 0\ , \tag{3.1.95}$$

as follows directly from (3.1.94), with the aid of

$$d\rho = \Omega_z{}^{-1}|\rho - \widehat{x}|^2 d\widehat{\Omega}\ .$$

To make the statistical averaging in (3.1.92), we introduce a surface area, S_0, defined so that S_0 is sufficiently small for the rough surface within S_0 to be approximated by one (and not two or more) tangent plane at a point, ρ_0, on S_0, for each given direction of the incident and reflected waves. But, S_0 is large enough to allow the statistical averaging, $\langle\cdots\rangle$, over the entire rough surface to be replaced by a local ensemble averaging, $\langle\cdots\rangle_{S_0}$ over S_0. Here, the local $\langle\cdots\rangle_{S_0}$ is given in terms of $\widetilde{R}(\boldsymbol{\lambda}'|\boldsymbol{\lambda})$ of (3.1.79), assuming that ζ_0 and ζ_0' are statistically independent at each point, ρ_0, on S_0, with a probability density function of ζ_0' given by $P(\zeta_0')$, and subject to

$$\int d\zeta'\,P(\zeta') = 1\ . \tag{3.1.96}$$

However, ζ_0 and ζ_0' at the neighborhood points are closely correlated within the correlation area of $\zeta(\rho)$, consistent with $P(\zeta_0')$. The shadow effect is necessarily involved in the $P(\zeta_0')$ since the tangent planes in the shadow area are to be entirely excluded (implying that $P(\zeta_0')$ is dependent also on the direction of each incident wave), and S_0 is large enough to enable the construction of the correct $P(\zeta_0')$ under this constraint. Hence, it follows from these conditions, that the size of S_0 is of the order of the correlation area of $\zeta(\rho)$.

By using (3.1.79), it is now straightforward to find that

$$\begin{aligned}\left\langle \widetilde{R}^*\widetilde{R}(\boldsymbol{u}',\boldsymbol{\lambda}'|\boldsymbol{u},\boldsymbol{\lambda})\right\rangle &\equiv \left\langle \widetilde{R}^*(\boldsymbol{\lambda}_1'|\boldsymbol{\lambda}_1)\widetilde{R}(\boldsymbol{\lambda}_2'|\boldsymbol{\lambda}_2)\right\rangle \\ &= (2\pi)^4|R_n|^2\delta(\boldsymbol{t}_2-\boldsymbol{t}_1)|\partial\boldsymbol{t}/\partial\boldsymbol{u}'|^2P(\boldsymbol{t}) \\ &\quad\times\left\langle \exp\left[\mathrm{i}(\zeta_0-\boldsymbol{t}\cdot\boldsymbol{\rho}_0)(q_{z_1}-q_{z_2})\right]\right\rangle_{S_0}\ .\end{aligned} \tag{3.1.97}$$

Here, the variables $\boldsymbol{t}_1$, q_{z_1}, and $\boldsymbol{t}_2$, q_{z_2} are defined by (3.1.84, 81), with the replacement of $\boldsymbol{\lambda}$, $\boldsymbol{\lambda}'$ by $\boldsymbol{\lambda}_1$, $\boldsymbol{\lambda}_1'$, and $\boldsymbol{\lambda}_2$, $\boldsymbol{\lambda}_2'$, respectively; wherein the relative variables $\boldsymbol{\lambda}$ and $\boldsymbol{\lambda}'$, defined by (2.3.26a), have been set equal to zero so that $\boldsymbol{t}_1=\boldsymbol{t}_2=\boldsymbol{t}$, in view of $|\boldsymbol{\lambda}|, |\boldsymbol{\lambda}'| \ll |\boldsymbol{u}|, |\boldsymbol{u}'| \sim k_0$, expect in the δ function and the phase term, which are given to the first order of $\boldsymbol{\lambda}$ and $\boldsymbol{\lambda}'$, by

$$\delta(\boldsymbol{t}_2-\boldsymbol{t}_1) = \delta(\partial\boldsymbol{t}/\partial\boldsymbol{u}'\cdot\boldsymbol{\lambda}' + \partial\boldsymbol{t}/\partial\boldsymbol{u}\cdot\boldsymbol{\lambda}) \tag{3.1.98a}$$

$$= \delta\left\{q_z^{-1}\left[\boldsymbol{\lambda}'-\boldsymbol{\lambda}+(q_{z_1}-q_{z_2})\boldsymbol{t}\right]\right\}\ , \tag{3.1.98b}$$

$$q_{z_1}-q_{z_2} = \boldsymbol{u}'\cdot\boldsymbol{\lambda}'\widetilde{h}(\boldsymbol{u}')^{-1} + \boldsymbol{u}\cdot\boldsymbol{\lambda}\widetilde{h}(\boldsymbol{u})^{-1}\ . \tag{3.1.98c}$$

Here, $\boldsymbol{t}$ is the same function of $\boldsymbol{u}'$ and $\boldsymbol{u}$ as it was of $\boldsymbol{\lambda}'$ and $\boldsymbol{\lambda}$ in (3.1.84). Hence, particularly when $\boldsymbol{\lambda}' = 0$,

$$\delta(\boldsymbol{t}_2 - \boldsymbol{t}_1)\,\big|_{\boldsymbol{\lambda}'=0} = |\partial \boldsymbol{t}/\partial \boldsymbol{u}|^{-1}\delta(\boldsymbol{\lambda}) \ . \qquad (3.1.99)$$

Thus the cross section, $\sigma^{(T)}(\widehat{\boldsymbol{\Omega}}'|\widehat{\boldsymbol{\Omega}})$ of (3.1.93), becomes, using (3.1.91, 92, 97, 99),

$$\sigma^{(T)}(\widehat{\boldsymbol{\Omega}}'|\widehat{\boldsymbol{\Omega}}) = |R_n|^2\tilde{h}(\boldsymbol{u}')\tilde{h}(\boldsymbol{u})|\partial \boldsymbol{t}/\partial \boldsymbol{u}'|P(\boldsymbol{t})\,\big|_{\Omega} \ , \qquad (3.1.100)$$

where use has been made of the relation in (3.1.85), the phase term by (3.1.98c), and the mark, $|_{\Omega}$, meaning to set

$$\boldsymbol{u}' = k_0\boldsymbol{\Omega}', \qquad \tilde{h}(\boldsymbol{u}') = k_0\Omega_z' \ ,$$
$$\boldsymbol{u} = k_0\boldsymbol{\Omega}, \qquad \tilde{h}(\boldsymbol{u}) = k_0\Omega_z \ . \qquad (3.1.101)$$

Note that the expression (3.1.100) is symmetrical with regard to $\boldsymbol{u}'$ and $\boldsymbol{u}$, except for a weighting function, $|\partial \boldsymbol{t}/\partial \boldsymbol{u}'|$.

The cross section (3.1.100) can be written in a compact form by introducing the Jacobian,

$$|\partial \boldsymbol{t}/\partial \widehat{\boldsymbol{\Omega}}'| = |\partial \boldsymbol{t}/\partial \boldsymbol{u}'|\,|\partial \boldsymbol{u}'/\partial \widehat{\boldsymbol{\Omega}}'| = k_0^2\Omega_z'|\partial \boldsymbol{t}/\partial \boldsymbol{u}'| \ . \qquad (3.1.102)$$

Hence,

$$\sigma^{(T)}(\widehat{\boldsymbol{\Omega}}'|\widehat{\boldsymbol{\Omega}}) = |R_n|^2\Omega_z|\partial \boldsymbol{t}/\partial \widehat{\boldsymbol{\Omega}}'|P(\boldsymbol{t}) \ , \qquad (3.1.103)$$

A great advantage of this result is that it leads directly to the optical relation (3.1.95) by changing the variable of integration, $\widehat{\boldsymbol{\Omega}}'$, to $\boldsymbol{t}$, and followed by use of the normalization (3.1.96).

Thus, the explicit expression of the cross section becomes, using (3.1.81–85),

$$\sigma^{(T)}(\widehat{\boldsymbol{\Omega}}'|\widehat{\boldsymbol{\Omega}}) = 2^{-1}k_0q^2q_z^{-3}\Omega_z|R_n|^2P(\boldsymbol{t}) \qquad (3.1.104a)$$
$$= (1 + \Omega_z'\Omega_z - \boldsymbol{\Omega}'\cdot\boldsymbol{\Omega})(\Omega_z' + \Omega_z)^{-3}\Omega_z$$
$$\times \left|R_n(\widehat{\boldsymbol{\Omega}}'|\widehat{\boldsymbol{\Omega}})\right|^2 P\left[(\boldsymbol{\Omega}' - \boldsymbol{\Omega})/(\Omega_z' + \Omega_z)\right] \ , \qquad (3.1.104b)$$

where $R_n(\widehat{\boldsymbol{\Omega}}'|\widehat{\boldsymbol{\Omega}})$ is given by (3.1.80) with

$$q = k_0 2^{1/2}[1 + \Omega_z'\Omega_z - \boldsymbol{\Omega}'\cdot\boldsymbol{\Omega}]^{1/2} \ . \qquad (3.1.104c)$$

In view of (3.1.94), $\sigma^{(T)}(\widehat{\boldsymbol{\Omega}}'|\widehat{\boldsymbol{\Omega}})$ means the resultant cross section, including the coherent one due to $\langle R\rangle$. But, (3.1.104a) suggests that $\langle R\rangle \sim 0$ in the present case of a large-scale rough surface; this can also be shown directly [3.1].

b) Shadowing and Multiple Reflection

In (3.1.103) for $\sigma^{(T)}(\hat{\Omega}'|\hat{\Omega})$, $P(\zeta'=t)$ is the probability density function of the tangent of the reflecting boundary for a given incident angle, $\hat{\Omega}$, under the condition of excluding the region of the shadow area (Fig. 3.2), and therefore it is necessarily dependent on $\hat{\Omega}$. That is, $P(\zeta')$ should be $P(\zeta', \hat{\Omega})$, subject to

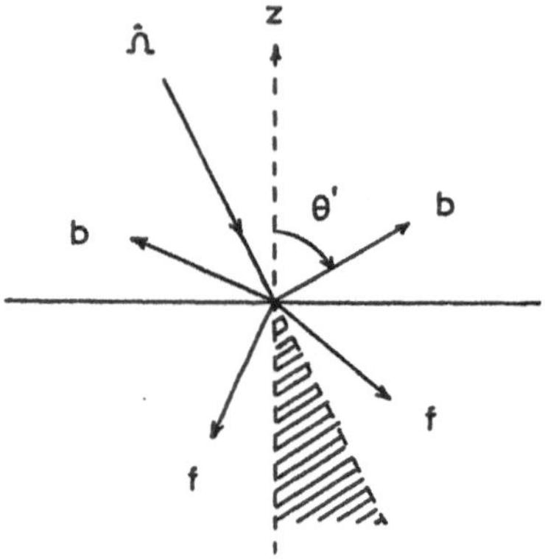

Fig. 3.2. The ζ' range of conditioned $P(\zeta', \hat{\Omega})$ for (3.1.115a,b). *b*: backward reflection; *f*: forward reflection; *shaded region* depicts the shadow range of tangent planes where $q_z < 0$

$$\int d\hat{\zeta}' \, P(\zeta', \hat{\Omega}) = 1 \ , \tag{3.1.105}$$

as will be given later, but, all the previous equations remain unchanged with this notational change. For example, the optical relation (3.1.95) holds as it is, with the remark that the $\hat{\Omega}'$ range of integration is not limited within the half-solid angle, 2π, but also includes the range of angles for the forward (downward) reflection wherein $\tilde{h}(\boldsymbol{\lambda}') = k_0 \Omega_z' < 0$, as illustrated in Fig. 3.2.

The function, $P(t)$, in (3.1.103) is hence replaced by $P(t, \hat{\Omega})$, which is given in terms of the unconditioned probability density function, say, $P_0(\zeta')$, by

$$P(t, \hat{\Omega}) = P_0(t) \Big/ \int_{q_z > 0} d\hat{\Omega}' \, |\partial t / \partial \hat{\Omega}'| P_0(t) \ .$$

Here, t is defined in the range, $q_z = k_0(\Omega_z' + \Omega_z) > 0$, excluding the shadow range where $q_z < 0$; and, as $q_z \to 0$, t tends to either $\Omega/\Omega_z = 0$ or ∞ as $q \to 0$ (grazing incidence to the surface) or $|q| \neq 0$ (vertical boundary), respectively, in view of (3.1.84).

To introduce the effects of the multiple reflection, we first observe that the resulting cross section should be able to be obtained as the solution of an integral equation similar to (3.2.78), to be derived quite generally as a basic equation for the incoherent cross section; i.e., the true cross section, $\sigma^{(TR)}(\hat{\Omega}'|\hat{\Omega})$, can be obtained, in terms of known $\sigma^{(T)}(\hat{\Omega}'|\hat{\Omega})$, as the solution of an integral equation of the form,

$$\sigma^{(TR)}(\hat{\Omega}'|\hat{\Omega}) = \sigma^{(T)}(\hat{\Omega}'|\hat{\Omega}) + \int_{2\pi} d\hat{\Omega}'' \sigma^{(T)}(\hat{\Omega}'|\hat{\Omega}'') C(\hat{\Omega}'') (\Omega_z'')^{-1} \sigma^{(TR)}(\hat{\Omega}''|\hat{\Omega}) \ . \tag{3.1.106}$$

Here, the last term means an effect of the waves reflected by the boundary itself, and the factor, $C(\hat{\Omega}'')$, is also unknown and must be obtained self-consistently. Note that the range of integration is limited within the 2π, and that the $\hat{\Omega}''$ integration range of $\sigma^{(T)}(\hat{\Omega}'|\hat{\Omega}'')$ in the second term is necessarily limited to those incident waves in the backward (upward) direction, whereas the $\hat{\Omega}'$ range is within the 2π (Fig. 3.2).

Here, $C(\hat{\Omega})$ and $P(t, \hat{\Omega})$ are not quite independent of each other, but are subjected to a relation resulting from the optical relation,

$$\int_{2\pi} d\hat{\Omega}' \sigma^{(TR)}(\hat{\Omega}'|\hat{\Omega}) = \Omega_z \ . \tag{3.1.107}$$

A formal solution of (3.1.106) for $\sigma^{(TR)}$ is obtained by introducing two $\hat{\Omega}$ matrices, $\sigma^{(T)}$ and $\sigma^{(TR)}$, defined by the matrix elements, $\sigma^{(T)}(\hat{\Omega}'|\hat{\Omega})$ and $\sigma^{(TR)}(\hat{\Omega}'|\hat{\Omega})$, respectively, so that

$$\sigma^{(TR)} = \sigma^{(T)}[1 + C\Omega_z^{-1}\sigma^{(TR)}] \ , \tag{3.1.108}$$

and hence,

$$\begin{aligned} \sigma^{(TR)} &= (1 - \sigma^{(T)} C \Omega_z^{-1})^{-1} \sigma^{(T)} \\ &= \sigma^{(T)} + \sigma^{(T)} C \Omega_z^{-1} \sigma^{(T)} \\ &\quad + \sigma^{(T)} C \Omega_z^{-1} \sigma^{(T)} C \Omega_z^{-1} \sigma^{(T)} + \cdots \ . \end{aligned} \tag{3.1.109}$$

To find another equation for $C(\hat{\Omega})$, we integrate (3.1.106), with respect to $\hat{\Omega}'$, over the range 2π, and use the condition (3.1.107), to obtain

$$\begin{aligned} &\Omega_z - \int_{2\pi} d\hat{\Omega}' \sigma^{(T)}(\hat{\Omega}'|\hat{\Omega}) \\ &= \int_{2\pi} d\hat{\Omega}'' \int_{2\pi} d\hat{\Omega}' \sigma^{(T)}(\hat{\Omega}'|\hat{\Omega}'') C(\hat{\Omega}'') (\Omega_z'')^{-1} \sigma^{(TR)}(\hat{\Omega}''|\hat{\Omega}) \ . \end{aligned} \tag{3.1.110}$$

Here, for the left-hand side, we can utilize optical relation (3.1.95) and write it as an integral extended only over the range of the forward reflection, say

F (Fig. 3.2), while, for the right-hand side, we observe that the forward (downward) reflection cannot occur for the incident wave in the backward (upward) directions. Hence, (3.1.110) is reduced to

$$\int_F d\hat{\Omega}' \sigma^{(T)}(\hat{\Omega}'|\hat{\Omega}) = \int_{2\pi} d\hat{\Omega}'' C(\hat{\Omega}'') \sigma^{(TR)}(\hat{\Omega}''|\hat{\Omega}) \qquad (3.1.111)$$

(note that $\Omega_z > 0$ on the right-hand side of (3.1.95), even for the incident wave from the backward direction).

Thus (3.1.106, 111) provide, in terms of given $\sigma^{(T)}$, a set of integral equations for the unknown $\sigma^{(TR)}(\hat{\Omega}'|\hat{\Omega})$ and $C(\hat{\Omega})$ that are defined only within the range 2π. In (3.1.111), the factor $C(\hat{\Omega}'')$ is expected to have a sharp distribution with the maximum in the horizontal direction. Hence, if both $\sigma^{(T)}(\hat{\Omega}'|\hat{\Omega})$ and $\sigma^{(TR)}(\hat{\Omega}''|\hat{\Omega})$ are regarded to be constant over the important integration range, use of $\sigma^{(TR)} \sim \sigma^{(T)}$ on the right-hand side leads to the solution,

$$C(\hat{\Omega}) \sim \delta(\Omega_z - 0) \int_F d\Omega'_z \ ,$$

which gives a rough estimate of $C(\hat{\Omega})$. Thus, $C(\hat{\Omega})$ is closely connected to the shadowing function, $S_s(\hat{\Omega})$, investigated in [3.6–9], suggesting that $C(\hat{\Omega}) \sim 1 - S_s(\hat{\Omega})$. The multiple reflection was also investigated in [3.10].

3.2 Statistical Green's Functions of First and Second Orders

The statistical Green's functions are obtained here first on the reference plane, $z = 0$, and are then continued to the outside space. The associated optical relations are also investigated in detail. The procedure is basically the same as for a random medium in Sect. 2.3.

The statistical surface Green's function of the first order, defined by $G = \langle g_b \rangle$, with the matrix elements,

$$G(\rho - \rho') = \langle g_b(\rho|\rho') \rangle \ , \qquad (3.2.1)$$

is governed by the average of (3.1.16a), which becomes, in terms of an effective matrix, $M^{(b)}$ of b, defined by

$$M^{(b)} G = \langle b\, g_b \rangle \ , \qquad (3.2.2a)$$

or, more precisely, by

$$\int d\rho'' M^{(b)}(\rho - \rho'') G(\rho'' - \rho') = \int d\rho'' \langle b(\rho|\rho'') g_b(\rho''|\rho') \rangle \ , \qquad (3.2.2b)$$

as

$$(\mathrm{i}h - B_0 - M^{(b)})G = 1 \ , \tag{3.2.3a}$$

$$(-\mathrm{i}h^* - B_0 - M^{(b)*})G^* = 1 \ , \tag{3.2.3b}$$

where the last is the complex-conjugate equation. Here, the procedure of evaluating $M^{(b)}$ is precisely the same as in the case of a random medium, and is given, for example, to the ladder approximation by,

$$M^{(b)} = \langle bGb \rangle, \qquad \langle b \rangle = 0 \ . \tag{3.2.4}$$

For another typical example of when discrete scatterers or bosses are randomly distributed over a smooth surface, the corresponding $M^{(b)}$ will be given by (3.2.36, 37).

The matrix, $M^{(b)}$, is connected to the averaged reflection coefficient, $\langle R \rangle$, through the equation,

$$G = \left(1 + \langle R \rangle\right)(2\mathrm{i}h)^{-1} \ , \tag{3.2.5}$$

from (3.1.33b), so that, upon the substitution into (3.2.3a), we find the relation,

$$M^{(B)} \equiv B_0 + M^{(b)} \tag{3.2.6a}$$

$$= -\mathrm{i}h\left(1 - \langle R \rangle\right)\left(1 + \langle R \rangle\right)^{-1} \ . \tag{3.2.6b}$$

Hence, conversely,

$$\langle R \rangle = (\mathrm{i}h - M^{(B)})^{-1}(\mathrm{i}h + M^{(B)}) \ . \tag{3.2.7}$$

When the boundary is rough enough that $\langle R \rangle \sim 0$, (3.2.5, 6) yield

$$G \sim G^{(0)} = (2\mathrm{i}h)^{-1}, \qquad M^{(B)} \sim -\mathrm{i}h \ , \tag{3.2.8}$$

which show that the first order Green's function becomes the same as that in free space, making no coherent wave reflection at all. Equation (3.2.7) shows that, even when $B_0 = 0$, $\langle R \rangle \rightarrow -1$ as $h \rightarrow 0$, as realized in the case of grazing wave incidence [3.4].

Alternatively, in terms of a scattering phase shift, δ_C, defined by

$$\langle R \rangle = \exp(-\mathrm{i}2\delta_C) \ , \tag{3.2.9a}$$

[where δ_C is commutable with h in view of the translational invariance; compare also to (3.1.35c)], (3.2.5, 6) are written as

$$G = (\mathrm{i}h)^{-1}\cos\delta_C \, \exp(-\mathrm{i}\delta_C) \ , \tag{3.2.9b}$$

$$M^{(B)} = h\tan\delta_C \ . \tag{3.2.9c}$$

For later convenience, we now introduce an attenuation coefficient of the coherent wave, A, defined by

$$G = 2AG^{(0)} = 2G^{(0)}A \ , \tag{3.2.10a}$$

similar to (3.1.24), so that, from (3.1.4),

$$A = \mathrm{i}hG = \mathrm{i}Gh = \left(1 + \langle R\rangle\right)/2 \tag{3.2.10b}$$

$$= h(h + \mathrm{i}M^{(B)})^{-1} \ , \tag{3.2.10c}$$

and is the same function of the coordinates as given by (3.1.25–28), with $B_0 \to M^{(B)}$.

To introduce the Green's function of second order, we attach the subscript 1 to the coordinates of quantities of the complex-conjugate wave function, e.g., $g_b^*(\rho_1|\rho_1')$ as the matrix elements of $g_b^*(1)$, and attach the subscript 2 to the coordinates of quantities of the original wave function, e.g., $g_b(\rho_2|\rho_2')$ as the matrix elements of $g_b(2)$. Thus the second order Green's function is defined by

$$I^{(B)}(1;2) = \langle g_b^*(1) g_b(2)\rangle \ , \tag{3.2.11}$$

in the same form as (2.1.10b) for a random medium, and has matrix elements of the form, $I^{(B)}(\rho_1;\rho_2|\rho_1';\rho_2')$.

The governing equation for $I^{(B)}(1;2)$ is also obtained in the same form as BS equation (2.2.21), i.e.,

$$I^{(B)}(1;2) = U^{(C)}(1;2)\left[1 + K^{(B)}(1;2)I^{(B)}(1;2)\right] \ , \tag{3.2.12}$$

$$U^{(C)}(1;2) = G^*(1)G(2) \ , \tag{3.2.13}$$

where $U^{(C)}(1;2)$ means the coherent part. That is, expressing the deterministic Green's function by

$$g_b^*(1) = G^*(1)\left[1 + \Delta b^*(1) g_b^*(1)\right] \ , \tag{3.2.14a}$$

$$g_b(2) = G(2)\left[1 + \Delta b(2) g_b(2)\right] \ , \tag{3.2.14b}$$

where

$$\Delta b = b - M^{(B)}, \qquad \langle \Delta b\, g_b\rangle = 0 \ , \tag{3.2.15a}$$

$$\Delta b^* = b - M^{(B)*}, \qquad \langle \Delta b^* g_b^*\rangle = 0 \ , \tag{3.2.15b}$$

(3.2.11) leads directly to the BS equation (3.2.12), with the factor, $K^{(B)}(1;2)$, defined by

$$K^{(B)}(1;2)I^{(B)}(1;2) = \langle \Delta b^*(1)\Delta b(2) g_b^*(1) g_b(2)\rangle \ , \tag{3.2.16}$$

and having the matrix elements of the form $K^{(B)}(\rho_1;\rho_2|\rho_1' ;\rho_2')$. In the case of a slightly random surface where $M^{(B)}$ is given by (3.2.4), then to the ladder approximation

$$K^{(B)}(1;2) = \langle b(1)\, b(2)\rangle \ . \tag{3.2.17}$$

3.2.1 Incoherent Scattering Matrix

A formal solution of BS equation (3.1.12) is

$$I^{(B)} = (1 - U^{(C)}K^{(B)})^{-1}U^{(C)} \tag{3.2.18a}$$

$$= U^{(C)}(1 - K^{(B)}U^{(C)})^{-1} \ , \tag{3.1.18b}$$

which leads to another version of the BS equation as

$$I^{(B)} = (1 + I^{(B)}K^{(B)})U^{(C)} \ . \tag{3.2.19}$$

More conveniently, in terms of the incoherent scattering matrix, $S^{(B)}$ of $K^{(B)}$, defined by

$$S^{(B)}U^{(C)} = K^{(B)}I^{(B)}, \qquad U^{(C)}S^{(B)} = I^{(B)}K^{(B)} \ , \tag{3.2.20}$$

the solution is written as

$$I^{(B)} = U^{(C)} + U^{(C)}S^{(B)}U^{(C)} \ , \tag{3.2.21}$$

with

$$S^{(B)} = K^{(B)}(1 + U^{(C)}S^{(B)}) \tag{3.2.22a}$$

$$= (1 - K^{(B)}U^{(C)})^{-1}K^{(B)} \tag{3.2.22b}$$

$$= K^{(B)} + K^{(B)}U^{(C)}K^{(B)} + K^{(B)}U^{(C)}K^{(B)}U^{(C)}K^{(B)} + \cdots \ , \tag{3.2.22c}$$

being of the same form as (2.3.11a,b).

3.2.2 Optical Condition

To investigate power equations associated with the BS equation, we first observe that the average of the vertical power flux on the plane $z = 0$, $\langle w_z(\rho)\rangle$, is obtained by replacing $\psi \to g_b$ in (3.1.52), in the form,

$$\langle w_z(\rho)\rangle = -\delta(\rho|1;2)(2i)^{-1}\Big\langle [B(1) - B(2)]\, g_b^*(1)g_b(2)\Big\rangle \tag{3.2.23a}$$

$$= -\frac{\partial}{\partial\rho}\cdot\Big\langle \beta^{(B)}(\rho|1;2)g_b^*(1)g_b(2)\Big\rangle \tag{3.2.23b}$$

$$= -\frac{\partial}{\partial\rho}\cdot\beta(\rho|1;2)I^{(B)}(1;2) \ . \tag{3.2.23c}$$

Here, the ρ matrix, $\delta(\rho|1;2)$, is defined in the same way as the $\hat{\boldsymbol{x}}$ matrix, $\delta(\hat{\boldsymbol{x}}|1;2)$, is defined by (2.3.14). And $\beta(\rho|1;2)$ is an effective matrix of $\beta^{(B)}(\rho|1;2)$, which is defined by the last equation, as $K^{(B)}(1;2)$ is defined by (3.2.16). The right-hand side of (3.2.23a) can be rewritten in terms of the matrix, $M^{(B)}$, of (3.2.6a), defined by

$$\langle Bg_b\rangle = M^{(B)}G, \qquad M^{(B)} = B_0 + M^{(b)} \ , \tag{3.2.24a}$$

by using the relations,

$$\langle B(1)g_b^*(1)g_b(2)\rangle = \left[M^{(B)*}(1) + G(2)K^{(B)}(1;2)\right]I^{(B)}(1;2) \ , \tag{3.2.24b}$$

$$\langle B(2)g_b^*(1)g_b(2)\rangle = \left[M^{(B)}(2) + G^*(1)K^{(B)}(1;2)\right]I^{(B)}(1;2) \ , \tag{3.2.24c}$$

which follow from (3.2.16, 15). Hence, with (3.2.23c) we find a basic relation between $M^{(B)}$, $K^{(B)}$ and $\beta(\rho)$, as

$$\begin{aligned}\frac{\partial}{\partial\rho}\cdot\beta(\rho|1;2) &= (2\mathrm{i})^{-1}\delta(\rho|1;2)\\ &\times\left\{M^{(B)*}(1) - M^{(B)}(2) - \left[G^*(1) - G(2)\right]K^{(B)}(1;2)\right\} \ ,\end{aligned} \tag{3.2.25}$$

which ensures power conservation at every point on the plane, $z = 0$, as will be shown below, and provides a local optical relation in contrast to the ρ-integrated relation to be given by (3.2.35).

The meaning of relation (3.2.25) becomes specific by rewriting it in terms of the scattering matrix, $S^{(B)}$, defined by (3.2.22), and the result is

$$\begin{aligned}\frac{\partial}{\partial\rho}\cdot\beta^{(e)}(\rho|1;2) = (2\mathrm{i})^{-1}\delta(\rho|1;2)\Big\{&M^{(B)*}(1) - M^{(B)}(2)\\ &-\mathrm{i}\left[h^*(1) + h(2)\right]U^{(C)}(1;2)S^{(B)}(1;2)\Big\} \ ,\end{aligned} \tag{3.2.26}$$

where

$$\beta^{(e)}(\rho|1;2) = \beta(\rho|1;2)\left[1 + U^{(C)}(1;2)S^{(B)}(1;2)\right] \ . \tag{3.2.27}$$

The proof is given by substituting the expression,

$$G^*(1) - G(2) = \left\{\mathrm{i}\left[h^*(1)+h(2)\right] + M^{(B)*}(1) - M^{(B)}(2)\right\}U^{(C)}(1;2) \ , \tag{3.2.28}$$

from (3.2.3) into (3.2.25), to rewrite the right-hand side as

$$\begin{aligned}(2\mathrm{i})^{-1}\delta(\rho|1;2)\Big\{&\left[M^{(B)*}(1) - M^{(B)}(2)\right]\left[1 - U^{(C)}(1;2)K^{(B)}(1;2)\right]\\ &-\mathrm{i}\left[h^*(1) + h(2)\right]U^{(C)}(1;2)K^{(B)}(1;2)\Big\} \ ;\end{aligned}$$

hence, multiplication of both sides to the right with

$$[1-U^{(C)}K^{(B)}]^{-1}=1+U^{(C)}S^{(B)}\ , \tag{3.2.29}$$

leads to (3.2.26). Thus, upon multiplication of both sides of (3.2.26) to the right with $U^{(C)}(1;2)$, the left-hand side becomes $-\langle w_z(\rho)\rangle$, in view of (3.2.23c, 27), while the right-hand side gives the sum of the vertical power fluxes of the coherent and incoherent waves, which are given, respectively, by [cf. (3.1.19a,b)]

$$w_z^{(C)}(\rho)=-\delta(\rho|1;2)(2\mathrm{i})^{-1}[M^{(B)*}(1)-M^{(B)}(2)]U^{(C)}(1;2)\ , \tag{3.2.30}$$

$$w_z^{(I)}(\rho)=\delta(\rho|1;2)2^{-1}[h^*(1)+h(2)]U^{(C)}S^{(B)}U^{(C)}(1;2)\ , \tag{3.2.31}$$

i.e.,

$$-\frac{\partial}{\partial\rho}\cdot\beta(\rho)I^{(B)}=\langle w_z(\rho)\rangle=w_z^{(C)}(\rho)+w_z^{(I)}(\rho)\ . \tag{3.2.32a}$$

Equation (3.2.32a) shows, explicitly, that the vertical power flux on the plane is exactly that which is converted to the power flux of a surface wave, given by the left-hand side; also ensured is

$$\int d\rho\,\langle w_z(\rho)\rangle=0\ , \tag{3.2.32b}$$

equivalent to the averaged version of (3.1.46b).

The average of the modified power equation (3.1.53) is hence guaranteed everywhere in the space and on the reference plane, $z=0$, and becomes, on replacing $\psi\rightarrow g_b$,

$$\frac{\partial}{\partial\widehat{\boldsymbol{x}}}\cdot[\widehat{\boldsymbol{\alpha}}(\widehat{\boldsymbol{x}}|1;2)+\delta(z)\boldsymbol{\beta}(\boldsymbol{\rho}|1;2)]I^{(B)}(1;2)=\delta(\widehat{\boldsymbol{x}}|1;2)\Delta G(1;2)\ , \tag{3.2.33a}$$

in terms of the operator, $\widehat{\boldsymbol{\alpha}}$, and the matrix, ΔG, defined by (2.3.19, 16b), respectively. The equation can be written more simply, according to the convention of (2.3.14), and $\boldsymbol{\beta}(\widehat{\boldsymbol{x}}|1;2)=\delta(z)\boldsymbol{\beta}(\boldsymbol{\rho}|1;2)$, as

$$\frac{\partial}{\partial\widehat{\boldsymbol{x}}}\cdot(\widehat{\boldsymbol{\alpha}}+\boldsymbol{\beta})I^{(B)}(\widehat{\boldsymbol{x}})=\Delta G(\widehat{\boldsymbol{x}})\ , \tag{3.2.33b}$$

and will later be reproduced based on the *BS* equation, by (3.2.55).

On the other hand, by the $\boldsymbol{\rho}$ integration of the local relation (3.2.25, 26), we obtain, in terms of the matrix,

$$\delta(1;2)=\int d\rho\,\delta(\rho|1;2)\ , \tag{3.2.34}$$

the integrated optical relations,

$$\delta(1;2)\Big\{M^{(B)*}(1) - M^{(B)}(2) - \big[G^*(1) - G(2)\big]K^{(B)}(1;2)\Big\} = 0 \ , \qquad (3.2.35a)$$

$$\delta(1;2)\Big\{M^{(B)*}(1) - M^{(B)}(2) - \mathrm{i}\big[h^*(1) + h(2)\big]U^{(C)}S^{(B)}(1;2)\Big\} = 0 \ . \qquad (3.2.35b)$$

Here, in view of

$$\delta(1;2)\big[B_0(1) - B_0(2)\big] = 0, \qquad B_0^T = B_0 \ , \qquad (3.2.35c)$$

$M^{(B)}$ and $M^{(B)*}$ can be replaced by $M^{(b)}$ and $M^{(b)*}$, respectively. In fact, relation (3.2.35a) is strictly satisfied by the $M^{(b)}$ and $K^{(B)}$ of (3.2.4, 17) for a slightly random surface, as shown by

$$\delta(1;2)\big[G^*(1) - G(2)\big]\langle b(1)b(2)\rangle = \langle b(G^* - G)b\rangle = \delta(1;2)\big[M^*(1) - M(2)\big] \ .$$

The same is also shown for a surface of randomly distributed bosses, to be treated in Sect. 3.2.3. Note that the Green's function involved in the $M^{(b)}$ is the statistical Green's function (not the free surface Green's function, g_0), and is essential for the strict satisfaction of the relation.

3.2.3 M and K for a Surface of Randomly Distributed Bosses

When small bosses are randomly distributed on the plane, $z = 0$, the term, b, is given by (3.1.43), whose mathematical situation is the same as when small discrete scatterers are randomly distributed in space. Thus, in reference to (2.2.14, 19), we obtain

$$M^{(b)}(\rho - \rho') = n\int d\rho_a \ \langle T_a^M(\rho - \rho_a|\rho' - \rho_a)\rangle' \qquad (3.2.36a)$$

or, in the ρ matrix form

$$M^{(b)}(2) = n\int d\rho_a \ \langle T_a^M(2)\rangle' \ . \qquad (3.2.36b)$$

Here, n is the density of bosses per unit area, ρ_a is the center coordinates of boss b_a, $\langle\cdots\rangle'$ means the statistical average over all possible characteristics of the bosses, including their size, shape, orientation, etc., and T_a^M is the scattering matrix of b_a, defined by

$$T_a^M = b_a(1 + GT_a^M) = (1 - b_aG)^{-1}b_a \ , \qquad (3.2.36c)$$

and differs from that of (3.1.38b) by $g_0 \to G \equiv G_M$. Equation (3.2.36a) provides a means to determine the still unknown M self-consistently [see also (3.2.88a)].

In the same way, from (2.2.19),

$$K^{(B)}(1;2) = n \int d\rho_a \langle T_a^{M*}(1) T_a^M(2) \rangle' \ , \tag{3.2.37}$$

which, with $M^{(b)}$ given by (3.2.36), strictly satisfies the optical relation (3.2.35a), in spite of the fact that $M^{(b)}$ involved in T_a^M is not Hermitian, as can be shown directly by using (2.2.10b).

3.2.4 Continuation to the Outside Space

So far the statistical Green's functions have been treated only on the reference plane, $z = 0$, but, the continuation of G into the outside space to obtain $G(z|z')$, for example, is straightforward by averaging (3.1.36b); hence,

$$G(z|z') = G^{(0)}(z - z') + G^{(0)}(z) \langle T \rangle G^{(0)}(-z') \ , \tag{3.2.38}$$

where, with $\langle R \rangle$ of (3.2.7),

$$\langle T \rangle = 2ih \langle R \rangle \ . \tag{3.2.39}$$

Here, when $z = 0$ and/or $z' = 0$ so that the interference of the direct and scattered waves becomes critically important, it is more convenient to use the z-continued version of expression (3.2.10a), given by

$$G(0|z) = 2AG^{(0)}(z) \ , \tag{3.2.40a}$$

$$G(z|0) = 2G^{(0)}(z)A \ . \tag{3.2.40b}$$

Thus it follows that the z-continuation of (3.2.21) for $I^{(B)}(1;2)$ can be made simply with the replacement of $G \to G(z|z')$; i.e., on using (3.2.40), we obtain an expression of the form,

$$\begin{aligned} I^{(B)}(z_1; z_2|z_1'; z_2') &= U^{(C)}(z_1; z_2|z_1'; z_2') \\ &+ U(z_1; z_2|0) \sigma^{(I)}(1;2) U(0|z_1'; z_2') \ . \end{aligned} \tag{3.2.41}$$

Here,

$$U^{(C)}(z_1; z_2|z_1'; z_2') = G^*(z_1|z_1') G(z_2|z_2') \ , \tag{3.2.42a}$$

$$U(z_1; z_2|0) = G^{(0)*}(z_1) G^{(0)}(z_2) = U(0|z_1; z_2) \ , \tag{3.2.42b}$$

being given by the free space Green's functions, and

$$U^{(C)}(z_1;z_2|0) = U(z_1;z_2|0)F^{(C)}(1;2) \ ,$$
$$U^{(C)}(0|z_1;z_2) = F^{(C)}(1;2)U(0|z_1;z_2) \ , \qquad (3.2.42c)$$

where

$$F^{(C)}(1;2) = 4A^*(1)A(2) \ , \qquad (3.2.42d)$$

so that the ρ matrix, $\sigma^{(I)}(1;2)$, is given by

$$\sigma^{(I)}(1;2) = F^{(C)}(1;2)S^{(B)}(1;2)F^{(C)}(1;2) \ , \qquad (3.2.43)$$

with the matrix elements, $\sigma^{(I)}(\rho_1;\rho_2|\rho_1';\rho_2')$, independent of the z-coordinate.

Equation (3.2.41) can also be directly derived by using the z-continued version of (3.2.14), i.e.,

$$g_b(z|z') = G(z|z') + G(z|0)\Delta b g_b(0|z') \ , \qquad (3.2.44a)$$
$$G(z|0) = \exp(-ihz)G, \qquad g_b(0|z') = g_b \exp(-ihz') \ , \qquad (3.2.44b)$$

which follows from (3.1.16b) upon substitution of (3.2.14). Hence the procedure becomes the same as when deriving the original BS equation (3.2.12) or, more generally, (2.3.7) for a random medium in space, on regarding Δb as a $\hat{x}$-coordinate matrix with the elements

$$\Delta b(\hat{x}|\hat{x}') = \delta(z)\Delta b(\rho|\rho')\delta(z') \ . \qquad (3.2.44c)$$

Equation (3.2.41) can be written as a $\hat{x}$ matrix equation, by

$$I^{(B)}(1;2) = U^{(C)}(1;2) + U^{(C)}(1;2)S^{(B)}(1;2)U^{(C)}(1;2) \quad (3.2.45a)$$
$$= U^{(C)}(1;2) + U(1;2)\sigma^{(I)}(1;2)U(1;2) \ , \qquad (3.2.45b)$$

in the same form as the original ρ matrix equation (3.2.21). Here, using (3.2.38), we can write $U^{(C)}(1;2)$ in the form,

$$U^{(C)}(1;2) = U(1;2) + U(1;2)V(1;2)U(1;2) \ , \qquad (3.2.46a)$$

with

$$V(1;2) = \langle T^*(1)\rangle\langle T(2)\rangle + \langle T^*(1)\rangle G^{-1}(2) + \langle T(2)\rangle\left[G^*(1)\right]^{-1} \ . \quad (3.2.46b)$$

Hence, (3.2.45b) can further be rewritten in the form,

$$I^{(B)}(1;2) = U(1;2) + U(1;2)\sigma^{(T)}(1;2)U(1;2) \ , \qquad (3.2.47)$$

in terms of a total scattering matrix, $\sigma^{(T)}$, defined by

$$\sigma^{(T)}(1;2) = \langle T^*(1)\rangle\langle T(2)\rangle + \sigma^{(I)}(1;2) \ , \qquad (3.2.48)$$

on neglect of the interference terms in (3.2.46b) which are negligible when the points z_j and z_j', $j = 1,2$, are separated far enough from the plane.

3.2.5 Rough Surface as an Effective Random Medium

Equation (3.1.16b) indicates that the deterministic Green's function, $g_b(z|z')$, is the solution of

$$[-\partial_z^2 - h^2]g_b(z|z') = \delta(z - z') \ , \qquad (3.2.49a)$$

subjected to

$$-\partial_z g_b(z|z') \big|_{z=0} = Bg_b(z{=}0|z') \ . \qquad (3.2.49b)$$

Here, the two equations can be unified to be written by one wave equation in the form,

$$[-\partial_z^2 - h^2 - B]g_b(z|z') = \delta(z - z') \ , \qquad (3.2.50a)$$

with the new boundary condition that $\partial_z g_b = 0$ at $z = -0$, and with the $\hat{x}$ matrix, B, defined in the same fashion as the $\hat{x}$ matrix, Δb, is defined by (3.2.44c). The proof is given by integrating (3.2.50a) over an infinitesimal range, $+0 \geq z \geq -0$; hence the boundary equation (3.2.49b) is reproduced. The full coordinate expression of (3,2.50a) is

$$(L - B)g_b(\hat{x}|\hat{x}') = \delta(\hat{x} - \hat{x}') \ , \qquad (3.2.50b)$$

which shows that B is involved in the equation in the same form as a medium, q, is involved in equation (2.1.8a) for G_q. The averaging of (3.2.50b) yields

$$(L - M^{(B)})G(\hat{x}|\hat{x}') = \delta(\hat{x} - \hat{x}') \ , \qquad (3.2.51)$$

in view of (3.2.24a); and the solution, G, and the complex-congugate, G^*, constitute $U^{(C)}$ of (3.2.42a).

Thus, power equation (2.3.18) for the coherent part, U, is now replaced by that equation for $U^{(C)}$, given by

$$\left(\frac{\partial}{\partial \hat{x}} \cdot \hat{\alpha} + \Gamma^{(B)}\right) U^{(C)}(\hat{x}) = \Delta G(\hat{x}) \ , \qquad (3.2.52)$$

where, on reference to (2.3.16a), the $\hat{x}$ matrix, $\Gamma^{(B)}$, is defined by

$$\Gamma^{(B)}(1;2) = (2\mathrm{i})^{-1}\left[M^{(B)*}(1) - M^{(B)}(2)\right] \ , \qquad (3.2.53)$$

and use has been made of the convention (2.3.14). The optical relation (3.2.25) is now rewritten as

$$\frac{\partial}{\partial \hat{x}} \cdot \beta(\hat{x}|1;2) = \delta(\hat{x}|1;2)\left[\Gamma^{(B)}(1;2) - \Delta G(1;2)K^{(B)}(1;2)\right] \ , \qquad (3.2.54a)$$

or, for short

$$\frac{\partial}{\partial \hat{x}} \cdot \beta(\hat{x}) = \Gamma^{(B)}(\hat{x}) - \Delta G(\hat{x})K^{(B)} \ . \qquad (3.2.54b)$$

Hence, power equation (3.2.33b) for $I^{(B)}(\hat{x})$ is reproduced directly by using the space-continued version of BS equation (3.2.12) and (3.2.52), i.e.,

$$\begin{aligned}\Big(\frac{\partial}{\partial\hat{x}}\cdot\hat{\alpha}+\Gamma^{(B)}\Big)I^{(B)}(\hat{x}) &= \Delta G(\hat{x})(1+K^{(B)}I^{(B)})\\ &= \Delta G(\hat{x})+\Big(\Gamma^{(B)}-\frac{\partial}{\partial\hat{x}}\cdot\beta\Big)I^{(B)}(\hat{x}) \ , \qquad (3.2.55)\end{aligned}$$

where the last term is obtained by using (3.2.54b).

The situation becomes more clear by rewriting (3.2.54b) in terms of the scattering matrix, $S^{(B)}$ of $K^{(B)}$: Multiplying the equation to the right with $(1-U^{(C)}K^{(B)})^{-1}$ and using (3.2.29), we obtain

$$\begin{aligned}\frac{\partial}{\partial\hat{x}}\cdot\beta^{(e)}(\hat{x}) &= \Gamma^{(B)}(\hat{x})(1+U^{(C)}S^{(B)})-\Delta G(\hat{x})S^{(B)}\\ &= \Gamma^{(B)}(\hat{x})+[\Gamma^{(B)}(\hat{x})U^{(C)}-\Delta G(\hat{x})]S^{(B)} \ , \qquad (3.2.56)\end{aligned}$$

and using (3.2.52) on the right-hand side gives

$$\frac{\partial}{\partial\hat{x}}\cdot[\beta^{(e)}(\hat{x})+\hat{\alpha}U^{(C)}S^{(B)}(\hat{x})] = \Gamma^{(B)}(\hat{x}) \ , \qquad (3.2.57)$$

which provides a space-continued version of (3.2.26). After multiplying both sides to the right with $U^{(C)}$, we obtain an equation with the meaning that the coherent power absorbed by the rough surface (3.2.30), as given by the right-hand side, is all spent to generate the surface wave and the incoherently scattered waves by the rough surface [cf. (2.3.21)].

3.2.6 Optical Expressions and Cross Sections per Unit Area

Equations (3.2.45, 47) for the Green's function of second order become simple in the asymptotic region far enough from the surface, by writing them in terms of optical expressions for various quantities involved. From (3.2.10a), we first observe that

$$G(\rho|\hat{x}) = 2\int d\rho' A(\rho-\rho')G^{(0)}(\rho'-\hat{x}) \ , \qquad (3.2.58a)$$

where, in asymptotic region of $k_0|\rho-\hat{x}|\gg 1$ and $|\rho-\rho'|\ll|\rho-\hat{x}|$,

$$G^{(0)}(\rho'-\hat{x}) \simeq \exp[-ik_0\Omega\cdot(\rho'-\rho)]\,G^{(0)}(\rho-\hat{x}) \ , \qquad (3.2.58b)$$

with the unit vector, $\hat{\Omega}=(\Omega,-\Omega_z)$ for $\Omega_z>0$, in the direction of $\rho-\hat{x}$. Hence (3.2.58a) can be given by an asymptotic expression,

$$G(\rho|\hat{x}) = 2A(\hat{\Omega})G^{(0)}(\rho-\hat{x}) \ , \qquad (3.2.59)$$

with the factor, $A(\widehat{\boldsymbol{\Omega}})$, which is given, in terms of the Fourier transform, $\widetilde{A}(\boldsymbol{u})$, from (3.2.10c),

$$\widetilde{A}(\boldsymbol{u}) = \widetilde{h}(\boldsymbol{u})\left[\widetilde{h}(\boldsymbol{u}) + \mathrm{i}\widetilde{M}^{(B)}(\boldsymbol{u})\right]^{-1} , \tag{3.2.60}$$

by

$$A(\widehat{\boldsymbol{\Omega}}) = \widetilde{A}(\boldsymbol{u})\Big|_{\boldsymbol{\Omega}} = \Omega_z\left[\Omega_z + \mathrm{i}k_0^{-1}M^{(B)}(\widehat{\boldsymbol{\Omega}})\right]^{-1} , \tag{3.2.61a}$$

$$M^{(B)}(\widehat{\boldsymbol{\Omega}}) = \widetilde{M}^{(B)}(\boldsymbol{u})\Big|_{\boldsymbol{\Omega}} ; \tag{3.2.61b}$$

and which agrees with the asymptotic expression (3.1.28b) for $A(\boldsymbol{\rho}, z)$ when $B_0 \to M^{(B)}$.

For evaluation of the second term in (3.2.45a), we can utilize (3.1.89) and (3.2.59) to obtain the asymptotic expression,

$$\begin{aligned} U^{(C)}(\boldsymbol{r}''', \boldsymbol{\rho}'''|\widehat{\boldsymbol{x}}') &= G^*(\boldsymbol{\rho}''' - \boldsymbol{r}'''/2|\widehat{\boldsymbol{x}}')G(\boldsymbol{\rho}''' + \boldsymbol{r}'''/2|\widehat{\boldsymbol{x}}') \\ &\simeq F^{(C)}(\widehat{\boldsymbol{\Omega}})\exp(-\mathrm{i}k_0\boldsymbol{\Omega}' \cdot \boldsymbol{r}''')\left|G^{(0)}(\boldsymbol{\rho}''' - \widehat{\boldsymbol{x}}')\right|^2 , \end{aligned} \tag{3.2.62a}$$

where

$$F^{(C)}(\widehat{\boldsymbol{\Omega}}) = |2A(\widehat{\boldsymbol{\Omega}})|^2 , \tag{3.2.62b}$$

in terms of the relative coordinates, $\boldsymbol{r}$ and $\boldsymbol{\rho}$, of (2.3.25-27). Here, $\widehat{\boldsymbol{\Omega}}' = (\boldsymbol{\Omega}', -\Omega_z')$ is in the direction of $\boldsymbol{\rho}''' - \widehat{\boldsymbol{x}}'$, being the incident direction. In the same way,

$$\begin{aligned} U^{(C)}(\widehat{\boldsymbol{x}}|\boldsymbol{r}'', \boldsymbol{\rho}'') &= G^*(\widehat{\boldsymbol{x}}|\boldsymbol{\rho}'' - \boldsymbol{r}''/2)G(\widehat{\boldsymbol{x}}|\boldsymbol{\rho}'' + \boldsymbol{r}''/2) \\ &\simeq F^{(C)}(\widehat{\boldsymbol{\Omega}})\exp(\mathrm{i}k_0\boldsymbol{\Omega} \cdot \boldsymbol{r}'')\left|G^{(0)}(\widehat{\boldsymbol{x}} - \boldsymbol{\rho}'')\right|^2 , \end{aligned} \tag{3.2.62c}$$

where $\widehat{\boldsymbol{\Omega}} = (\boldsymbol{\Omega}, +\Omega_z)$, is in the direction of $\widehat{\boldsymbol{x}} - \boldsymbol{\rho}''$, or of the scattered wave.

Thus, performing the $\boldsymbol{r}''$ and $\boldsymbol{r}'''$ integrations in the second term of (3.2.45a), the result is obtained when $\widehat{\boldsymbol{x}}_1 = \widehat{\boldsymbol{x}}_2 = \widehat{\boldsymbol{x}}$ and $\widehat{\boldsymbol{x}}_1' = \widehat{\boldsymbol{x}}_2' = \widehat{\boldsymbol{x}}'$, in the form (Fig. 3.3),

$$\begin{aligned} I^{(B)}(\widehat{\boldsymbol{x}}|\widehat{\boldsymbol{x}}') &\equiv I^{(B)}(\widehat{\boldsymbol{x}}; \widehat{\boldsymbol{x}}|\widehat{\boldsymbol{x}}'; \widehat{\boldsymbol{x}}') \\ &= |G(\widehat{\boldsymbol{x}}|\widehat{\boldsymbol{x}}')|^2 + \int d\boldsymbol{\rho}'' d\boldsymbol{\rho}''' |\widehat{\boldsymbol{x}} - \boldsymbol{\rho}''|^{-2} \\ &\quad \times \sigma^{(I)}(\widehat{\boldsymbol{\Omega}}|\boldsymbol{\rho}'' - \boldsymbol{\rho}'''|\widehat{\boldsymbol{\Omega}}')\left|G^{(0)}(\boldsymbol{\rho}''' - \widehat{\boldsymbol{x}}')\right|^2 . \end{aligned} \tag{3.2.63}$$

Here,

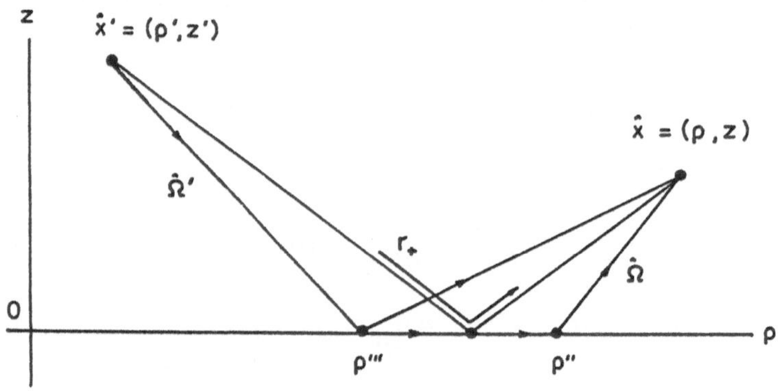

Fig. 3.3. Geometry and notations for (3.2.63, 71)

$$\sigma^{(I)}(\widehat{\boldsymbol{\Omega}}|\boldsymbol{\rho}''-\boldsymbol{\rho}'''|\widehat{\boldsymbol{\Omega}}')=F^{(C)}(\widehat{\boldsymbol{\Omega}})S^{(B)}(\widehat{\boldsymbol{\Omega}}|\boldsymbol{\rho}''-\boldsymbol{\rho}'''|\widehat{\boldsymbol{\Omega}}')F^{(C)}(\widehat{\boldsymbol{\Omega}}') \ , \qquad (3.2.64)$$

where, in terms of the matrix elements, $S^{(B)}(\boldsymbol{\rho}_1'';\boldsymbol{\rho}_2''|\boldsymbol{\rho}_1''';\boldsymbol{\rho}_2''')=S^{(B)}(\boldsymbol{r}''|\boldsymbol{\rho}''-\boldsymbol{\rho}'''|\boldsymbol{r}''')$ of $S^{(B)}(1;2)$,

$$S^{(B)}(\widehat{\boldsymbol{\Omega}}|\boldsymbol{\rho}''-\boldsymbol{\rho}'''|\widehat{\boldsymbol{\Omega}}')=(4\pi)^{-2}\int d\boldsymbol{r}''d\boldsymbol{r}'''$$
$$\times\exp\left[\mathrm{i}k_0(\boldsymbol{\Omega}\cdot\boldsymbol{r}''-\boldsymbol{\Omega}'\cdot\boldsymbol{r}''')\right]S^{(B)}(\boldsymbol{r}''|\boldsymbol{\rho}''-\boldsymbol{\rho}'''|\boldsymbol{r}''') \ , \qquad (3.2.65)$$

which is a short-range function of $\boldsymbol{\rho}''-\boldsymbol{\rho}'''$, appreciable only within an effective range of multiple scattering (3.2.22c). In terms of the Fourier transform, $\widetilde{S}^{(B)}(\boldsymbol{u}|\boldsymbol{\lambda}|\boldsymbol{u}')$ of $S^{(B)}$, defined by [cf. (2.3.26–29)]

$$\widetilde{S}^{(B)}(\boldsymbol{\lambda}_1;\boldsymbol{\lambda}_2|\boldsymbol{\lambda}_1';\boldsymbol{\lambda}_2')=(2\pi)^2\delta(\boldsymbol{\lambda}-\boldsymbol{\lambda}')\widetilde{S}^{(B)}(\boldsymbol{u}|\boldsymbol{\lambda}|\boldsymbol{u}') \ , \qquad (3.2.66)$$

(3.2.65) can be expressed as

$$S^{(B)}(\widehat{\boldsymbol{\Omega}}|\boldsymbol{\rho}''-\boldsymbol{\rho}'''|\widehat{\boldsymbol{\Omega}}')=(2\pi)^{-2}\int d\boldsymbol{\lambda}\exp\left[-\mathrm{i}\boldsymbol{\lambda}\cdot(\boldsymbol{\rho}''-\boldsymbol{\rho}''')\right]$$
$$\times(4\pi)^{-2}\widetilde{S}^{(B)}(\boldsymbol{u}=k_0\boldsymbol{\Omega}|\boldsymbol{\lambda}|\boldsymbol{u}'=k_0\boldsymbol{\Omega}') \ , \qquad (3.2.67)$$

and hence, also,

$$S^{(B)}(\widehat{\boldsymbol{\Omega}}|\widehat{\boldsymbol{\Omega}}')\equiv\int d\boldsymbol{\rho}''S^{(B)}(\widehat{\boldsymbol{\Omega}}|\boldsymbol{\rho}''-\boldsymbol{\rho}'''|\widehat{\boldsymbol{\Omega}}')$$
$$=(4\pi)^{-2}\widetilde{S}^{(B)}(k_0\boldsymbol{\Omega}|\boldsymbol{\lambda}=0|k_0\boldsymbol{\Omega}') \ . \qquad (3.2.68)$$

Thus, when the points, $\widehat{\boldsymbol{x}}$ and $\widehat{\boldsymbol{x}}'$, are separated far enough from the surface, expression (3.2.63) can be approximated by

$$I^{(B)}(\hat{\boldsymbol{x}}|\hat{\boldsymbol{x}}') = \left|G(\hat{\boldsymbol{x}}|\hat{\boldsymbol{x}}')\right|^2 + \int d\rho''|\hat{\boldsymbol{x}} - \rho''|^{-2}\sigma^{(I)}(\hat{\boldsymbol{\Omega}}|\hat{\boldsymbol{\Omega}}')\left|G^{(0)}(\rho'' - \hat{\boldsymbol{x}}')\right|^2 \ , \quad (3.2.69)$$

in terms of an incoherent scattering cross section per unit area, $\sigma^{(I)}(\hat{\boldsymbol{\Omega}}|\hat{\boldsymbol{\Omega}}')$, given in view of (3.2.68, 64), by

$$\sigma^{(I)}(\hat{\boldsymbol{\Omega}}|\hat{\boldsymbol{\Omega}}') = F^{(C)}(\hat{\boldsymbol{\Omega}})S^{(B)}(\hat{\boldsymbol{\Omega}}|\hat{\boldsymbol{\Omega}}')F^{(C)}(\hat{\boldsymbol{\Omega}}') \ . \quad (3.2.70)$$

In the same way, also, for the coherent term in (3.2.69), we obtain a similar expression of the form,

$$\left|G(\hat{\boldsymbol{x}}|\hat{\boldsymbol{x}}')\right|^2 = \left|G^{(0)}(\hat{\boldsymbol{x}} - \hat{\boldsymbol{x}}')\right|^2 + \int d\rho''|\hat{\boldsymbol{x}} - \rho''|^{-2} \times \Omega_z\left|\langle R(\hat{\boldsymbol{\Omega}})\rangle\right|^2\delta^2(\hat{\boldsymbol{\Omega}} - \hat{\boldsymbol{\Omega}}')\left|G^{(0)}(\rho'' - \hat{\boldsymbol{x}}')\right|^2 \ , \quad (3.2.71a)$$

as will be shown later, separately. Here, the second term differs from zero only when making the specular reflection, with the factor,

$$\langle R(\hat{\boldsymbol{\Omega}})\rangle = \langle \widetilde{R}(u = k_0\boldsymbol{\Omega})\rangle \ , \quad (3.2.71b)$$

and the whole expression gives a specific version of the matrix equation (3.2.46a) when neglecting the interference terms. Thus, by substituting (3.2.71a) in (3.2.69), we obtain the entire asymptotic expression as

$$I^{(B)}(\hat{\boldsymbol{x}}|\hat{\boldsymbol{x}}') = \left|G^{(0)}(\hat{\boldsymbol{x}} - \hat{\boldsymbol{x}}')\right|^2 + \int d\rho''|\hat{\boldsymbol{x}} - \rho''|^{-2}\sigma^{(T)}(\hat{\boldsymbol{\Omega}}|\hat{\boldsymbol{\Omega}}')\left|G^{(0)}(\rho'' - \hat{\boldsymbol{x}}')\right|^2 \ . \quad (3.2.72)$$

Here,

$$\sigma^{(T)}(\hat{\boldsymbol{\Omega}}|\hat{\boldsymbol{\Omega}}') = \left|\Omega_z\langle R(\hat{\boldsymbol{\Omega}})\rangle^2\right|\delta^2(\hat{\boldsymbol{\Omega}} - \hat{\boldsymbol{\Omega}}') + \sigma^{(I)}(\hat{\boldsymbol{\Omega}}|\hat{\boldsymbol{\Omega}}') \ , \quad (3.2.73)$$

and provides the resultant cross section per unit area, including the coherent one, whose original version was given in matrix form by (3.2.48).

To derive (3.2.71a) from (3.2.46a,b) on neglect of the interference terms, we observe that the second term can be written by following the same procedure as in (3.2.63–67), in the form,

$$\int d\rho'' d\rho'''|\hat{\boldsymbol{x}} - \rho''|^{-2}\sigma^{(C)}(\hat{\boldsymbol{\Omega}}|\rho'' - \rho'''|\hat{\boldsymbol{\Omega}}')|G^{(0)}(\rho''' - \hat{\boldsymbol{x}}')|^2 \ .$$

Here, on reference to (3.2.67),

$$\sigma^{(C)}(\widehat{\Omega}|\rho''-\rho'''|\widehat{\Omega}') = (4\pi)^{-2}\delta(u-u')\int d\lambda$$
$$\times \exp\left[-i\lambda\cdot(\rho''-\rho''')\right]\left\langle\widetilde{T}^*(u-\lambda/2)\right\rangle\left\langle\widetilde{T}(u+\lambda/2)\right\rangle$$
$$\simeq 4^{-1}\left|\left\langle\widetilde{T}(u)\right\rangle\right|^2\delta(u-u')\Big|_{\Omega}\,\delta(\rho''-\rho''')$$
$$= \left|\widetilde{h}(u)\left\langle\widetilde{R}(u)\right\rangle\right|^2\delta(u-u')\Big|_{\Omega}\,\delta(\rho''-\rho''') ,$$

where use has been made of $|\lambda| \ll |u| \sim k_0$, (3.2.39), and

$$\delta(u-u')\Big|_{\Omega} = k_0^{-2}\delta(\Omega-\Omega') = (k_0^2\Omega_z)^{-1}\delta^2(\Omega-\Omega') .$$

Thus, in terms of the notation, $\left\langle R(\widehat{\Omega})\right\rangle$, by (3.2.71b), we obtain

$$\sigma^{(C)}(\widehat{\Omega}|\rho''-\rho'''|\widehat{\Omega}') = \Omega_z\left|\left\langle R(\widehat{\Omega})\right\rangle\right|^2\delta^2(\widehat{\Omega}-\widehat{\Omega}')\delta(\rho''-\rho''') ,$$

to yield the second term of (3.2.71a), which after the ρ'' integration, becomes (Fig. 3.3)

$$|\langle R(\widehat{\Omega})\rangle|^2(4\pi r_+)^{-2} ,$$
$$r^+ = \left[(x-x')^2+(z+z')^2\right]^{1/2} .$$

Here, the last derivation is straightforward by changing the variable of integration, ρ'' to $\widehat{\Omega}'$, $d\rho'' = (\Omega_z')^{-1}|\rho''-\widehat{x}'|^2 d\widehat{\Omega}'$, and using (3.1.76) with the Jacobian,

$$\left|\partial(\widehat{\Omega}-\widehat{\Omega}')/\partial\widehat{\Omega}'\right| = r_+^2|\widehat{x}-\rho''|^{-2} .$$

The incoherent cross section, $\sigma^{(I)}(\widehat{\Omega}|\widehat{\Omega}')$, is given by (3.2.70) in terms of $S^{(B)}(\widehat{\Omega}|\widehat{\Omega}')$, defined by (3.2.68) and derived, therefore, from the solution of integral equation (3.2.22a) whose Fourier transform is

$$\widetilde{S}^{(B)}(u|\lambda|u') = \widetilde{K}^{(B)}(u|\lambda|u') + (2\pi)^{-2}\int du''\widetilde{K}^{(B)}(u|\lambda|u'')$$
$$\times\,\widetilde{U}^{(C)}(u'',\lambda)\widetilde{S}^{(B)}(u''|\lambda|u') , \tag{3.2.74}$$
$$\widetilde{U}^{(C)}(u,\lambda) = \widetilde{G}^*(u-\lambda/2)\widetilde{G}(u+\lambda/2) , \tag{3.2.75}$$

(which should not be confused with $\widetilde{U}(\widehat{u},\widehat{\lambda})$ of (2.3.30)). Here, $\lambda = 0$ in the present case, and, using (3.2.10a, 8),

$$\widetilde{U}^{(C)}(u,\lambda{=}0)\Big|_{\Omega} = F^{(C)}(\widehat{\Omega})\widetilde{U}(u,\lambda{=}0)\Big|_{\Omega}$$
$$= F^{(C)}(\widehat{\Omega})(2k_0\Omega_z)^{-2} . \tag{3.2.76}$$

Hence, with $du'' = k_0^2 \Omega_z'' d\hat{\Omega}''$, and a Ω matrix, $K^{(B)}(\hat{\Omega}|\hat{\Omega}')$, defined by

$$K^{(B)}(\hat{\Omega}|\hat{\Omega}') = (4\pi)^{-2}\widetilde{K}^{(B)}(u{=}k_0\Omega|\lambda{=}0|u'{=}k_0\Omega') \ , \qquad (3.2.77)$$

(3.2.74) is rewritten as an integral equation for $S^{(B)}(\hat{\Omega}|\hat{\Omega}')$,

$$\begin{aligned} S^{(B)}(\hat{\Omega}|\hat{\Omega}') = K^{(B)}(\hat{\Omega}|\hat{\Omega}') + \int d\hat{\Omega}'' K^{(B)}(\hat{\Omega}|\hat{\Omega}'') \\ \times F^{(C)}(\hat{\Omega}'')(\Omega_z'')^{-1} S^{(B)}(\hat{\Omega}''|\hat{\Omega}') \ . \end{aligned} \qquad (3.2.78)$$

Here, the integration range is not limited within the half-solid angle, 2π, but also includes the nonoptical range, $|\Omega| > 1$, so that $\Omega_z = \widetilde{h}(u)/k_0$ goes from $-\mathrm{i}\infty$ to 1 via 0.

3.2.7 Optical Relations

The incoherent cross section, $\sigma^{(I)}(\hat{\Omega}|\hat{\Omega}')$, of (3.2.70), is subject to an optical relation to ensure power conservation. It is derived from the Fourier transform of the integrated optical relation (3.2.35b), i.e.,

$$\begin{aligned} (\widetilde{M}^{(B)*} - \widetilde{M}^{(B)})(u) - \mathrm{i}(2\pi)^{-2} \int du'(\widetilde{h}^* + \widetilde{h})(u') \\ \times \widetilde{U}^{(C)}(u',\lambda)\widetilde{S}^{(B)}(u'|\lambda|u)\Big|_{\lambda=0} = 0 \ , \end{aligned} \qquad (3.2.79)$$

and can be written, using (3.2.76) and a new quantity, $\gamma^{(B)}(\hat{\Omega})$, defined by

$$\begin{aligned} \gamma^{(B)}(\hat{\Omega}) &= (2\mathrm{i}k_0)^{-1}[\widetilde{M}^{(B)*}(u) - \widetilde{M}^{(B)}(u)]\Big|_{\Omega} \\ &\equiv k_0^{-1}\Gamma^{(B)}(\hat{\Omega}) \ , \end{aligned} \qquad (3.2.80)$$

as

$$\gamma^{(B)}(\hat{\Omega}) = \int_{2\pi} d\hat{\Omega}' F^{(C)}(\hat{\Omega}')S^{(B)}(\hat{\Omega}'|\hat{\Omega}) \ . \qquad (3.2.81)$$

Note that the integration is made only over the optical range, 2π. From (3.2.70), $\sigma^{(I)}(\hat{\Omega}|\hat{\Omega}')$ is thus found to be subject to

$$\gamma^{(B)}(\hat{\Omega})F^{(C)}(\hat{\Omega}) = \int_{2\pi} d\hat{\Omega}' \sigma^{(I)}(\hat{\Omega}'|\hat{\Omega}) \ . \qquad (3.2.82)$$

Also, for $\langle R(\hat{\Omega})\rangle$, constituting the coherent cross section, there exists an optical relation resulting from the identity from (3.2.6), i.e.,

$$M^{(B)} = -\mathrm{i}h(1 - \langle R\rangle)(1 + \langle R\rangle)^{-1} \ . \qquad (3.2.83)$$

Hence, taking the anti-Hermitian parts of both sides, we find a relation between $\langle R\rangle$ and $\Gamma^{(B)}$, defined by (3.2.53), as

$$(1+\langle R^*\rangle)\Gamma^{(B)}(1+\langle R\rangle) = 2^{-1}(h^*+h) - \langle R\rangle^* 2^{-1}(h^*+h)\langle R\rangle \\ +2^{-1}(h^*-h)\langle R\rangle + \langle R\rangle^* 2^{-1}(h-h^*) \ . \tag{3.2.84}$$

Here, in the optical range where $h^* = h$ (3.1.8), the last two terms make no contribution. Hence, by using (3.2.5, 10a, 62b), the $\widehat{\Omega}$ version of (3.2.84) can be written in terms of $F^{(C)}(\widehat{\Omega})$ and $\gamma^{(B)}(\widehat{\Omega})$ (3.2.80), as

$$\gamma^{(B)}(\widehat{\Omega})F^{(C)}(\widehat{\Omega}) = \Omega_z\left[1-\left|\langle R(\widehat{\Omega})\rangle\right|^2\right] \ , \tag{3.2.85a}$$

with the same left-hand side as that of (3.2.82), or,

$$\Omega_z\left|\langle R(\widehat{\Omega})\rangle\right|^2 = \Omega_z - \gamma^{(B)}(\widehat{\Omega})F^{(C)}(\widehat{\Omega}) \ , \tag{3.2.85b}$$

which can be used as a convenient alternative expression of the coherent term in (3.2.73).

Thus, the resultant $\sigma^{(T)}(\widehat{\Omega}|\widehat{\Omega}')$ of (3.2.73) is confirmed to satisfy

$$\int_{2\pi} d\widehat{\Omega}'\sigma^{(T)}(\widehat{\Omega}'|\widehat{\Omega}) = \Omega_z \ , \tag{3.2.86}$$

as required for the scattered waves given by (3.2.72).

In the integral equation (3.2.78) for $S^{(B)}(\widehat{\Omega}|\widehat{\Omega}')$, the factors $K^{(B)}(\widehat{\Omega}|\widehat{\Omega}')$ and $F^{(C)}(\widehat{\Omega})$ can be given to an appropriate approximation, according to the respective definitions (3.2.77, 62b), but, in order to ensure optical relation (3.2.81) for $S^{(B)}(\widehat{\Omega}|\widehat{\Omega}')$, they should be consistent with condition (3.2.35a) whose $\widehat{\Omega}$ expression is given by

$$\gamma^{(B)}(\widehat{\Omega}) = \int d\widehat{\Omega}'[\Omega_z' + \gamma^{(B)}(\widehat{\Omega}')]\,F^{(C)}(\widehat{\Omega}')(\Omega_z')^{-1}K^{(B)}(\widehat{\Omega}'|\widehat{\Omega}) \ , \tag{3.2.87}$$

in consequence of (3.2.28, 80). Here, it may be remarked that, if the term $\gamma^{(B)}(\widehat{\Omega}')$ in the integrand is negligible, then $K^{(B)}(\widehat{\Omega}|\widehat{\Omega}')$ satisfies the same relation as (3.2.81), implying that $S^{(B)}(\widehat{\Omega}|\widehat{\Omega}') \sim K^{(B)}(\widehat{\Omega}|\widehat{\Omega}')$; and, also, that for the $\gamma^{(B)}(\widehat{\Omega}')$ integrand term, the range of integration is not limited within the optical range, 2π, as remarked for the integral equation (3.2.78), in contrast with the range of (3.2.81) which is limited strictly within the 2π.

3.2.8 Examples

a) Surface of Randomly Distributed Bosses

The ρ matrices, $M^{(B)}$ and $K^{(B)}$, in this case, are given by (3.2.36, 37), respectively, and we obtain

$$\widetilde{M}^{(B)}(\boldsymbol{u}) = n\langle \widetilde{T}_a^M(\boldsymbol{u}|\boldsymbol{u})\rangle' \tag{3.2.88a}$$

$$\widetilde{K}^{(B)}(\boldsymbol{u}|\boldsymbol{\lambda}=0|\boldsymbol{u}') = n\langle \widetilde{T}_a^{M*}(\boldsymbol{u}|\boldsymbol{u}')\widetilde{T}_a^M(\boldsymbol{u}|\boldsymbol{u}')\rangle' \ , \tag{3.2.88b}$$

where $\widetilde{T}_a^M(\boldsymbol{u}|\boldsymbol{u}')$ is the Fourier transform of the scattering matrix, $T_a^M(\rho|\rho')$, for one boss, b_a (3.2.36c). Hence, from (3.2.61b, 77)

$$M^{(B)}(\boldsymbol{\Omega}) = n\langle \widetilde{T}_a^M(\boldsymbol{u}=k_0\boldsymbol{\Omega}|\boldsymbol{u}=k_0\boldsymbol{\Omega})\rangle' \ ,$$

$$K^{(B)}(\boldsymbol{\Omega}|\boldsymbol{\Omega}') = (4\pi)^{-2} n\left\langle |\widetilde{T}_a^M(\boldsymbol{u}=k_0\boldsymbol{\Omega}|\boldsymbol{u}'=k_0\boldsymbol{\Omega}')|^2\right\rangle' \ ,$$

and $A(\widehat{\boldsymbol{\Omega}})$ and $F^{(C)}(\widehat{\boldsymbol{\Omega}})$ are given by (3.2.61a, 62b), respectively. Thus the incoherent cross section, $\sigma^{(I)}(\widehat{\boldsymbol{\Omega}}|\widehat{\boldsymbol{\Omega}}')$, is given by (3.2.70) after getting $S^{(B)}(\widehat{\boldsymbol{\Omega}}|\widehat{\boldsymbol{\Omega}}') \sim K^{(B)}(\widehat{\boldsymbol{\Omega}}|\widehat{\boldsymbol{\Omega}}')$ as the solution of integral equation (3.2.78), while, for the coherent part, expression (3.2.85b) can be conveniently utilized with $\gamma^{(B)}(\widehat{\boldsymbol{\Omega}})$, given by (3.2.80). As to the optical relations, the basic $M^{(B)}$ and $K^{(B)}$ satisfy the condition (3.2.35a), strictly, and therefore, so do optical relations (3.2.82, 86); whereas the still undetermined $M^{(B)}$ is to be obtained as a self-consistent solution of (3.2.88a).

b) Slightly Random Surface

When the surface displacement, $\zeta(\rho)$, is small enough, the random part of the surface impedance, b, is given by (3.1.14), and the resulting matrices, $M^{(b)}$ and $K^{(B)}$, are given by (3.2.4, 17), respectively. Hence, $S^{(B)}(\widehat{\boldsymbol{\Omega}}|\widehat{\boldsymbol{\Omega}}')$ is given (as will be shown separately later) by

$$\begin{aligned} S^{(B)}(\widehat{\boldsymbol{\Omega}}|\widehat{\boldsymbol{\Omega}}') &\sim K^{(B)}(\widehat{\boldsymbol{\Omega}}|\widehat{\boldsymbol{\Omega}}') = (4\pi)^{-2}\widetilde{K}^{(B)}(\boldsymbol{u}|\boldsymbol{\lambda}=0|\boldsymbol{u}')\,\big|_{\boldsymbol{\Omega}} \\ &= 2^{-4}k_0^4\Phi\left[k_0(\boldsymbol{\Omega}'-\boldsymbol{\Omega})\right]\left|1+(B_0/k_0)^2-\boldsymbol{\Omega}\cdot\boldsymbol{\Omega}'\right|^2 \ , \end{aligned} \tag{3.2.89}$$

where $\Phi(\boldsymbol{u})$ is the "power" spectrum density of $\zeta(\rho)$, defined by

$$\Phi(\boldsymbol{u}) = 4(2\pi)^{-2}\int d\rho \exp\left[i\boldsymbol{u}\cdot(\rho-\rho')\right]\langle\zeta(\rho)\zeta(\rho')\rangle \ , \tag{3.2.90}$$

so that, from (3.2.70, 62b) $\sigma^{(I)}(\widehat{\boldsymbol{\Omega}}|\widehat{\boldsymbol{\Omega}}')$ is obtained in the form,

$$\sigma^{(I)}(\widehat{\boldsymbol{\Omega}}|\widehat{\boldsymbol{\Omega}}') = k_0^4\Phi\left[k_0(\boldsymbol{\Omega}'-\boldsymbol{\Omega})\right]\left|f(\widehat{\boldsymbol{\Omega}}|\widehat{\boldsymbol{\Omega}}')\right|^2 \ . \tag{3.2.91}$$

Here,

$$f(\widehat{\boldsymbol{\Omega}}|\widehat{\boldsymbol{\Omega}}') = A(\widehat{\boldsymbol{\Omega}})[1+(B_0/k_0)^2 - \boldsymbol{\Omega}\cdot\boldsymbol{\Omega}']A(\widehat{\boldsymbol{\Omega}}') \ , \qquad (3.2.92)$$

where $A(\widehat{\boldsymbol{\Omega}})$ is defined by (3.2.61), and, when neglecting the term, $M^{(B)}(\widehat{\boldsymbol{\Omega}})$,

$$f(\widehat{\boldsymbol{\Omega}}|\widehat{\boldsymbol{\Omega}}') \simeq \begin{cases} 1-\boldsymbol{\Omega}\cdot\boldsymbol{\Omega}', & B_0 = 0 \qquad (3.2.93a)\\ -\Omega_z\Omega_z', & B_0 = \infty \ . \qquad (3.2.93b)\end{cases}$$

Hence, in the case of backscattering where $\widehat{\boldsymbol{\Omega}}+\widehat{\boldsymbol{\Omega}}'=0$,

$$\sigma^{(I)}(-\widehat{\boldsymbol{\Omega}}|\widehat{\boldsymbol{\Omega}}) = k_0^4\Phi(2k_0\boldsymbol{\Omega})\times\begin{cases}(1+\Omega^2)^2, & B_0=0 \qquad (3.2.94a)\\ \Omega_z^4, & B_0=\infty \ , \qquad (3.2.94b)\end{cases}$$

excluding the range $\Omega_z \le \left|M^{(B)}(\widehat{\boldsymbol{\Omega}})/k_0\right|$ when $B_0=0$. In the last case,

$$A(\widehat{\boldsymbol{\Omega}}) = \Omega_z\left[\Omega_z + \mathrm{i}M^{(B)}(\widehat{\boldsymbol{\Omega}})/k_0\right]^{-1}, \qquad B_0 = 0 \ , \qquad (3.2.95)$$

resulting in $\sigma^{(I)}(-\widehat{\boldsymbol{\Omega}}|\widehat{\boldsymbol{\Omega}}) \to 0$ as $\Omega_z \to 0$, in contrast with the uncorrected expression (3.2.94a).

Physically, this means that the waves scattered in the grazing directions of $\Omega_z \sim 0$ are severely rescattered by the rough surface so that no wave is left for the angle $\Omega_z = 0$. Mathematically, this is a consequence of using the statistical Green's function, G, in the BS equation (3.2.12), instead of using the unperturbed g_0. The situation is the same also for $M^{(b)}$, given by (3.2.4), which satisfies the optical condition (3.2.35a) exactly, by virtue of G involved.

On the other hand, when $B_0=0$, (3.2.4) gives

$$\begin{aligned}\mathrm{i}\widetilde{M}^{(b)}(k_0\boldsymbol{\Omega}) &\simeq (2\pi)^{-2}\int d\boldsymbol{u}'\,\left[h(\boldsymbol{u}')\right]^{-1}\widetilde{K}^{(B)}(\boldsymbol{u}'|\boldsymbol{\lambda}{=}0|\boldsymbol{u})\,\Big|_{\boldsymbol{\Omega}}\\ &= 4k_0\int d\widehat{\boldsymbol{\Omega}}'\,K^{(B)}(\widehat{\boldsymbol{\Omega}}'|\widehat{\boldsymbol{\Omega}}) \ , \qquad (3.2.96)\end{aligned}$$

in view of (3.2.89). Hence, when

$$\langle\zeta(\boldsymbol{\rho})\zeta(\boldsymbol{\rho}')\rangle = \langle\zeta^2\rangle\exp\left[-(\boldsymbol{\rho}-\boldsymbol{\rho}')^2/2l^2\right] \ , \qquad (3.2.97a)$$

so that

$$\Phi(\boldsymbol{u}) = 2\pi^{-1}l^2\langle\zeta^2\rangle\exp\left[-(\boldsymbol{u}l)^2/2\right] \ , \qquad (3.2.97b)$$

then, when $\Omega_z=0$ and $B_0=0$, use of (3.2.89) shows that

$$\begin{aligned}\widetilde{M}^{(b)}(k_0) &\sim 2^{-1/4}\pi^{-1/2}\mathrm{e}^{-\mathrm{i}\pi/4}\Gamma(5/4)(l^{-1}k_0)^{3/2}\langle\zeta^2\rangle, \ k_0 l \gg 1 \ , \qquad (3.2.98a)\\ &\sim 2^{-3/2}\pi^{1/2}l^{-1}k_0^2\langle\zeta^2\rangle, \qquad k_0 l \ll 1 \ . \qquad (3.2.98b)\end{aligned}$$

Hence, $\widetilde{M}^{(b)}(k_0)$, given by (3.2.98b), has a real positive value, confirming that, when the correlation distance of the surface change is small enough compared with the wavelength, the rough surface supports a surface wave in the conventional sense, as given by (3.1.31) with $B_0 \to \widetilde{M}^{(b)}(k_0)$.

With the notation $b(j)$, $j = 1, 2$, to designate b of (3.1.14) as a ρ_j-coordinate operator, (3.2.17) gives the Fourier transform, $\widetilde{K}^{(B)}$, according to definition (2.3.28b), as

$$\begin{aligned}\widetilde{K}^{(B)}(\lambda_1; \lambda_2 | \lambda_1'; \lambda_2') &= \int d\rho_1 d\rho_2 \exp\left[-\mathrm{i}(\lambda_1 \cdot \rho_1 - \lambda_2 \cdot \rho_2)\right] \\ &\times \left\langle b(1) b(2) \right\rangle \exp\left[\mathrm{i}(\lambda_1' \cdot \rho_1' - \lambda_2' \cdot \rho_2')\right] \\ &= (k_0^2 + B_0^2 - \lambda_1 \cdot \lambda_1')(k_0^2 + B_0^2 - \lambda_2 \cdot \lambda_2') \int d\rho_1 d\rho_2 \\ &\times \exp\left\{\mathrm{i}\left[(\lambda_1' - \lambda_1) \cdot \rho_1 - (\lambda_2' - \lambda_2) \cdot \rho_2\right]\right\} \left\langle \zeta(\rho_1) \zeta(\rho_2) \right\rangle .\end{aligned}$$

Here, from definition (3.2.90),

$$\left\langle \zeta(\rho_1) \zeta(\rho_2) \right\rangle = 4^{-1} \int dv \exp\left[-\mathrm{i}v \cdot (\rho_1 - \rho_2)\right] \Phi(v) ,$$

hence, upon the substitution, we obtain [cf. (3.2.66)]

$$\begin{aligned}\widetilde{K}^{(B)}(u|\lambda|u') &= 4^{-1}(2\pi)^2 (k_0^2 + B_0^2 - \lambda_1 \cdot \lambda_1') \\ &\times (k_0^2 + B_0^2 - \lambda_2 \cdot \lambda_2') \Phi(u' - u) ,\end{aligned}$$

with

$$\begin{aligned}\lambda_1 &= u - \lambda/2, \qquad & \lambda_2 &= u + \lambda/2 , \\ \lambda_1' &= u' - \lambda/2, \qquad & \lambda_2' &= u' + \lambda/2 ,\end{aligned}$$

yielding the optical expression (3.2.89).

3.3 Transmissible (Two-Sided) Rough Boundary

As illustrated in Fig. 3.4, the rough boundary, S, is assumed to be described by $z = -\zeta(\rho) < 0$, and also to be non-dissipative, as in the previous sections; the unit vector, normal to S, is $\hat{n}^{(S)} = (n^{(S)}, n_z)$, $n_z > 0$. Hence, the space is divided into two parts, R_1 of $z > -\zeta$, and R_2 of $z < -\zeta$, with the propagation constants k_1 and k_2, respectively. Two reference boundary planes, S_1 at $z = 0$, and S_2 at $z = -d_2$, are introduced so that S is completely within a boundary space enclosed by S_1 and S_2. The unit vectors directed as outward normals to

S_1 and S_2, are denoted by $\widehat{\boldsymbol{n}}^{(j)}$, $j = 1, 2$, and the notation $\partial_n^{(j)} = \widehat{\boldsymbol{n}}^{(j)} \cdot \partial / \partial \widehat{\boldsymbol{x}}$ will hereafter be used.

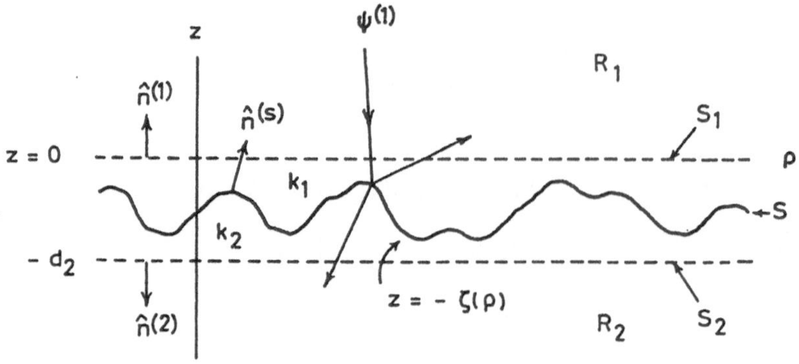

Fig. 3.4. Boundary profile and notations for (3.3.3, 10)

With the scalar wave equations (2.1.7), or (3.1.1) when suppressing the ρ-coordinates, the boundary condition on S can generally be reduced to a continuity of

$$\psi\,|_S , \qquad \eta^{-1} \widehat{\boldsymbol{n}}^{(S)} \cdot \frac{\partial}{\partial \widehat{\boldsymbol{x}}} \psi\,|_S \;, \tag{3.3.1}$$

with some real constant, η, characterizing the medium; which may be the density (sound wave), the dielectric constant ε, or the magnetic susceptibility (electro-magnetic waves of vertical or horizontal polarization, respectively, in two-dimensional space). Consistent with the boundary condition, the power flux vector is defined by

$$\widehat{\boldsymbol{w}} = (2\mathrm{i}\eta)^{-1} \psi^* \Big(\frac{\overleftarrow{\partial}}{\partial \widehat{\boldsymbol{x}}} - \frac{\overrightarrow{\partial}}{\partial \widehat{\boldsymbol{x}}} \Big) \psi \;, \tag{3.3.2}$$

so that continuity of the component, normal to S, is guaranteed.

In spite of the quite different situation compared with that of a rough surface in Sect. 3.1, many basic equations can be formulated in the same form, including those of surface impedance, Green's functions, and optical conditions [3.11, 12].

3.3.1 Basic Equations

The boundary condition can be transferred onto the two reference planes, S_1 and S_2, as we will show in Sect. 3.3.2, in the form,

$$-\eta_1^{-1}\partial_n^{(1)}\psi_1 = B_{11}\psi_1 + B_{12}\psi_2 \ ,$$
$$-\eta_2^{-1}\partial_n^{(2)}\psi_2 = B_{21}\psi_1 + B_{22}\psi_2 \ . \tag{3.3.3}$$

Here, ψ_j and η_j denote ψ on S_j, and η in space R_j, respectively; B_{ij}, $i, j = 1, 2$, are $\boldsymbol{\rho}$ matrices representing $\boldsymbol{\rho}$ operators (3.1.9, 10) and have the matrix elements $B_{ij}(\boldsymbol{\rho}|\boldsymbol{\rho}')$, so that (3.3.3) can be written explicitly as

$$-\eta_j^{-1}\partial_n^{(j)}\psi_j(\boldsymbol{\rho}) = \sum_{i=1,2}\int d\boldsymbol{\rho}'\, B_{ji}(\boldsymbol{\rho}|\boldsymbol{\rho}')\psi_i(\boldsymbol{\rho}') \ . \tag{3.3.4}$$

The boundary equation (3.3.3) can be written in a compact form by introducing 2×2 matrices $\boldsymbol{B}$, $\boldsymbol{\eta}$, and a two vector $\boldsymbol{\psi}$, written by boldface letters and defined by the elements

$$\boldsymbol{B} = \begin{pmatrix} B_{11} & B_{12} \\ B_{21} & B_{22} \end{pmatrix}, \qquad \boldsymbol{\eta} = \begin{pmatrix} \eta_1 & 0 \\ 0 & \eta_2 \end{pmatrix}, \qquad \boldsymbol{\psi} = \begin{pmatrix} \psi_1 \\ \psi_2 \end{pmatrix} \ , \tag{3.3.5}$$

as

$$-\boldsymbol{\eta}^{-1}\partial_n\boldsymbol{\psi} = \boldsymbol{B}\boldsymbol{\psi}, \qquad -\boldsymbol{\eta}^{-1}\partial_n\boldsymbol{\psi}^* = \boldsymbol{\psi}^*\boldsymbol{B}^\dagger \ . \tag{3.3.6}$$

Here, the last expression is the complex-conjugate equation, and $\boldsymbol{B}^\dagger$ denotes the Hermitian-conjugate matrix of $\boldsymbol{B}$, with respect to both the coordinates and the indices, i.e., $B_{ij}^\dagger(\boldsymbol{\rho}|\boldsymbol{\rho}') = B_{ji}^*(\boldsymbol{\rho}'|\boldsymbol{\rho})$. Here, since the integrated power w_z emitted away from the boundaries, $S_1 + S_2$, is given by the integrals,

$$w_z \equiv \sum_{i=1,2}\int_{S_j} d\boldsymbol{\rho}\,(2\mathrm{i}\eta_j)^{-1}\psi_j^*(\overleftarrow{\partial_n^{(j)}} - \overrightarrow{\partial_n^{(j)}})\psi_j \tag{3.3.7a}$$
$$= \sum_{i,j=1,2}(2\mathrm{i})^{-1}\int d\boldsymbol{\rho}\,d\boldsymbol{\rho}'\,\psi_j^*(\boldsymbol{\rho})$$
$$\times\left[B_{ji}(\boldsymbol{\rho}|\boldsymbol{\rho}') - B_{ji}^\dagger(\boldsymbol{\rho}|\boldsymbol{\rho}')\right]\psi_i(\boldsymbol{\rho}') \ , \tag{3.3.7b}$$

expressed in terms of $\boldsymbol{B}$, and should be zero for arbitrary $\psi_j(\boldsymbol{\rho})$, we find that $\boldsymbol{B}$ is a Hermitian matrix, subject to the condition,

$$\boldsymbol{B}^\dagger = \boldsymbol{B} \ . \tag{3.3.8}$$

To obtain the general solution of homogeneous wave equation (3.1.1) for $j(z) = 0$, we introduce an operator solution, $\psi^{(1)}(z)$, with the components $\psi_1^{(1)}(z)$, $z \geq 0$, and $\psi_2^{(1)}(z)$, $z \leq -d_2$, given, in terms of notations,

$$h_1 = \left[k_1^2 + (\partial/\partial\boldsymbol{\rho})^2\right]^{1/2}, \qquad h_2 = \left[k_2^2 + (\partial/\partial\boldsymbol{\rho})^2\right]^{1/2} \ , \tag{3.3.9}$$

by

$$\psi_1^{(1)}(z) = \left[\exp(ih_1 z) - \exp(-ih_1 z)\right](2ih_1)^{-1}\eta_1 + \exp(-ih_1 z)g_{11}, \qquad z \geq 0 \; , \tag{3.3.10a}$$

$$\psi_2^{(1)}(z) = \exp\left[ih_2(z+d_2)\right]g_{21}, \qquad z \leq -d_2 \; . \tag{3.3.10b}$$

Here, g_{11} and g_{21} are ρ operators independent of z, and give the boundary values of $\psi^{(1)}(z)$ at $z=0$ and $-d_2$, respectively; $\psi_1^{(1)}(z)$ has the same form as (3.1.15) except for the factor η_1. Now, substitution of (3.3.10) into (3.3.3) yields

$$(ih_1\eta_1^{-1} - B_{11})g_{11} - B_{12}g_{21} = 1 \; , \tag{3.3.11a}$$

$$-B_{21}g_{11} + (ih_2\eta_2^{-1} - B_{22})g_{21} = 0 \; , \tag{3.3.11b}$$

which provides g_{11} and g_{21} as the solutions, in terms of B_{ij}.

In the same way, by introducing another operator solution, $\psi^{(2)}(z)$, with the expressions,

$$\psi_1^{(2)}(z) = \exp(-ih_1 z)g_{12}, \qquad z \geq 0 \; , \tag{3.3.12a}$$

$$\psi_2^{(2)}(z) = \left\{\exp\left[-ih_2(z+d_2)\right] - \exp\left[ih_2(z+d_2)\right]\right\}(2ih_2)^{-1}\eta_2 + \exp\left\{ih_2(z+d_2)\right\}g_{22}, \qquad z \leq -d_2 \tag{3.3.12b}$$

(which represents waves incident from the R_2 space side), we obtain equations for g_{12} and g_{22}, similar to (3.3.11).

The four equations thus obtained can be unified and written as one 2×2 matrix equation,

$$\begin{pmatrix} ih_1\eta_1^{-1} - B_{11} & -B_{12} \\ -B_{21} & ih_2\eta_2^{-1} - B_{22} \end{pmatrix} \begin{pmatrix} g_{11} & g_{12} \\ g_{21} & g_{22} \end{pmatrix} = \begin{pmatrix} 1 & 0 \\ 0 & 1 \end{pmatrix} \; , \tag{3.3.13}$$

or,

$$(i\boldsymbol{h}\boldsymbol{\eta}^{-1} - \boldsymbol{B})\boldsymbol{g} = 1 \; , \tag{3.3.14}$$

where the matrices $\boldsymbol{h}$, $\boldsymbol{g}$, and 1 are defined by the elements $h_i\delta_{ij}$, g_{ij}, and δ_{ij}, respectively. Hence, $\boldsymbol{g}$ is obtained as the solution,

$$\boldsymbol{g} = (i\boldsymbol{h}\boldsymbol{\eta}^{-1} - \boldsymbol{B})^{-1} \; ; \tag{3.3.15a}$$

wherein $\boldsymbol{h}$ and $\boldsymbol{B}$ are not commutable, i.e.,

$$[\boldsymbol{h}, \boldsymbol{\eta}] = 0, \qquad [\boldsymbol{h}, \boldsymbol{B}] \neq 0 \; . \tag{3.3.15b}$$

When $\boldsymbol{\eta} = 1$, (3.3.14) has the same form as (3.1.16a) for the surface Green's function, g_b, on a one-sided boundary, exactly, and $\boldsymbol{g}$ will hence be referred to as the surface Green's function, too. In fact, the elements, g_{ij}, provide the boundary values on S_1 and S_2 of the z-continued Green's function, $\boldsymbol{g}(z|z')$, of Sect. 3.3.5.

Reflection-transmission coefficient matrix, $\boldsymbol{R}$, can also be defined in the same form as (3.1.33b), according to

$$\boldsymbol{g} = (1+\boldsymbol{R})(2\mathrm{i}\boldsymbol{h})^{-1}\boldsymbol{\eta} \tag{3.3.16a}$$

$$= \begin{pmatrix} (1+R_{11})(2\mathrm{i}h_1)^{-1}\eta_1 & R_{12}(2\mathrm{i}h_2)^{-1}\eta_2 \\ R_{21}(2\mathrm{i}h_1)^{-1}\eta_1 & (1+R_{22})(2\mathrm{i}h_2)^{-1}\eta_2 \end{pmatrix} , \tag{3.3.16b}$$

so that expressions (3.3.10) for $\psi^{(1)}(z)$ can be written as

$$\psi_1^{(1)}(z) = \left[\exp(\mathrm{i}h_1 z) + \exp(-\mathrm{i}h_1 z)R_{11}\right](2\mathrm{i}h_1)^{-1}\eta_1, \quad z \geq 0 , \tag{3.3.17a}$$

$$\psi_2^{(1)}(z) \doteq \exp\left[\mathrm{i}h_2(z+d_2)\right]R_{21}(2\mathrm{i}h_1)^{-1}\eta_1, \quad z \leq -d_2 ; \tag{3.3.17b}$$

similar expressions are also obtained for $\psi^{(2)}(z)$ from (3.3.12). From (3.3.16a, 14), $\boldsymbol{R}$ is given in terms of $\boldsymbol{B}$, by

$$\boldsymbol{R} = \boldsymbol{g}\, 2\mathrm{i}\boldsymbol{h}\boldsymbol{\eta}^{-1} - 1 \tag{3.3.18a}$$

$$= (\mathrm{i}\boldsymbol{h}\boldsymbol{\eta}^{-1} - \boldsymbol{B})^{-1}(\mathrm{i}\boldsymbol{h}\boldsymbol{\eta}^{-1} + \boldsymbol{B}) , \tag{3.3.18b}$$

and, conversely,

$$\boldsymbol{B} = \mathrm{i}\boldsymbol{h}\boldsymbol{\eta}^{-1} - \boldsymbol{g}^{-1} \tag{3.3.19a}$$

$$= \mathrm{i}\boldsymbol{h}\boldsymbol{\eta}^{-1}(\boldsymbol{R}-1)(\boldsymbol{R}+1)^{-1} . \tag{3.3.19b}$$

Note that both (3.3.18, 19) have the same form as (3.1.34a,b), respectively, with $\boldsymbol{h} \rightarrow \boldsymbol{h}\boldsymbol{\eta}^{-1}$, and also that, once $\boldsymbol{g}$ or $\boldsymbol{R}$ is known, $\boldsymbol{B}$ is determined, thereby, according to (3.3.19).

The matrix elements of $\boldsymbol{B}$ or $\boldsymbol{g}$ are not quite independent of one another, and are subjected to the symmetries which for $\boldsymbol{B}$ are given by

$$\boldsymbol{B}^T = \boldsymbol{B} = \boldsymbol{B}^* = \boldsymbol{B}^\dagger , \tag{3.3.20}$$

where the superscript ,T, denotes the transposed matrix with respect to both the coordinates and the subscripts; the first equality holds even in the case of a dissipative medium. The proof is given first by applying the Green's theorem to the boundary space enclosed by $S_1 + S_2$ (where $j(\widehat{\boldsymbol{x}}) = 0$) to find that, for arbitrary solutions ψ' and ψ'' of wave equation (3.1.1) subjected to (3.3.1),

$$\int_{S_1+S_2} d\rho\, \eta^{-1}\psi'(\overleftarrow{\partial}_n - \overrightarrow{\partial}_n)\psi'' = 0 , \tag{3.3.21a}$$

in consequence of the vanishing surface integral around S by virtue of the boundary condition, and then by utilizing boundary Eq. (3.3.4). Hence,

$$\sum_{i,j} \int d\rho \, d\rho' \, \psi_i'(\rho) [-B_{ji}(\rho'|\rho) + B_{ij}(\rho|\rho')] \psi_j''(\rho') = 0 \ , \tag{3.3.21b}$$

showing that $B_{ji}(\rho'|\rho) = B_{ij}(\rho|\rho')$.

Since $\boldsymbol{h}^T = \boldsymbol{h}$, it follows from (3.3.14, 18a) that

$$\boldsymbol{g}^T = \boldsymbol{g} \ , \tag{3.3.22a}$$

$$\boldsymbol{h}\boldsymbol{\eta}^{-1}\boldsymbol{R} = (\boldsymbol{h}\boldsymbol{\eta}^{-1}\boldsymbol{R})^T = \boldsymbol{R}^T \boldsymbol{h}\boldsymbol{\eta}^{-1} \ , \tag{3.3.22b}$$

so that

$$\widetilde{h}_1(\boldsymbol{\lambda})\eta_1^{-1}\widetilde{R}_{12}(\boldsymbol{\lambda}|\boldsymbol{\lambda}') = \widetilde{R}_{21}(\boldsymbol{\lambda}'|\boldsymbol{\lambda})\widetilde{h}_2(\boldsymbol{\lambda}')\eta_2^{-1} \ ;$$

while, from (3.3.18b),

$$\boldsymbol{R}^{-1} = (\mathrm{i}\boldsymbol{h}\boldsymbol{\eta}^{-1} + \boldsymbol{B})^{-1}(\mathrm{i}\boldsymbol{h}\boldsymbol{\eta}^{-1} - \boldsymbol{B}) \ , \tag{3.3.23a}$$

$$\boldsymbol{R}^* = (\mathrm{i}\boldsymbol{h}^*\boldsymbol{\eta}^{-1} + \boldsymbol{B})^{-1}(\mathrm{i}\boldsymbol{h}^*\boldsymbol{\eta}^{-1} - \boldsymbol{B}) \ , \tag{3.3.23b}$$

in view of $\boldsymbol{B}^* = \boldsymbol{B}$, showing that $\boldsymbol{R}^{-1} = \boldsymbol{R}^*$, if all the processes are taken place in the optical range where $\boldsymbol{h}^* = \boldsymbol{h}$.

The Hermitian condition for $\boldsymbol{B}$ imposes another constraint on $\boldsymbol{g}$, which eventually leads to the optical relation. This can be simply derived by using the invariance against the Hermitian conjugation of the expression (3.3.19a), i.e.,

$$\mathrm{i}\boldsymbol{h}\boldsymbol{\eta}^{-1} - \boldsymbol{g}^{-1} = -\mathrm{i}\boldsymbol{h}^\dagger\boldsymbol{\eta}^{-1} - (\boldsymbol{g}^\dagger)^{-1} \ , \tag{3.3.24}$$

which leads to [cf.(3.1.44b)]

$$\boldsymbol{w}_n \equiv 2^{-1}\boldsymbol{g}^\dagger(\boldsymbol{h}^\dagger + \boldsymbol{h})\boldsymbol{\eta}^{-1}\boldsymbol{g} - (2\mathrm{i})^{-1}(\boldsymbol{g}^\dagger - \boldsymbol{g}) = 0 \ . \tag{3.3.25}$$

Here, the diagonal elements, say w_{n1} and w_{n2}, are found to be exactly the integrated powers out of $S_1 + S_2$ of $\psi^{(1)}(z)$ and $\psi^{(2)}(z)$, respectively, as can be directly shown by using (3.3.7a) with (3.3.10, 12).

In the same way, the corresponding constraint for $\boldsymbol{R}$ is found to be, using the expression (3.3.19b),

$$\begin{aligned} &\boldsymbol{R}^\dagger(\boldsymbol{h}^\dagger + \boldsymbol{h})\boldsymbol{\eta}^{-1}\boldsymbol{R} - (\boldsymbol{h}^\dagger + \boldsymbol{h})\boldsymbol{\eta}^{-1} \\ &+\boldsymbol{R}^\dagger(\boldsymbol{h}^\dagger - \boldsymbol{h})\boldsymbol{\eta}^{-1} - \boldsymbol{\eta}^{-1}(\boldsymbol{h}^\dagger - \boldsymbol{h})\boldsymbol{R} = 0 \ . \end{aligned} \tag{3.3.26}$$

Here, in the optical range, the last two terms make no contribution, so that, in terms of the Fourier transform, $\widetilde{R}_{ij}(\boldsymbol{\lambda}|\boldsymbol{\lambda}')$ of $R_{ij}(\rho|\rho')$ defined by (3.1.70), the relation (3.3.26) can be written as

$$\sum_{j=1}^{2}(2\pi)^{-2}\int_{|\boldsymbol{\lambda}|\le k_j} d\boldsymbol{\lambda}\,\eta_j^{-1}\tilde{h}_j(\boldsymbol{\lambda})\tilde{R}^*_{ja}(\boldsymbol{\lambda}|\boldsymbol{\lambda}_1)\tilde{R}_{jb}(\boldsymbol{\lambda}|\boldsymbol{\lambda}_2)$$
$$= (2\pi)^2\delta(\boldsymbol{\lambda}_1-\boldsymbol{\lambda}_2)\delta_{ab}\tilde{h}_a(\boldsymbol{\lambda}_1)\eta_a^{-1} \ ; \qquad (3.3.27)$$

which corresponds to (3.1.50) for a rough surface, under the same condition.

3.3.2 Evaluation of the Surface Green's Function

The method is essentially the same as employed in Sect. 3.1 for a rough surface, although the actual procedure is not quite the same, by first deriving the Green's function, $\boldsymbol{g}$, directly independent of $\boldsymbol{B}$ [3.11].

a) First Method

We first observe that $\psi_j^{(i)}(z)$, $j = 1, 2$, of (3.3.17) is a solution of the inhomogeneous wave (3.1.1) in the space k_j, and that, on S, the boundary value is $\psi_j^{(1)}(-\zeta)$ when ζ is a constant independent of $\boldsymbol{\rho}$; even when $\zeta = \zeta(\boldsymbol{\rho})$, the boundary value can be written in the same form as

$$\psi^{(1)}\,|_S = \psi_1^{(1)}(-\zeta) = \psi_2^{(1)}(-\zeta) \ . \qquad (3.3.28)$$

From (3.3.17) we have

$$\psi_1^{(1)}(-\zeta) = \left[\exp^{(N)}(-ih_1\zeta) + \exp^{(N)}(ih_1\zeta)R_{11}\right](2ih_1)^{-1}\eta_1 \ , \qquad (3.3.29a)$$
$$\psi_2^{(1)}(-\zeta) = \exp^{(N)}(ih_2\bar{\zeta})R_{21}(2ih_1)^{-1}\eta_1 \ , \qquad (3.3.29b)$$

where $\bar{\zeta} = d_2 - \zeta$, and, with the ordering symbol $\mathcal{N}$ defined by (3.1.60b),

$$\exp^{(N)}(ih\zeta) \equiv \mathcal{N}\left[\exp(ih\zeta)\right] = \sum_{n=0}^{\infty}(n!)^{-1}\zeta^n(ih)^n \ , \qquad (3.3.30)$$

and it hold the relations

$$\mathcal{N}[f] \equiv f^{(N)}, \qquad \mathcal{N}[f+g] = f^{(N)} + g^{(N)} \ ,$$
$$\mathcal{N}[fg] = \mathcal{N}[gf] = \mathcal{N}[g^{(N)}f^{(N)}] \neq g^{(N)}f^{(N)} \ . \qquad (3.3.31)$$

The symbol $\mathcal{N}$ will be redefined in (3.3.38) with a minor additional condition.

Thus, from (3.3.28, 29), we find a simple relation between R_{21} and R_{11}, as

$$R_{21} = \left[\exp^{(N)}(ih_2\bar{\zeta})\right]^{-1}\left[\exp^{(N)}(-ih_1\zeta) + \exp^{(N)}(ih_1\zeta)R_{11}\right] \ . \qquad (3.3.32)$$

Also, for g_{21} and g_{11}, there exists a corresponding relation given from (3.3.10),

$$g_{21} = \left[\exp^{(N)}(ih_2\overline{\zeta})\right]^{-1}\left[-\sin^{(N)}(h_1\zeta)h_1^{-1}\eta_1 + \exp^{(N)}(ih_1\zeta)g_{11}\right] \ . \quad (3.3.33)$$

The situation is the same also for the second boundary condition, i.e., the continuity of

$$\eta^{-1}\widehat{\boldsymbol{n}}^{(S)} \cdot \frac{\partial}{\partial \widehat{\boldsymbol{x}}}\psi^{(1)} \mid_S \equiv \eta^{-1}(\boldsymbol{n}^{(S)} \cdot \partial_\rho + n_z^{(S)}\partial_z)\psi^{(1)} \mid_S \ , \quad (3.3.34)$$

where $\partial/\partial\widehat{\boldsymbol{x}} = (\partial_\rho, \partial_z)$ and the unit vector normal to S, $\widehat{\boldsymbol{n}}^{(S)} = (\boldsymbol{n}^{(S)}, n_z^{(S)})$, $n_z^{(S)} > 0$, is given by (3.1.54) in terms of $\zeta(\boldsymbol{\rho})$. Hence, for $\psi_2^{(1)}(z)$, $z \le -d_2$, defined in (3.3.17), for example, the quantity (3.3.34) can be written as an ordered equation,

$$\begin{aligned} &\eta_2^{-1}\widehat{\boldsymbol{n}}^{(S)} \cdot \frac{\partial}{\partial \widehat{\boldsymbol{x}}}\psi_2^{(1)} \mid_S \\ &= \eta_2^{-1}\mathcal{N}\left[(\boldsymbol{n}^{(S)} \cdot \partial_\rho + n_z^{(S)}ih_2)\exp^{(N)}(ih_2\overline{\zeta})\right]R_{21}(2ih_1)^{-1}\eta_1 \ . \end{aligned} \quad (3.3.35)$$

The following equations become simple by introduction of a surface impedance, $B_S^{(1)}(\boldsymbol{\rho}, \partial_\rho)$, on the real boundary, S, defined by

$$-\eta^{-1}\widehat{\boldsymbol{n}}^{(S)} \cdot \frac{\partial}{\partial \widehat{\boldsymbol{x}}}\psi^{(1)} \mid_S = B_S^{(1)}\psi^{(1)} \mid_S \ . \quad (3.3.36)$$

Here, $B_S^{(1)}$ must be the same for either of the solutions $\psi_1^{(1)}(z)$ and $\psi_2^{(1)}(z)$ on the different sides of S, in view of the boundary conditions (3.3.1) on S; and, for the $\psi_2^{(1)}(z)$, substitution of (3.3.29b, 35) in (3.3.36) yields

$$\begin{aligned} -\eta_2^{-1}&\mathcal{N}\left[(\boldsymbol{n}^{(S)} \cdot \partial_\rho + n_z^{(S)}ih_2)\exp^{(N)}(-ih_2\zeta)\right] \\ &= B_S^{(1)}\exp^{(N)}(-ih_2\zeta) \ , \end{aligned} \quad (3.3.37)$$

where the common factor has been dropped out from both sides. Equation (3.3.37) is independent of R_{21}, and the unknown $B_S^{(1)}$ can thereby be determined.

Also, for $\psi_1^{(1)}(z)$, we obtain, on using expression (3.3.29a) in both sides of (3.3.36), a similar equation as

$$\begin{aligned} &\mathcal{N}\left\{\left[-\boldsymbol{n}^{(S)} \cdot \partial_\rho + n_z^{(S)}ih_1 - \eta_1 B_S^{(1)}\right]\exp^{(N)}(ih_1\zeta)\right\}R_{11} \\ &= \mathcal{N}\left[(\boldsymbol{n}^{(S)} \cdot \partial_\rho + n_z^{(S)}ih_1 + \eta_1 B_S^{(1)})\exp^{(N)}(-ih_1\zeta)\right] \ . \end{aligned} \quad (3.3.38)$$

Here, $B_S^{(1)}$ should be to the left of all factors, and, hereafter, the symbol $\mathcal{N}$ will be redefined with this additional condition. Note that (3.3.38) exactly agrees with (3.1.69) for R by the replacement of $R_{11} \to R$ and $\eta_1 B_S^{(1)} \to B_0$;

hence the integral equation for $\widetilde{R}_{11}(\boldsymbol{\lambda}|\boldsymbol{\lambda}')$ also becomes the same as (3.1.71) with the same replacement.

By using (3.3.16), (3.3.38) can be rewritten as an independent equation to find g_{11} by

$$\begin{aligned}&\mathcal{N}[(-\eta_1^{-1}\boldsymbol{n}^{(S)}\cdot\partial_\rho+\eta_1^{-1}n_z^{(S)}\mathrm{i}h_1-B_S^{(1)})\exp^{(N)}(\mathrm{i}h_1\zeta)]g_{11}\\&=\mathcal{N}[-(\eta_1^{-1}\boldsymbol{n}^{(S)}\cdot\partial_\rho+B_S^{(1)})\sin^{(N)}(h_1\zeta)h_1^{-1}\eta_1+n_z^{(S)}\cos^{(N)}(h_1\zeta)]\ ,\end{aligned}\quad(3.3.39)$$

and g_{21} is obtained in terms of g_{11} from (3.3.33)

b) Slightly Random Boundary

To the first order of $\zeta(\rho)$, $\boldsymbol{n}^{(S)}=\zeta'\equiv\partial\zeta/\partial\rho$ and $n_z^{(S)}=1$; hence, (3.3.37) is reduced to

$$-\eta_2^{-1}(\mathrm{i}h_2+\zeta h_2^2+\zeta'\cdot\partial_\rho)=B_S^{(1)}(1-\mathrm{i}\zeta h_2)\ ,\qquad(3.3.40\mathrm{a})$$

showing that

$$B_S^{(1)}=-\eta_2^{-1}[\mathrm{i}h_2+(\zeta h_2-h_2\zeta)h_2+\zeta'\cdot\partial_\rho]\ .\qquad(3.3.40\mathrm{b})$$

In the same way, (3.3.39) is reduced to

$$[\mathrm{i}h_1\eta_1^{-1}-B_S^{(1)}-\eta_1B_S^{(1)}\zeta B_S^{(1)}-\eta_1^{-1}(\zeta h_1^2+\zeta'\cdot\partial_\rho)]g_{11}=1\ ,\qquad(3.3.41)$$

Thus the solution of (3.3.41) is obtained in the form,

$$g_{11}=g_0+g_0b^{(11)}g_0\ .\qquad(3.3.42)$$

Here,

$$g_0=[\mathrm{i}(h_1\eta_1^{-1}+h_2\eta_2^{-1})]^{-1}\ ,\qquad(3.3.43)$$

and, to the first order of ζ,

$$\begin{aligned}b^{(11)}=&(\eta_1^{-1}-\eta_2^{-1})\partial_\rho\cdot\zeta\partial_\rho+(\eta_2-\eta_1)\eta_2^{-2}h_2\zeta h_2\\&+(k_1^2\eta_1^{-1}-k_2^2\eta_2^{-1})\zeta\ ,\end{aligned}\qquad(3.3.44)$$

where use has been made of

$$h_1^2\eta_1^{-1}-h_2^2\eta_2^{-1}=k_1^2\eta_1^{-1}-k_2^2\eta_2^{-1}+(\eta_1^{-1}-\eta_2^{-1})\partial_\rho^2\ .$$

On the other hand, (3.3.33) for g_{21} becomes, on putting $d_2=0$ so that $\overline{\zeta}=-\zeta$,

$$(1-\mathrm{i}\zeta h_2)g_{21}=-\eta_1\zeta+(1+\mathrm{i}\zeta h_1)g_{11}\ .\qquad(3.3.45)$$

Hence, g_{21} is also obtained in the form,

$$g_{21} = g_0 + g_0 b^{(21)} g_0 \ , \tag{3.3.46a}$$

$$b^{(21)} = (\eta_1^{-1} - \eta_2^{-1})(\partial_\rho \cdot \zeta \partial_\rho - h_1 \zeta h_2) + (k_1^2 \eta_1^{-1} - k_2^2 \eta_2^{-1})\zeta \ . \tag{3.3.46b}$$

To the first order of ζ, (3.3.45) can also be written as

$$g_{21} - g_{11} = -\mathrm{i}(\eta_1/\eta_2 - 1)\zeta h_2 g_{21} \ , \tag{3.3.47a}$$

in view of $g_{21} = g_{11} = g_0$ to the zero order. Hence, (3.3.42) leads to

$$\mathrm{i}h_1\eta_1^{-1} g_{11} + \mathrm{i}h_2\eta_2^{-1} g_{21} = 1 + b_1 g_{11} \ , \tag{3.3.47b}$$

$$b_1 = (\eta_1^{-1} - \eta_2^{-1})\partial_\rho \cdot \zeta \partial_\rho + (k_1^2\eta_1^{-1} - k_2^2\eta_2^{-1})\zeta \ , \tag{3.3.47c}$$

so that (3.3.47a,b) provide a set of equations for the unknown g_{11} and g_{21}. Here, the solutions are shown to be strictly consistent with power conservation, as follows: With a new Hermitean operator, b_2, defined by

$$b_2 = (\eta_1 - \eta_2)\eta_2^{-2} h_2^\dagger \zeta h_2 \ , \tag{3.3.48a}$$

we first rewrite (3.3.47a) as

$$\mathrm{i}h_2^\dagger \eta_2^{-1}(-g_{11} + g_{21}) = b_2 g_{21} \ , \tag{3.3.48b}$$

and then, together with (3.3.47b), write the two equations by one 2×2 matrix equation of the form [3.11],

$$\boldsymbol{\pi}\vec{g} = \vec{1} + \boldsymbol{b}\vec{g} \ . \tag{3.3.49}$$

Here,

$$\vec{g} = \begin{pmatrix} g_{11} \\ g_{21} \end{pmatrix}, \quad \vec{1} = \begin{pmatrix} 1 \\ 0 \end{pmatrix} \ , \tag{3.3.50a}$$

$$\boldsymbol{\pi} = \begin{pmatrix} \mathrm{i}h_1\eta_1^{-1} & \mathrm{i}h_2\eta_2^{-1} \\ -\mathrm{i}h_2^\dagger\eta_2^{-1} & \mathrm{i}h_2^\dagger\eta_2^{-1} \end{pmatrix}, \quad \boldsymbol{b} = \begin{pmatrix} b_1 & 0 \\ 0 & b_2 \end{pmatrix} \ . \tag{3.3.50b}$$

Hence, since $\boldsymbol{b}^\dagger = \boldsymbol{b}$, the Hermitian conjugation of (3.3.49) yields

$$\vec{g}^\dagger \boldsymbol{\pi}^\dagger = \vec{1} + \vec{g}^\dagger \boldsymbol{b} \ . \tag{3.3.51}$$

Here,

$$\boldsymbol{\pi} - \boldsymbol{\pi}^\dagger = \mathrm{i}(\boldsymbol{h}^\dagger + \boldsymbol{h})\boldsymbol{\eta}^{-1} \ . \tag{3.3.52a}$$

Hence the total integrated power out of $S_1 + S_2$, as given by (3.3.25), can be written as

$$w_{n1} = (2\mathrm{i})^{-1}\left[\vec{g}^{\dagger}(\pi - \pi^{\dagger})\vec{g} + \vec{1}^{\dagger}\vec{g} - \vec{g}^{\dagger}\vec{1}\right] , \tag{3.3.52b}$$

which becomes exactly zero in consequence of (3.3.49, 51).

Summarizing, the original equations (3.3.47a,b) can be regarded as a generalized versions of (3.1.14, 16a) for g_b when the boundary is a one-sided (surface), and the solution is significant for all orders of ζ, in the sense of selected summation as when treating light wave in turbulent air (Chap. 7); where the medium fluctuation is slight, but still, its accumulated effect can be quite large.

c) Alternative Method

In the previous method, the two wave functions, $\psi_1^{(1)}(z)$ and $\psi_2^{(1)}(z)$ of (3.3.10), were continued directly on the real boundary through the boundary conditions (3.3.1), and R_{11} and R_{21} were obtained therefrom in terms of the surface impedance $B_S^{(1)}$, defined by (3.3.36), and determined by (3.3.37). Therefore, the method is essentialy the same as that employed in Sect. 3.1, with the only difference of B_0 being replaced by $B_S^{(1)}\eta_1$, which depends also on $\zeta(\rho)$. There exists an alternative method which enables R_{21} to be directly obtained, and R_{11} to be obtained from the known R_{21}; in this method, the continuation of the wave functions is performed rather indirectly with the aid of the Green's theorem.

By applying the Green's theorem to the boundary space enclosed by S_1 and S (Fig. 3.4), we find, for arbitrary solutions, ψ' and ψ'', of wave equation (3.1.1), the relation,

$$[\psi';\psi''] \equiv \int_{z=0} d\rho\, \eta_1^{-1}\psi'(\hat{x})\left(\frac{\overleftarrow{\partial}}{\partial z} - \frac{\overrightarrow{\partial}}{\partial z}\right)\psi''(\hat{x}) \tag{3.3.53a}$$

$$= \int_{z=-\zeta} ds\, \eta_1^{-1}\psi'(\hat{x})\left(\frac{\overleftarrow{\partial}}{\partial \hat{x}}\cdot \hat{n}^{(S)} - \hat{n}^{(S)}\cdot\frac{\overrightarrow{\partial}}{\partial \hat{x}}\right)\psi''(\hat{x}) , \tag{3.3.53b}$$

where $ds = d\rho/n_z^{(S)}$ is the two-dimensional element of surface S. Here, the relation can be rewritten in a compact form by regarding $\psi'(\hat{x})$ and $\psi''(\hat{x})$ as components of ρ vectors, $\psi'(z)$ and $\psi''(z)$ (which, more generally, may be ρ matrices), respectively [cf.(3.3.34)],

$$[\psi',\psi''] \equiv \eta_1^{-1}\psi'(z)\left(\frac{\overleftarrow{\partial}}{\partial z} - \frac{\overrightarrow{\partial}}{\partial z}\right)\psi''(z)\bigg|_{z=0} \tag{3.3.54a}$$

$$= \overline{\mathcal{N}}\left[\psi'(z)\frac{\overleftarrow{\partial}}{\partial \hat{x}}\cdot\hat{n}^{(S)}(n_z^{(S)}\eta)^{-1}\right]\mathcal{N}[\psi''(z)]\bigg|_{z=-\zeta}$$

$$- \overline{\mathcal{N}}[\psi'(z)]\mathcal{N}\left[(n_z^{(S)}\eta)^{-1}\hat{n}^{(S)}\cdot\frac{\overrightarrow{\partial}}{\partial \hat{x}}\psi''(z)\right]\bigg|_{z=-\zeta} . \tag{3.3.54b}$$

Here, the ρ integration is involved as a natural consequence of the inner product, i.e., for any ρ matrices, A and B,

$$AB(\rho'|\rho'') = \int d\rho\, A(\rho'|\rho)B(\rho|\rho'') \ ;$$

and the simbol $\overline{\mathcal{N}}$ designates an ordering similar to $\mathcal{N}$; however, the ordering is inverse such that, for example,

$$\begin{aligned} \overline{\mathcal{N}}\left[\exp(ih\zeta)\right] &\equiv \exp^{(\overline{\mathcal{N}})}(ih\zeta) \\ &= \sum_{n=0}^{\infty}(n!)^{-1}(ih)^n\zeta^n \ , \end{aligned} \tag{3.3.55}$$

wherein all the ζ variables (and/or ρ functions involved, if any) are understood to be placed to the right of all the h's (and/or $\partial_\rho \equiv \partial/\partial\rho$, if any).

Here, we set $\psi''(z) = \psi_1^{(1)}(z)$ of (3.3.10a) and

$$\psi'(z) = \psi^{(+)}(z) \equiv \exp(-ih_1 z) \ ; \tag{3.3.56}$$

hence, the substitution into (3.3.54a) simply yields

$$[\psi^{(+)}, \psi^{(1)}] = -1 \ . \tag{3.3.57}$$

For the substitution into (3.3.54b), on the other hand, we observe that, by virtue of boundary condition (3.3.1), all the factors associated with $\psi_1^{(1)}(z)$ can be replaced with those of $\psi_2^{(1)}(z)$, given by (3.3.10b); and, since, by (3.1.54),

$$\hat{\boldsymbol{n}}^{(S)}/n_z^{(S)} = (\zeta', 1), \qquad \zeta' = \partial\zeta/\partial\rho \ , \tag{3.3.58}$$

also that

$$\begin{aligned} &\mathcal{N}\left[\hat{\boldsymbol{n}}^{(S)}(n_z^{(S)}\eta)^{-1} \cdot \frac{\overrightarrow{\partial}}{\partial\hat{\boldsymbol{x}}}\psi_2^{(1)}(z(=-\zeta))\right] \\ &= \mathcal{N}\left[\eta_2^{-1}(\zeta' \cdot \partial_\rho + ih_2)\exp^{(N)}(ih_2\overline{\zeta})\right] g_{21} \ , \end{aligned} \tag{3.3.59a}$$

$$\begin{aligned} &\mathcal{N}\left[\psi^{(+)}(z(=-\zeta))\frac{\overleftarrow{\partial}}{\partial\hat{\boldsymbol{x}}} \cdot \hat{\boldsymbol{n}}^{(S)}(n_z^{(S)}\eta)^{-1}\right] \\ &= \mathcal{N}\left[\eta_1^{-1}\exp^{(\overline{N})}(ih_1\zeta)(\overleftarrow{\partial}_\rho \cdot \zeta' - ih_1)\right] \ , \end{aligned} \tag{3.3.59b}$$

where we can set $\overleftarrow{\partial}_\rho = -\overrightarrow{\partial}_\rho$ by partial integration. Thus, with (3.3.57), the result from (3.3.54b) can be written in the form,

$$T^{(+)}g_{21} = 1 \ . \tag{3.3.60}$$

Here,

$$\begin{aligned} T^{(+)} =& \eta_1^{-1}\overline{\mathcal{N}}\left[\exp^{(\overline{N})}(ih_1\zeta)(\partial_\rho\cdot\zeta' + ih_1)\right]\exp^{(N)}(ih_2\overline{\zeta}) \\ &+ \eta_2^{-1}\exp^{(\overline{N})}(ih_1\zeta)\mathcal{N}\left[(\zeta'\cdot\partial_\rho + ih_2)\exp^{(N)}(ih_2\overline{\zeta})\right] \ . \end{aligned} \tag{3.3.61}$$

In the same way, by replasing $\psi^{(+)}(z)$ with

$$\psi^{(-)}(z) \equiv \exp(ih_1 z) \ , \tag{3.3.62}$$

(3.3.54a) becomes, on using (3.3.17a),

$$[\psi^{(-)}, \psi^{(1)}] = (2ih_1)\eta_1^{-1}R_{11}(2ih_1)^{-1}\eta_1 \ , \tag{3.3.63}$$

whereas (3.3.54b) becomes what would be obtained by replacing $h_1 \to -h_1$ in (3.3.59a,b). Thus, we obtain another equation for R_{11} of the form,

$$(G_1^{(0)})^{-1}R_{11}G_1^{(0)} = -T^{(-)}g_{21} \ . \tag{3.3.64}$$

Here,

$$G_1^{(0)} = (2ih_1)^{-1}\eta_1 \ , \tag{3.3.65}$$

meaning the free space Green's function in the space of $\boldsymbol{k}_1$ [cf. (3.1.4)], and $T^{(-)}$ is obtained from $T^{(+)}$ with $h_1 \to -h_1$. Hence, with (3.3.60), g_{11} can be written in terms of g_{21}, as

$$g_{11} = (1 + R_{11})G_1^{(0)} = G_1^{(0)}[T^{(+)} - T^{(-)}]g_{21} \ . \tag{3.3.66}$$

This second method is particularly convenient for electro-magnetic waves and other waves of a multi-component, in view of several conditions to be fulfilled on the real boundary for the first method [3.12].

3.3.3 Power Equations

To derive related power equations, we first observe that $w_z(\rho)$ of (3.1.52a) for a rough surface is now replaced by the total power flux density away from $S_1 + S_2$ at a point, ρ, i.e.,

$$\begin{aligned} \sum_{a=1}^{2}\widehat{\boldsymbol{n}}^{(a)}\cdot\widehat{\boldsymbol{w}}_a(\rho) = \sum_{i,j=1}^{2}\int d\rho_1 d\rho_2 \\ \times(-)\frac{\partial}{\partial\rho}\cdot\beta_{ij}^{(B)}(\rho|\rho_1;\rho_2)\psi_i^*(\rho_1)\psi_j(\rho_2) \ , \end{aligned} \tag{3.3.67a}$$

with

$$\frac{\partial}{\partial \rho} \cdot \beta_{ij}^{(B)}(\rho|\rho_1;\rho_2) = (2\mathrm{i})^{-1}$$
$$\times \left[B_{ji}^*(\rho|\rho_1)\delta(\rho-\rho_2) - B_{ij}(\rho|\rho_2)\delta(\rho-\rho_1)\right] , \qquad (3.3.67\mathrm{b})$$

which results from the Hermitian condition, $\boldsymbol{B}^\dagger = \boldsymbol{B}$ (3.3.20), similar to (3.1.20b). To write power equations, however, it is more convenient to express the equations in terms of the full coordinates $\widehat{\boldsymbol{x}}$, by introducing the $\widehat{\boldsymbol{x}}$-coordinate matrix, $\boldsymbol{B}$, defined by the matrix elements,

$$B_{ij}(\boldsymbol{x}|\boldsymbol{x}') = \delta(z-d_i)B_{ij}(\rho|\rho')\delta(z'-d_j) , \qquad (3.3.68)$$

where $i, j = 1, 2$, and $d_1 = 0$, and also the corresponding $\beta_{ij}^{(B)}(\widehat{\boldsymbol{x}}|\widehat{\boldsymbol{x}}_1;\widehat{\boldsymbol{x}}_2)$, defined by

$$\frac{\partial}{\partial \rho} \cdot \beta_{ij}^{(B)}(\widehat{\boldsymbol{x}}|\widehat{\boldsymbol{x}}_1;\widehat{\boldsymbol{x}}_2) = (2\mathrm{i})^{-1}$$
$$\times \left[B_{ji}^*(\widehat{\boldsymbol{x}}|\widehat{\boldsymbol{x}}_1)\delta(\widehat{\boldsymbol{x}}-\widehat{\boldsymbol{x}}_2) - B_{ij}(\widehat{\boldsymbol{x}}|\widehat{\boldsymbol{x}}_2)\delta(\widehat{\boldsymbol{x}}-\widehat{\boldsymbol{x}}_1)\right] , \qquad (3.3.69)$$

in exactly the same form as (3.3.67b). Hence, (3.3.67a) is also replaced by

$$\sum_{a=1}^{2} \widehat{\boldsymbol{n}}^{(a)} \cdot \widehat{\boldsymbol{w}}_a(\widehat{\boldsymbol{x}}) = \sum_{i,j=1}^{2} \int d\widehat{\boldsymbol{x}}_1 d\widehat{\boldsymbol{x}}_2$$
$$\times(-)\frac{\partial}{\partial \rho} \cdot \beta_{ij}^{(B)}(\widehat{\boldsymbol{x}}|\widehat{\boldsymbol{x}}_1;\widehat{\boldsymbol{x}}_2)\psi_i^*(\widehat{\boldsymbol{x}}_1)\psi_j(\widehat{\boldsymbol{x}}_2) , \qquad (3.3.70)$$

which differs from zero only on S_1 and S_2.

Equation (3.3.69) can be written in a simple matrix form as [cf. (3.2.23a)]

$$\frac{\partial}{\partial \rho} \cdot \boldsymbol{\beta}^{(B)}(\widehat{\boldsymbol{x}}|1;2) = (2\mathrm{i})^{-1}\boldsymbol{\delta}(\widehat{\boldsymbol{x}}|1;2)\left[\boldsymbol{B}^*(1) - \boldsymbol{B}(2)\right] . \qquad (3.3.71)$$

Here, like $\delta(\widehat{\boldsymbol{x}}|1;2)$ as defined by (2.3.14), the matrix $\boldsymbol{\delta}(\widehat{\boldsymbol{x}}|1;2)$ is defined by the matrix elements,

$$\delta_{ij}(\widehat{\boldsymbol{x}}|\widehat{\boldsymbol{x}}_1;\widehat{\boldsymbol{x}}_2) = \delta_{ij}\delta(\widehat{\boldsymbol{x}}-\widehat{\boldsymbol{x}}_1)\delta(\widehat{\boldsymbol{x}}-\widehat{\boldsymbol{x}}_2) , \qquad (3.3.72\mathrm{a})$$

so that, for any $\widehat{\boldsymbol{x}}$ 2×2 matrices $A^*(1)$ and $B(2)$, the product,

$$\boldsymbol{\delta}(\widehat{\boldsymbol{x}}|1;2)\boldsymbol{A}^*(1)\boldsymbol{B}(2) \equiv \boldsymbol{A}^*\boldsymbol{B}(\widehat{\boldsymbol{x}}|1;2) \equiv \boldsymbol{A}^*\boldsymbol{B}(\widehat{\boldsymbol{x}}) , \qquad (3.3.72\mathrm{b})$$

represents

$$\sum_{a,b=1}^{2} \int d\hat{\boldsymbol{x}}_1 d\hat{\boldsymbol{x}}_2\, \delta_{ab}(\hat{\boldsymbol{x}}|\hat{\boldsymbol{x}}_1;\hat{\boldsymbol{x}}_2) A^*_{ac}(\hat{\boldsymbol{x}}_1|\hat{\boldsymbol{x}}'_1) B_{bd}(\hat{\boldsymbol{x}}_2|\hat{\boldsymbol{x}}'_2)$$
$$= \sum_a A^*_{ac}(\hat{\boldsymbol{x}}|\hat{\boldsymbol{x}}'_1) B_{ad}(\hat{\boldsymbol{x}}|\hat{\boldsymbol{x}}'_2) \ . \tag{3.3.72c}$$

Hence, (3.3.70) can be written as

$$\sum_a \hat{\boldsymbol{n}}^{(a)} \cdot \hat{\boldsymbol{w}}_a(\hat{\boldsymbol{x}}) = -\frac{\partial}{\partial \boldsymbol{\rho}} \cdot \boldsymbol{\beta}^{(B)}(\hat{\boldsymbol{x}}) \boldsymbol{\psi}^* \boldsymbol{\psi} \ . \tag{3.3.73}$$

Thus the deterministic power equation corresponding to (3.1.53) can be written as

$$\frac{\partial}{\partial \hat{\boldsymbol{x}}} \cdot \left[\hat{\boldsymbol{\alpha}}(\hat{\boldsymbol{x}}) + \boldsymbol{\beta}^{(B)}(\hat{\boldsymbol{x}})\right] \boldsymbol{\psi}^* \boldsymbol{\psi} = (2\mathrm{i})^{-1} \left[\boldsymbol{\psi}^* \boldsymbol{j}(\hat{\boldsymbol{x}}) - \boldsymbol{\psi} \boldsymbol{j}^*(\hat{\boldsymbol{x}})\right] \ , \tag{3.3.74}$$

by writting $\hat{\boldsymbol{w}}$ of (3.3.2) in the form [cf.(2.3.19)],

$$\hat{\boldsymbol{w}}(\hat{\boldsymbol{x}}) = \hat{\boldsymbol{\alpha}}(\hat{\boldsymbol{x}}|1;2) \boldsymbol{\psi}^*(1) \boldsymbol{\psi}(2) \ , \tag{3.3.75a}$$

in terms of the matrix, $\hat{\boldsymbol{\alpha}}(\hat{\boldsymbol{x}}|1;2)$, defined by the elements,

$$\hat{\boldsymbol{\alpha}}_{ij}(\hat{\boldsymbol{x}}|\hat{\boldsymbol{x}}_1;\hat{\boldsymbol{x}}_2) = \delta_{ij}(\hat{\boldsymbol{x}}|\hat{\boldsymbol{x}}_1;\hat{\boldsymbol{x}}_2)(2\mathrm{i}\eta_j)^{-1} \left(\frac{\partial}{\partial \hat{\boldsymbol{x}}_1} - \frac{\partial}{\partial \hat{\boldsymbol{x}}_2}\right) \ . \tag{3.3.75b}$$

Here it may be remarked that the $\boldsymbol{\rho}$ equation (3.3.67b) can also be written in the same form as (3.3.71),

$$\frac{\partial}{\partial \boldsymbol{\rho}} \cdot \boldsymbol{\beta}^{(B)}(\boldsymbol{\rho}|1;2) = (2\mathrm{i})^{-1} \boldsymbol{\delta}(\boldsymbol{\rho}|1;2) \left[\boldsymbol{B}^*(1) - \boldsymbol{B}(2)\right] \ . \tag{3.3.76}$$

Here $\boldsymbol{B}^*$ and $\boldsymbol{B}$ are the original $\boldsymbol{\rho}$ matrices, $\boldsymbol{\delta}(\boldsymbol{\rho}|1;2)$ is defined by the same equation as (3.3.72a,b) for $\boldsymbol{\delta}(\hat{\boldsymbol{x}}|1;2)$, except that all the $\hat{\boldsymbol{x}}$-coordinates are replaced by the corresponding $\boldsymbol{\rho}$ coordinates. However, involved in the power equation is $\boldsymbol{\beta}(\hat{\boldsymbol{x}})$ and not $\boldsymbol{\beta}(\boldsymbol{\rho})$, the latter being given by the z integration of the former.

3.3.4 Statistical Green's Functions of First and Second Orders

The procedure of deriving equations for the statistical Green's functions is almost the same as for a rough surface; we only need to replace the lightface letters in Sect. 3.2 by those of boldface letters and then $\boldsymbol{h} \to \boldsymbol{h}\eta^{-1}$. That is, on writing $\boldsymbol{B} = \boldsymbol{B}_0 + \boldsymbol{b}$ with the random part $\boldsymbol{b}$ in (3.3.14), we obtain, as a basic equation for the deterministic Green's function, $\boldsymbol{g}$,

$$(ih\eta^{-1} - B_0 - b)g = 1 \ , \tag{3.3.77}$$

which has the same form as the corresponding equation (3.1.16a) for g_b.

Hence, to find the first order Green's function, $G = \langle g \rangle$, we introduce an effective 2×2 matrix operator, $M^{(b)}$ of b, defined by

$$\langle bg \rangle = M^{(b)} G, \qquad M^{(b)T} = M(b) \ , \tag{3.3.78a}$$

or, writing explicitly,

$$\sum_j \langle b_{ij} g_{jk} \rangle = \sum_j M_{ij}^{(b)} G_{jk} \ , \tag{3.3.78b}$$

so that the average of (3.3.77) is written by

$$(ih\eta^{-1} - B_0 - M^{(b)})G = 1 \ . \tag{3.3.79a}$$

In the same way, since $B_0^* = B_0$, we obtain

$$(-ih^*\eta^{-1} - B_0 - M^{(b)*})G^* = 1 \ . \tag{3.3.79b}$$

The second order Green's function is defined, in matrix form, by

$$I^{(B)}(1;2) = \langle g^*(1) g(2) \rangle \ , \tag{3.3.80a}$$

with the elements

$$I_{ij;kl}^{(B)}(\rho_1;\rho_2|\rho_1';\rho_2') = \langle g_{ik}^*(\rho_1|\rho_1') g_{jl}(\rho_2|\rho_2') \rangle \ . \tag{3.3.80b}$$

Here, as in the previous sections, the coordinates having indices 1 and 2 are used for quantities of the complex-conjugate wave function and the original wave function, respectively, in the manner of (3.3.80). To find a governing equation for $I^{(B)}(1;2)$, we introduce a quantity, Δb, defined by

$$\Delta b = b - M^{(b)}, \qquad \langle \Delta b g \rangle = 0 \ , \tag{3.3.81}$$

and, using (3.3.79), exhibit g in terms of G, by

$$g = G(1 + \Delta b g), \tag{3.3.82a}$$

$$g^* = G^*(1 + \Delta b^* g^*), \qquad \langle \Delta b^* g^* \rangle = 0 \ , \tag{3.3.82b}$$

Hence, by the substitution into the right-hand side of (3.3.80a), the resulting equation can be written as

$$I^{(B)}(1;2) = U^{(C)}(1;2)[1 + K^{(B)}(1;2)I^{(B)}(1;2)] \ , \qquad (3.3.83)$$

in exactly the same form as the BS equation (3.2.12). Here,

$$U^{(C)}(1;2) = G^*(1)G(2) \ , \qquad (3.3.84)$$

and the matrix, $K^{(B)}(1;2)$, is defined, like $M^{(b)}$, in (3.3.78), according to

$$K^{(B)}(1;2)I^{(B)}(1;2) = \langle \Delta b^*(1)\Delta b(2)g^*(1)g(2)\rangle \ . \qquad (3.3.85)$$

Here, to the first order of b,

$$M^{(b)} = \langle bGb\rangle, \qquad \langle b\rangle = 0 \ , \qquad (3.3.86a)$$

$$K^{(B)}(1;2) = \langle b(1)b(2)\rangle \qquad (3.3.86b)$$

(when setting $d_2 = 0$ (Fig. 3.4), however, the BS equation (3.3.83) should be modified based on (3.3.47a,b) or (3.3.49, 51) [3.11]).

In another typical case of when small bosses are randomly distributed on S_2 at $z = -d_2$, b is given, in the same form as (3.1.43), by

$$b = \sum_a b_a \ , \qquad (3.3.87a)$$

where b_a denotes the contribution from the a-th boss and differs from zero only over a small area of the boss. Thus, refering to (3.2.36, 37) we obtain

$$M^{(b)}(2) = n\int d\rho_a \langle T_a^M(2)\rangle' \ , \qquad (3.3.87b)$$

$$K^{(B)}(1;2) = n\int d\rho_a \langle T_a^{M*}(1)T_a^M(2)\rangle' \ , \qquad (3.3.87c)$$

in terms of the scattering matrix, T_a^M of b_a, with the center at ρ_a.

a) Incoherent Scattering Matrix and Optical Condition

The solution of the BS equation (3.3.83) can also be written in the same form as (3.2.21) by

$$I^{(B)} = U^{(C)} + U^{(C)}S^{(B)}U^{(C)} \ , \qquad (3.3.88)$$

in terms of the scattering matrix, $S^{(B)}$ of $K^{(B)}$, defined by

$$S^{(B)} = K^{(B)}(1 + U^{(C)}S^{(B)}) \qquad (3.3.89a)$$

$$= (1 - K^{(B)}U^{(C)})^{-1}K^{(B)} \ . \qquad (3.3.89b)$$

Here, the matrix elements of $\boldsymbol{S}^{(B)}$ will be denoted by $S_{ij;kl}(\boldsymbol{\rho}_1;\boldsymbol{\rho}_2|\boldsymbol{\rho}_1';\boldsymbol{\rho}_2')$ or, simply, $S_{ij;kl}$, in the same fashion as those for $\boldsymbol{I}^{(B)}$ and $\boldsymbol{U}^{(C)}$. For example, we obtain, as a specific expression of (3.3.88),

$$\begin{aligned} I_{11;11}(\boldsymbol{\rho}_1;\boldsymbol{\rho}_2|\boldsymbol{\rho}_1';\boldsymbol{\rho}_2') &= U^{(C)}_{11;11}(\boldsymbol{\rho}_1;\boldsymbol{\rho}_2|\boldsymbol{\rho}_1';\boldsymbol{\rho}_2') \\ &+ \sum_{i,j,k,l}\int d\boldsymbol{\rho}_i'' d\boldsymbol{\rho}_j'' d\boldsymbol{\rho}_k'' d\boldsymbol{\rho}_l'' U^{(C)}_{11;ij}(\boldsymbol{\rho}_1;\boldsymbol{\rho}_2|\boldsymbol{\rho}_i'';\boldsymbol{\rho}_j'') \\ &\times S^{(B)}_{ij;kl}(\boldsymbol{\rho}_i'';\boldsymbol{\rho}_j''|\boldsymbol{\rho}_k'';\boldsymbol{\rho}_l'')U^{(C)}_{kl;11}(\boldsymbol{\rho}_k'';\boldsymbol{\rho}_l''|\boldsymbol{\rho}_1';\boldsymbol{\rho}_2')\ , \end{aligned} \tag{3.3.90}$$

with

$$U^{(C)}_{ij;kl}(\boldsymbol{\rho}_1;\boldsymbol{\rho}_2|\boldsymbol{\rho}_1';\boldsymbol{\rho}_2') = G^*_{ik}(\boldsymbol{\rho}_1-\boldsymbol{\rho}_1')G_{jl}(\boldsymbol{\rho}_2-\boldsymbol{\rho}_2')\ . \tag{3.3.91}$$

Equation (3.3.90) contains the terms of $S_{22;22}$ and the other elements having index 2, showing that the backscattered wave, $I^{(B)}_{11;11}$, is necessarily affected by a multiple scattering on both sides of S.

The local optical relations to ensure power conservation in the scattering are also obtained in the same form as (3.2.25, 26); i.e., with a resultant effective matrix $\boldsymbol{M}^{(B)} = \boldsymbol{M}^{(B)T}$, defined by

$$\boldsymbol{M}^{(B)} = \boldsymbol{B}_0 + \boldsymbol{M}^{(b)}, \qquad \langle \boldsymbol{B}\boldsymbol{g}\rangle = \boldsymbol{M}^{(B)}\boldsymbol{G}\ , \tag{3.3.92a}$$

we obtain, as an $\hat{\boldsymbol{x}}$ matrix equation based on (3.3.68, 71),

$$\begin{aligned} &\frac{\partial}{\partial\boldsymbol{\rho}}\cdot\boldsymbol{\beta}(\hat{\boldsymbol{x}}|1;2) = (2\mathrm{i})^{-1}\boldsymbol{\delta}(\hat{\boldsymbol{x}}|1;2) \\ &\times\left\{\boldsymbol{M}^{(B)*}(1) - \boldsymbol{M}^{(B)}(2) - [\boldsymbol{G}^*(1)-\boldsymbol{G}(2)]\boldsymbol{K}^{(B)}(1;2)\right\}\ , \end{aligned} \tag{3.3.92b}$$

similar to (3.2.54a). Hence, from (3.2.26), the relation (3.3.92b) can be written in terms of $\boldsymbol{S}^{(B)}$ as

$$\begin{aligned} &\frac{\partial}{\partial\boldsymbol{\rho}}\cdot\boldsymbol{\beta}^{(e)}(\hat{\boldsymbol{x}}|1;2) = \boldsymbol{\delta}(\hat{\boldsymbol{x}}|1;2) \\ &\times\left\{\boldsymbol{\Gamma}^{(B)}(1;2) - 2^{-1}[\boldsymbol{h}^*(1)+\boldsymbol{h}(2)]\eta^{-1}\boldsymbol{U}^{(C)}\boldsymbol{S}^{(B)}(1;2)\right\}\ , \end{aligned} \tag{3.3.93}$$

where

$$\boldsymbol{\beta}^{(e)}(\hat{\boldsymbol{x}}|1;2) = \boldsymbol{\beta}(\hat{\boldsymbol{x}}|1;2)[1+\boldsymbol{U}^{(C)}\boldsymbol{S}^{(B)}(1;2)]\ , \tag{3.3.94a}$$

$$\boldsymbol{\Gamma}^{(B)}(1;2) = (2\mathrm{i})^{-1}[\boldsymbol{M}^{(B)*}(1) - \boldsymbol{M}^{(B)}(2)]\ . \tag{3.3.94b}$$

The integrated optical relation is obtained therefrom, as

$$\begin{aligned}\boldsymbol{\delta}(1;2)\Big\{&\boldsymbol{\Gamma}^{(B)}(1;2)-2^{-1}\left[\boldsymbol{h}^*(1)+\boldsymbol{h}(2)\right]\\&\times\boldsymbol{\eta}^{-1}\boldsymbol{U}^{(C)}\boldsymbol{S}^{(B)}(1;2)\Big\}=0\ ,\end{aligned}\tag{3.3.94c}$$

with the matrix element $\boldsymbol{\delta}(1;2)$, similar to $\delta(1;2)$ of (3.2.34).

Equation (3.3.93) can be written in another form of (3.2.57), with $U^{(C)}\to\boldsymbol{U}^{(C)}$, $S^{(B)}\to\boldsymbol{S}^{(B)}$, and $\Gamma^{(B)}\to\boldsymbol{\Gamma}^{(B)}$, i.e.,

$$\frac{\partial}{\partial\widehat{\boldsymbol{x}}}\cdot\left[\boldsymbol{\beta}^{(e)}(\widehat{\boldsymbol{x}})+\widehat{\boldsymbol{\alpha}}\boldsymbol{U}^{(C)}\boldsymbol{S}^{(B)}(\widehat{\boldsymbol{x}})\right]=\boldsymbol{\Gamma}^{(B)}(\widehat{\boldsymbol{x}})\ .\tag{3.3.95}$$

Alternatively, we can also write (3.3.93), on multiplying to the right with $\boldsymbol{U}^{(C)}$, as

$$-\frac{\partial}{\partial\boldsymbol{\rho}}\cdot\boldsymbol{\beta}(\widehat{\boldsymbol{x}})\boldsymbol{I}^{(B)}=\left\langle\boldsymbol{w}_n(\widehat{\boldsymbol{x}})\right\rangle=\boldsymbol{w}_n^{(C)}(\widehat{\boldsymbol{x}})+\boldsymbol{w}_n^{(I)}(\widehat{\boldsymbol{x}})\ ,\tag{3.3.96a}$$

in terms of the vertical power fluxes of coherent and incoherent waves on S_{12}, $\boldsymbol{w}_n^{(C)}$ and $\boldsymbol{w}_n^{(I)}$, respectively, given by

$$\boldsymbol{w}_n^{(C)}(\widehat{\boldsymbol{x}})=-\boldsymbol{\delta}(\widehat{\boldsymbol{x}}|1;2)\boldsymbol{\Gamma}^{(B)}\boldsymbol{U}^{(C)}(1;2)\ ,\tag{3.3.96b}$$

$$\boldsymbol{w}_n^{(I)}(\widehat{\boldsymbol{x}})=\boldsymbol{\delta}(\widehat{\boldsymbol{x}}|1;2)2^{-1}\left[\boldsymbol{h}^*(1)+\boldsymbol{h}(2)\right]\boldsymbol{\eta}^{-1}\boldsymbol{U}^{(C)}\boldsymbol{S}^{(B)}\boldsymbol{U}^{(C)}(1;2)\ ,\tag{3.3.96c}$$

in exactly the same form as (3.2.32a) for a rough surface. Hence, power equation (3.2.33b) also formally remains unchanged with $I^{(B)}\to\boldsymbol{I}^{(B)}$ and $\Delta G\to\Delta\boldsymbol{G}$. The relation $\int d\boldsymbol{\rho}\left\langle\boldsymbol{w}_n(\widehat{\boldsymbol{x}})\right\rangle=0$, which results from (3.3.96a), leads to the average version of relation (3.3.25) in view of

$$\boldsymbol{G}^\dagger-\boldsymbol{G}=\boldsymbol{G}^\dagger\left[\mathrm{i}(\boldsymbol{h}^\dagger+\boldsymbol{h})\boldsymbol{\eta}^{-1}+\boldsymbol{M}^{(B)\dagger}-\boldsymbol{M}^{(B)}\right]\boldsymbol{G}\ .\tag{3.3.97}$$

Finally, as in the case of a rough surface (Sect 3.2), various equations associated with the incoherent wave (including cross sections and optical relations) can be written in terms of $\boldsymbol{S}^{(B)}$, in such a way to enable a straightforward physical interpretation. Here, $\boldsymbol{S}^{(B)}$ is obtained as the solution of integral equation (3.3.89a), or of its optical version to be given by (3.3.141).

3.3.5 Continuation to the Outside Spaces

Continuation of the deterministic Green's function defined on the boundary planes, S_1 and S_2, into the outside spaces can be made on following the procedure of Sect. 3.2.4, as follows: Equations (3.3.10a,b) are replaced by a set of equations for the space Green's function, say $g_{ij}(z|z')$, as

$$g_{11}(z|z') = \Big\{\exp\big[-ih_1|z-z'|\big] - \exp\big[-ih_1|z+z'|\big]\Big\}(2ih_1)^{-1}\eta_1 \\ + \exp(-ih_1 z) g_{11} \exp(-ih_1 z'), \quad z, z' \geq 0 \ , \tag{3.3.98a}$$

$$g_{21}(z|z') = \exp\big[ih_2(z+d_2)\big] g_{21} \exp(-ih_1 z'), \\ z \leq -d_2, \quad z' \geq 0 \ ; \tag{3.3.98b}$$

similar equations for $g_{22}(z|z')$ and $g_{12}(z|z')$ are also obtained from (3.3.12a,b). In fact, since the g_{ij}'s are the same as those defined in Sect. 3.3.1, $g_{ij}(z|z')$ is found to be the solution of

$$(-\partial_z^2 - h_i^2) g_{ij}(z|z') = \eta_i \delta_{ij} \delta(z-z') \ , \tag{3.3.99}$$

subjected to boundary condition (3.3.3) on S_1 and S_2.

Thus the Green's function of first order in space, say $G_{ij}(z|z')$, is obtained simply by replacing all the coefficient g_{ij}'s in (3.3.98) with G_{ij} of (3.3.79a). The situation is also the same for the Green's function of second order, resulting in most of the previous equations remaining unchanged with the redefinition of $\boldsymbol{G}^*(1)$, $\boldsymbol{G}(2)$, and $\boldsymbol{I}^{(B)}(1;2)$ by those continued into the outside spaces, in just the same way as in the case of a rough surface. Hence, $\boldsymbol{I}^{(B)}(1;2)$ now represents the $\widehat{\boldsymbol{x}}$ matrix having the elements $I^{(B)}_{ij;kl}(\widehat{\boldsymbol{x}}_1;\widehat{\boldsymbol{x}}_2|\widehat{\boldsymbol{x}}'_1;\widehat{\boldsymbol{x}}'_2)$, with the possible abbreviations $I^{(B)}_{ij;kl}(z_1;z_2|z'_1;z'_2)$ and $I^{(B)}_{ij;kl}$. Here, since $\boldsymbol{K}^{(B)}$ and $\boldsymbol{S}^{(B)}$ are ρ matrices, $\boldsymbol{S}^{(B)}$ as an $\widehat{\boldsymbol{x}}$ matrix, for example, has the matrix elements given by

$$S^{(B)}_{ij;kl}(\widehat{\boldsymbol{x}}_1;\widehat{\boldsymbol{x}}_2|\widehat{\boldsymbol{x}}'_1;\widehat{\boldsymbol{x}}'_2) = \delta_i(z_1)\delta_j(z_2) \\ \times S^{(B)}_{ij;kl}(\rho_1;\rho_2|\rho'_1;\rho'_2)\delta_k(z'_1)\delta_l(z'_2) \ , \tag{3.3.100}$$

in terms of notation,

$$\delta_j(z) = \delta(z+d_j), \qquad d_1 = 0 \ . \tag{3.3.101}$$

For later convenience, we introduce here several 2×2 matrices, as follows

$$\boldsymbol{G}^{(0)} = (2i\boldsymbol{h})^{-1}\boldsymbol{\eta} = \begin{pmatrix} G_1^{(0)} & 0 \\ 0 & G_2^{(0)} \end{pmatrix} \\ = \begin{pmatrix} (2ih_1)^{-1}\eta_1 & 0 \\ 0 & (2ih_2)^{-1}\eta_2 \end{pmatrix} \ , \tag{3.3.102}$$

where $G_j^{(0)}$ means the Green's function in a homogeneous medium of k_j, except for the constant factor, η_j (which therefore differs from $G^{(0)}$ in (3.1.4) and (3.2.8), and also should not be confused with g_0 of (3.3.43) for the smooth boundary of $\zeta = 0$); also two "attenuation" coefficient matrices, $\boldsymbol{A}$ and $\overline{\boldsymbol{A}}$, for the coherent wave, defined by

$$\boldsymbol{G} = 2\boldsymbol{A}\boldsymbol{G}^{(0)} = 2\boldsymbol{G}^{(0)}\overline{\boldsymbol{A}} \ , \tag{3.3.103}$$

and therefore given by

$$\begin{aligned} \boldsymbol{A} &= \mathrm{i}\boldsymbol{G}\boldsymbol{h}\boldsymbol{\eta}^{-1} = (\boldsymbol{h} + \mathrm{i}\boldsymbol{\eta}\boldsymbol{M}^{(B)})^{-1}\boldsymbol{h} \ , \\ \overline{\boldsymbol{A}} &= \mathrm{i}\boldsymbol{\eta}^{-1}\boldsymbol{h}\boldsymbol{G} = \boldsymbol{A}^T \ . \end{aligned} \tag{3.3.104}$$

Hence, from (3.3.18a)

$$\langle \boldsymbol{R} \rangle = 2\boldsymbol{A} - 1 = (\mathrm{i}\boldsymbol{h}\boldsymbol{\eta}^{-1} - \boldsymbol{M}^{(B)})^{-1}(\mathrm{i}\boldsymbol{h}\boldsymbol{\eta}^{-1} + \boldsymbol{M}^{(B)}) \ . \tag{3.3.105}$$

Alternatively,

$$\langle \boldsymbol{R} \rangle = \boldsymbol{G}^{(0)} \langle \boldsymbol{T} \rangle \ , \tag{3.3.106}$$

where $\langle \boldsymbol{T} \rangle$ is a scattering matrix, defined by

$$\langle \boldsymbol{T} \rangle = 2\mathrm{i}\boldsymbol{h}\boldsymbol{\eta}^{-1} \langle \boldsymbol{R} \rangle = \langle \boldsymbol{T} \rangle^T \ , \tag{3.3.107}$$

and has symmetrical matrix elements, in contrast with $\langle \boldsymbol{R} \rangle$ (3.3.22b). Finally, we introduce a z-continued version of $\boldsymbol{G}^{(0)}$, say $\boldsymbol{G}^{(0)}(z - z')$, defined by the diagonal elements,

$$G_j^{(0)}(z - z') = G_j^{(0)} \exp\left(-\mathrm{i}h_j|z - z'|\right), \qquad j = 1, 2 \ , \tag{3.3.108}$$

and, particularly when the point, z', is placed on S_j at $z = -d_j$,

$$G_j^{(0)}(z|0) = G_j^{(0)}(0|z) \equiv G_j^{(0)}(z + d_j) \ . \tag{3.3.109}$$

Hence,

$$\boldsymbol{G} = \left(1 + \langle \boldsymbol{R} \rangle\right) \boldsymbol{G}^{(0)} = \boldsymbol{G}^{(0)} + \boldsymbol{G}^{(0)} \langle \boldsymbol{T} \rangle \boldsymbol{G}^{(0)} \ , \tag{3.3.110}$$

and is continued to the outside spaces as

$$\boldsymbol{G}(z|z') = \boldsymbol{G}^{(0)}(z - z') + \boldsymbol{G}^{(0)}(z|0) \langle \boldsymbol{T} \rangle \boldsymbol{G}^{(0)}(0|z') \ , \tag{3.3.111}$$

which represents the averaged version of the whole set of equations (3.3.98). Note that $\boldsymbol{G}^{(0)}(z - z') = 0$ when the points, z and z', are not in the same space of either $\boldsymbol{k}_1$ or $\boldsymbol{k}_2$. In the same way, the continued version of (3.3.103) is

$$\begin{aligned} \boldsymbol{G}(0|z) &= 2\boldsymbol{A}\boldsymbol{G}^{(0)}(0|z) \ , \\ \boldsymbol{G}(z|0) &= 2\boldsymbol{G}^{(0)}(z|0)\overline{\boldsymbol{A}} \ , \end{aligned} \tag{3.3.112}$$

which, as well as $\boldsymbol{G}(z|z')$, are non-diagonal 2×2 matrices, also having the elements for the points z and z' existing in separate spaces.

Now, the continued version of (3.3.88) for the second order Green's function is written as

$$\begin{aligned} \boldsymbol{I}^{(B)}(z_1; z_2|z_1'; z_2') &= \boldsymbol{U}^{(C)}(z_1; z_2|z_1'; z_2') \\ &+ \boldsymbol{U}(z_1; z_2|0)\boldsymbol{\sigma}^{(I)}(1;2)\boldsymbol{U}(0|z_1'; z_2') \ , \end{aligned} \tag{3.3.113}$$

in the same form as (3.2.41). Here $\boldsymbol{\sigma}^{(I)}(1;2)$ is the $\boldsymbol{\rho}$ matrix defined by

$$\boldsymbol{\sigma}^{(I)}(1;2) = \overline{\boldsymbol{F}}^{(C)}(1;2)\boldsymbol{S}^{(B)}(1;2)\boldsymbol{F}^{(C)}(1;2) \ , \tag{3.3.114}$$

where $\overline{\boldsymbol{F}}^{(C)}$ and $\boldsymbol{F}^{(C)}$ are the "attenuation" factors of $\boldsymbol{U}^{(C)}$, defined by

$$\boldsymbol{U}^{(C)} = \boldsymbol{U}\overline{\boldsymbol{F}}^{(C)} = \boldsymbol{F}^{(C)}\boldsymbol{U} \ , \tag{3.3.115}$$

$$\boldsymbol{U}(1;2) = \boldsymbol{G}^{(0)*}(1)\boldsymbol{G}^{(0)}(2) \ , \tag{3.3.116}$$

and, with expressions (3.3.104) for $\boldsymbol{A}$ and $\overline{\boldsymbol{A}}$,

$$\overline{\boldsymbol{F}}^{(C)}(1;2) = 2^2\overline{\boldsymbol{A}}^*(1)\overline{\boldsymbol{A}}(2), \qquad \boldsymbol{F}^{(C)}(1;2) = 2^2\boldsymbol{A}^*(1)\boldsymbol{A}(2) \ . \tag{3.3.117}$$

Equation (3.3.113) can be written as an $\widehat{\boldsymbol{x}}$ matrix equation

$$\boldsymbol{I}^{(B)}(1;2) = \boldsymbol{U}^{(C)}(1;2) + \boldsymbol{U}(1;2)\boldsymbol{\sigma}^{(I)}(1;2)\boldsymbol{U}(1;2) \ , \tag{3.3.118}$$

and, using (3.3.111), further written as

$$\boldsymbol{I}^{(B)}(1;2) = \boldsymbol{U}(1;2) + \boldsymbol{U}(1;2)\boldsymbol{\sigma}^{(T)}(1;2)\boldsymbol{U}(1;2) \ , \tag{3.3.119a}$$

in terms of a resultant scattering matrix, $\boldsymbol{\sigma}^{(T)}(1;2)$, defined by

$$\boldsymbol{\sigma}^{(T)}(1;2) = \langle \boldsymbol{T}^*(1)\rangle\langle \boldsymbol{T}(2)\rangle + \boldsymbol{\sigma}^{(I)}(1;2) \ ; \tag{3.3.119b}$$

which is a consequence of using

$$\boldsymbol{U}^{(C)}(1;2) = \boldsymbol{U}(1;2) + \boldsymbol{U}(1;2)\boldsymbol{V}^{(C)}(1;2)\boldsymbol{U}(1;2) \ , \tag{3.3.120a}$$

$$\begin{aligned} \boldsymbol{V}^{(C)}(1;2) &= \langle \boldsymbol{T}^*(1)\rangle\langle \boldsymbol{T}(2)\rangle \\ &+ \langle \boldsymbol{T}^*(1)\rangle \boldsymbol{G}^{-1}(2) + \langle \boldsymbol{T}(2)\rangle \left[\boldsymbol{G}^*(1)\right]^{-1} \ , \end{aligned} \tag{3.3.120b}$$

similar to (3.2.46a,b), upon neglecting the interference terms of $\boldsymbol{V}^{(C)}(1;2)$.

3.3.6 Scattering Cross Sections and Optical Relations

The procedure of deriving the optical expression of (3.3.113) is essentially the same as in the case of a rough surface (Sect. 3.2.6) and the result is obtained, when the points $\hat{\boldsymbol{x}}_1 = \hat{\boldsymbol{x}}_2 = \hat{\boldsymbol{x}}$ and $\hat{\boldsymbol{x}}_1' = \hat{\boldsymbol{x}}_2' = \hat{\boldsymbol{x}}'$ are in the spaces of $\boldsymbol{k}_i$ and $\boldsymbol{k}_j$, respectively, in the form [3.11]

$$I_{ij}^{(B)}(\hat{\boldsymbol{x}}|\hat{\boldsymbol{x}}') \equiv I_{ii;jj}^{(B)}(\hat{\boldsymbol{x}};\hat{\boldsymbol{x}}|\hat{\boldsymbol{x}}';\hat{\boldsymbol{x}}') = \left|G_{ij}(\hat{\boldsymbol{x}}|\hat{\boldsymbol{x}}')\right|^2 + \int d\boldsymbol{\rho}''d\boldsymbol{\rho}'''$$
$$\times|\hat{\boldsymbol{x}} - \boldsymbol{\rho}''|^{-2}\eta_i^2\sigma_{ij}^{(I)}(\hat{\boldsymbol{\Omega}}|\boldsymbol{\rho}'' - \boldsymbol{\rho}'''|\hat{\boldsymbol{\Omega}}')\left|G_j^{(0)}(\boldsymbol{\rho}''' - \hat{\boldsymbol{x}}')\right|^2 \ . \quad (3.3.121)$$

Here, $\hat{\boldsymbol{\Omega}}$ and $\hat{\boldsymbol{\Omega}}'$ are the unit vectors in the directions of $\hat{\boldsymbol{x}} - \boldsymbol{\rho}''$ and $\boldsymbol{\rho}''' - \hat{\boldsymbol{x}}'$, respectively, and (Fig. 3.5)

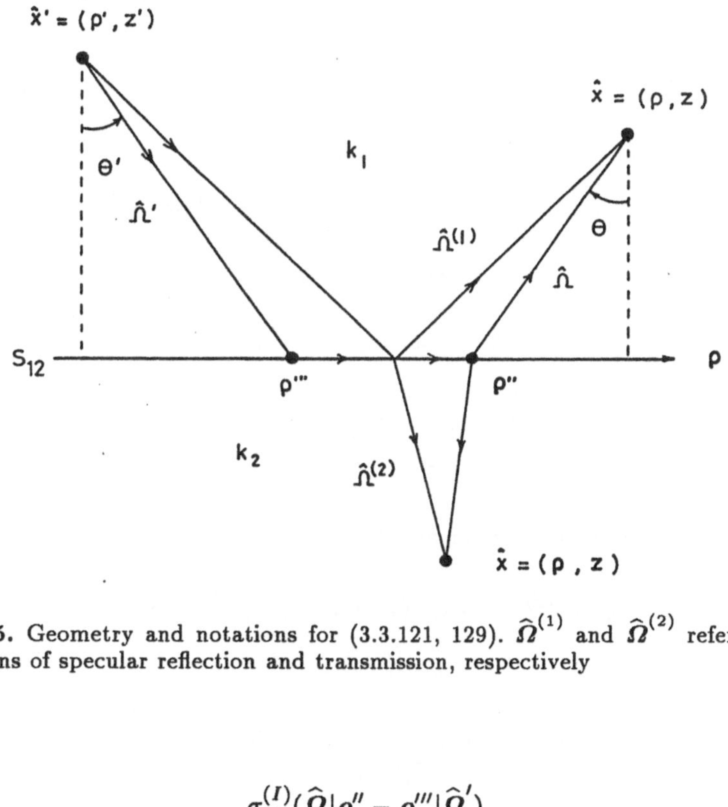

Fig. 3.5. Geometry and notations for (3.3.121, 129). $\hat{\boldsymbol{\Omega}}^{(1)}$ and $\hat{\boldsymbol{\Omega}}^{(2)}$ refer to the directions of specular reflection and transmission, respectively

$$\sigma_{ij}^{(I)}(\hat{\boldsymbol{\Omega}}|\boldsymbol{\rho}'' - \boldsymbol{\rho}'''|\hat{\boldsymbol{\Omega}}')$$
$$= \sum_{a,b,c,d} \overline{F}_{ii;ab}^{(C)}(\hat{\boldsymbol{\Omega}})S_{ab;cd}^{(B)}(\hat{\boldsymbol{\Omega}}|\boldsymbol{\rho}'' - \boldsymbol{\rho}'''|\hat{\boldsymbol{\Omega}}')F_{cd;jj}^{(C)}(\hat{\boldsymbol{\Omega}}') \ , \quad (3.3.122)$$

where

$$\overline{F}^{(C)}_{ii;ab}(\widehat{\boldsymbol{\Omega}}) = 4\overline{A}^*_{ia}(\widehat{\boldsymbol{\Omega}})\overline{A}_{ib}(\widehat{\boldsymbol{\Omega}}) \ , \tag{3.3.123}$$

$$\overline{A}_{ib}(\widehat{\boldsymbol{\Omega}}) = \widetilde{\overline{A}}_{ib}(\boldsymbol{u}) \ |_{\boldsymbol{\Omega}} = \mathrm{i}\widetilde{h}_i(\boldsymbol{u})\eta_i^{-1}\widetilde{G}_{ib}(\boldsymbol{u}) \ |_{\boldsymbol{\Omega}} \ ; \tag{3.3.124}$$

$F^{(C)}_{cd;jj}$ is similarly given in terms of $A_{cj}(\widehat{\boldsymbol{\Omega}})$'s, with the mark $|_{\boldsymbol{\Omega}}$ meaning to set

$$\boldsymbol{u} = k_i\boldsymbol{\Omega}, \qquad \widetilde{h}_i(\boldsymbol{u}) = k_i\Omega_z \ . \tag{3.3.125}$$

For given $\widehat{\boldsymbol{\Omega}}$, however, $\boldsymbol{u}$ is undetermined by the factor k_i, and therefore, whenever there is confusion, the notation $\widehat{\boldsymbol{\Omega}}^{(i)}$ will be used for $\widehat{\boldsymbol{\Omega}}$, to mean that $\boldsymbol{u} = k_i\boldsymbol{\Omega}^{(i)}$. In (3.3.124), $\widetilde{G}_{ij}(\boldsymbol{u})$ is obtained from (3.3.79a) as the solution of

$$\sum_j \left[\mathrm{i}\widetilde{h}_i(\boldsymbol{u})\eta_i^{-1}\delta_{ij} - \widetilde{M}^{(B)}_{ij}(\boldsymbol{u})\right]\widetilde{G}_{jk}(\boldsymbol{u}) = \delta_{ik} \ , \tag{3.3.126}$$

and $\widetilde{G}_{jk}(\boldsymbol{u}) = \widetilde{G}_{kj}(-\boldsymbol{u})$, from (3.3.22a), so that $\overline{A}_{ij}(\widehat{\boldsymbol{\Omega}}) = \overline{A}_{ji}(-\widehat{\boldsymbol{\Omega}})$;

$$\begin{aligned} &S^{(B)}_{ab;cd}(\widehat{\boldsymbol{\Omega}}|\rho|\widehat{\boldsymbol{\Omega}}') \\ &= (2\pi)^{-2}\int d\boldsymbol{\lambda}\, \mathrm{e}^{-\mathrm{i}\boldsymbol{\lambda}\cdot\boldsymbol{\rho}}(4\pi)^{-2}\widetilde{S}_{ab;cd}(\boldsymbol{u}|\boldsymbol{\lambda}|\boldsymbol{u}') \ |_{\boldsymbol{\Omega}} \ , \end{aligned} \tag{3.3.127}$$

which corresponds to (3.2.67) for a rough surface.

Thus, when the points $\widehat{\boldsymbol{x}}$ and $\widehat{\boldsymbol{x}}'$ are both separated far enough from the boundary, (3.3.121) leads to the asymptotic expression

$$\begin{aligned} I^{(B)}_{ij}(\widehat{\boldsymbol{x}}|\widehat{\boldsymbol{x}}') &= \left|G^{(0)}_i(\widehat{\boldsymbol{x}} - \widehat{\boldsymbol{x}}')\right|^2\delta_{ij} \\ &+ \int d\boldsymbol{\rho}'' \, |\widehat{\boldsymbol{x}} - \boldsymbol{\rho}''|^{-2}\sigma^{(T)}_{ij}(\widehat{\boldsymbol{\Omega}}|\widehat{\boldsymbol{\Omega}}')\left|G^{(0)}_j(\boldsymbol{\rho}'' - \widehat{\boldsymbol{x}}')\right|^2 \ . \end{aligned} \tag{3.3.128}$$

Here, from (3.3.120, 107) [refer also to (3.2.71a)]

$$\begin{aligned} \sigma^{(T)}_{ij}(\widehat{\boldsymbol{\Omega}}|\widehat{\boldsymbol{\Omega}}') &= \left|\Omega^{(i)}_z\langle R_{ij}(\widehat{\boldsymbol{\Omega}})\rangle^2\right|\delta^2_s(\widehat{\boldsymbol{\Omega}}^{(i)} - \widehat{\boldsymbol{\Omega}}^{(i)'}) \\ &\quad + \eta_i^2\sigma^{(I)}_{ij}(\widehat{\boldsymbol{\Omega}}|\widehat{\boldsymbol{\Omega}}') \ , \end{aligned} \tag{3.3.129}$$

where $\widehat{\boldsymbol{\Omega}}^{(i)'} = \widehat{\boldsymbol{\Omega}}^{(i)}(\boldsymbol{\Omega}')$ means the unit vector in the direction of wave propagation when the incident wave of given $\boldsymbol{u} = k_j\boldsymbol{\Omega}'$ is observed in the medium of k_i (Fig. 3.5);

$$\langle R_{ij}(\widehat{\boldsymbol{\Omega}})\rangle = \langle \widetilde{R}_{ij}(\boldsymbol{u})\rangle \ |_{\boldsymbol{\Omega}} \neq \langle R_{ji}(-\widehat{\boldsymbol{\Omega}})\rangle \qquad (i \neq j) \ , \tag{3.3.130}$$

$$\sigma^{(I)}_{ij}(\widehat{\boldsymbol{\Omega}}|\widehat{\boldsymbol{\Omega}}') = \int d\boldsymbol{\rho}\, \sigma^{(I)}_{ij}(\widehat{\boldsymbol{\Omega}}|\boldsymbol{\rho}|\widehat{\boldsymbol{\Omega}}') \ , \tag{3.3.131}$$

which is given in view of (3.3.122), by $\sigma_{ij}^{(I)} = \sigma_{ii;jj}^{(I)}$ from

$$\sigma_{ij;kl}^{(I)}(\widehat{\Omega}|\widehat{\Omega}') = \sum_{a,b,c,d} \overline{F}_{ij;ab}^{(C)}(\widehat{\Omega}) S_{ab;cd}^{(B)}(\widehat{\Omega}|\widehat{\Omega}') F_{cd;kl}^{(C)}(\widehat{\Omega}') \ , \quad (3.3.132a)$$

$$S_{ab;cd}^{(B)}(\widehat{\Omega}|\widehat{\Omega}') = (4\pi)^{-2} \widetilde{S}_{ab;cd}^{(B)}(\boldsymbol{u}|\boldsymbol{\lambda}=0|\boldsymbol{u}') \,|_{\Omega} \ , \quad (3.3.132b)$$

[cf.(3.2.66, 67)].

a) Optical Relations

When the medium is non-dissipative, as has been assumed, the cross section $\boldsymbol{\sigma}^{(T)}(\widehat{\Omega}|\widehat{\Omega}')$ of (3.3.129), is subject to an optical relation which follows from the Fourier transform of relation (3.3.94c), i.e.,

$$\widetilde{\Gamma}_{cd}^{(B)}(\boldsymbol{u}') = \sum_{j,a,b} (2\pi)^{-2} \int_{|\boldsymbol{u}|\le k_j} d\boldsymbol{u}\, \widetilde{h}_j(\boldsymbol{u}) \eta_j^{-1} \times \widetilde{U}_{jj;ab}^{(C)}(\boldsymbol{u}, \boldsymbol{\lambda}=0) \widetilde{S}_{ab;cd}^{(B)}(\boldsymbol{u}|\boldsymbol{\lambda}=0|\boldsymbol{u}') \ , \quad (3.3.133)$$

where

$$\widetilde{\Gamma}_{cd}^{(B)}(\boldsymbol{u}) = (2\mathrm{i})^{-1} [\widetilde{M}_{dc}^{(B)*}(\boldsymbol{u}) - \widetilde{M}_{cd}^{(B)}(\boldsymbol{u})] = \widetilde{\Gamma}_{dc}^{(B)*}(-\boldsymbol{u}) \ . \quad (3.3.134a)$$

Hence, changing the variable of integration $\boldsymbol{u}$ to $\widehat{\Omega}$, with $d\boldsymbol{u} = k_j^2 \Omega_z^{(j)} d\widehat{\Omega}^{(j)}$ and the notation,

$$\Gamma_{cd}^{(B)}(\widehat{\Omega}) = \widetilde{\Gamma}_{cd}^{(B)}(\boldsymbol{u}) \,|_{\Omega} = \Gamma_{dc}^{(B)*}(-\widehat{\Omega}) \ , \quad (3.3.134b)$$

we obtain

$$\Gamma_{cd}^{(B)}(\widehat{\Omega}') = \sum_{j,a,b} k_j \eta_j \int_{2\pi} d\widehat{\Omega}^{(j)} \overline{F}_{jj;ab}^{(C)}(\widehat{\Omega}) S_{ab;cd}^{(B)}(\widehat{\Omega}|\widehat{\Omega}') \ , \quad (3.3.135)$$

in consequence of (3.3.132, 115, 102). Equation (3.3.135) can be rewritten in terms of $\boldsymbol{\sigma}^{(I)}$ of (3.3.132) as

$$\sum_{c,d} \Gamma_{cd}^{(B)}(\widehat{\Omega}') F_{cd;kl}^{(C)}(\widehat{\Omega}') = \sum_{j} k_j \eta_j \int_{2\pi} d\widehat{\Omega}\, \sigma_{jj;kl}^{(I)}(\widehat{\Omega}|\widehat{\Omega}') \ , \quad (3.3.136)$$

which provides the optical relation connecting the attenuation coefficient for the coherent wave on the left-hand side to the total incoherent cross section on the right-hand side.

Another relation exists connecting the attenuation coefficient to the reflection coefficient, $\langle \boldsymbol{R} \rangle$, for the coherent wave; i.e., (3.3.79a,b) lead to a relation similar to (3.3.26),

$$\mathrm{i}^{-1}4\boldsymbol{A}^{\dagger}(\boldsymbol{M}^{\dagger}-\boldsymbol{M})\boldsymbol{A}+\langle\boldsymbol{R}^{\dagger}\rangle(\boldsymbol{h}^{\dagger}+\boldsymbol{h})\boldsymbol{\eta}^{-1}\langle\boldsymbol{R}\rangle-(\boldsymbol{h}^{\dagger}+\boldsymbol{h})\boldsymbol{\eta}^{-1}=0\ , \tag{3.3.137}$$

which is valid only within the optical range, $\boldsymbol{h}^{\dagger}=\boldsymbol{h}$. Hence, by the Fourier transformation, we obtain

$$\sum_{i,j}4\widetilde{\Gamma}_{ij}^{(B)}(\boldsymbol{u})\widetilde{A}_{ia}^{*}\widetilde{A}_{jb}(\boldsymbol{u})+\sum_{j}\widetilde{h}_{j}(\boldsymbol{u})\eta_{j}^{-1}\langle\widetilde{R}_{ja}^{*}(\boldsymbol{u})\rangle\langle\widetilde{R}_{jb}(\boldsymbol{u})\rangle$$
$$-\widetilde{h}_{a}(\boldsymbol{u})\eta_{a}^{-1}\delta_{ab}=0\ , \tag{3.3.138}$$

whose optical expression becomes

$$\sum_{i,j}\Gamma_{ij}^{(B)}(\widehat{\boldsymbol{\Omega}})F_{ij;ab}^{(C)}(\widehat{\boldsymbol{\Omega}})=k_{a}\eta_{a}^{-1}\Omega_{z}^{(a)}\delta_{ab}$$
$$-\sum_{j}k_{j}\eta_{j}^{-1}\Omega_{z}^{(j)}\langle R_{ja}^{*}(\widehat{\boldsymbol{\Omega}})\rangle\langle R_{jb}(\widehat{\boldsymbol{\Omega}})\rangle\ , \tag{3.3.139}$$

with the same left-hand side as (3.3.136).

Hence, two relations (3.3.136, 139), can be combined to be written in terms of $\boldsymbol{\sigma}^{(T)}$ of (3.3.129), as a simple relation:

$$\sum_{j}k_{j}\eta_{j}^{-1}\int_{2\pi}d\widehat{\boldsymbol{\Omega}}\,\sigma_{jj;ab}^{(T)}(\widehat{\boldsymbol{\Omega}}|\widehat{\boldsymbol{\Omega}}')=k_{a}\eta_{a}^{-1}\Omega_{z}^{(a)}\delta_{ab}\ , \tag{3.3.140}$$

which is similar in form to (3.2.86), and ensures power conservation in scattering by the two-sided boundary. It may be remarked that in both optical relations (3.3.136, 140), the $\widehat{\boldsymbol{\Omega}}$ range of integration is limited strictly within the optical range, 2π, and that this is a consequence of using the expression (3.3.133) in terms of the scattering matrix, $\boldsymbol{S}^{(B)}$ (instead of $\boldsymbol{K}^{(B)}$).

Equation (3.3.89a) for $\boldsymbol{S}^{(B)}$ leads to an integral equation for the factor $S_{ab;cd}^{(B)}(\widehat{\boldsymbol{\Omega}}|\widehat{\boldsymbol{\Omega}}')$ involved in $\sigma_{ij;kl}^{(I)}(\widehat{\boldsymbol{\Omega}}|\widehat{\boldsymbol{\Omega}}')$ of (3.3.132), as [cf.(3.2.78, 77)]

$$S_{ab;cd}^{(B)}(\widehat{\boldsymbol{\Omega}}|\widehat{\boldsymbol{\Omega}}')=K_{ab;cd}^{(B)}(\widehat{\boldsymbol{\Omega}}|\widehat{\boldsymbol{\Omega}}')+\sum_{i,j,k,l}\eta_{i}\eta_{j}\int(d\widehat{\boldsymbol{\Omega}}''\Omega_{z}''^{-1})_{ij}$$
$$\times K_{ab;ij}^{(B)}(\widehat{\boldsymbol{\Omega}}|\widehat{\boldsymbol{\Omega}}'')F_{ij;kl}^{(C)}(\widehat{\boldsymbol{\Omega}}'')S_{kl;cd}^{(B)}(\widehat{\boldsymbol{\Omega}}''|\widehat{\boldsymbol{\Omega}}')\ . \tag{3.3.141}$$

Here,

$$(d\widehat{\boldsymbol{\Omega}}\Omega_{z}^{-1})_{ij}=d\boldsymbol{u}\left[\widetilde{h}_{i}^{*}(\boldsymbol{u})\widetilde{h}_{j}(\boldsymbol{u})\right]^{-1} \tag{3.3.142}$$
$$=(k_{i}/k_{j})(\Omega_{z}^{(j)})^{-1}d\widehat{\boldsymbol{\Omega}}^{(i)}$$
$$=(k_{j}/k_{i})(\Omega_{z}^{(i)})^{-1}d\widehat{\boldsymbol{\Omega}}^{(j)}\ , \tag{3.3.143}$$

and the path of integration is not limited within the optical range, 2π, unlike those in (3.3.136, 140), instead, it is over the entire $\boldsymbol{u}$-range of integration so that $\Omega_{z}^{(j)}$ changes from $-\mathrm{i}\infty$ to 1 via 0.

3.3.7 Case of a Slightly Random Boundary

Based on Eqs. (3.3.49, 51) for the deterministic Green's functions, g_{11} and g_{21}, an equation of the statistical Green's function of second order can be obtained in the form of the BS equation (3.3.83), so that the incoherent cross sections are obtained in the form of (3.3.132) (Appendix C of [3.11]). To obtain the cross sections to the first order approximation, however, we can directly derive the results by using (3.3.42, 46): We first introduce attenuation factors, A_1 and A_2, defined by

$$g_0 = 2A_jG_j^{(0)} = 2G_j^{(0)}A_j, \qquad j = 1,2 , \tag{3.3.144}$$

and rewrite (3.3.42, 46) by

$$g_{11} = g_0 + 2^2G_1^{(0)}A_1b^{(11)}A_1G_1^{(0)} , \tag{3.3.145}$$

$$g_{21} = g_0 + 2^2G_2^{(0)}A_2b^{(21)}A_1G_1^{(0)} . \tag{3.3.146}$$

Hence, on reference to (3.3.113, 114), we obtain

$$\sigma_{11}^{(I)}(1;2) = 2^4A_1^*(1)A_1(2)\langle b^{(11)*}(1)b^{(11)}(2)\rangle A_1^*(1)A_1(2) , \tag{3.3.147}$$

$$\sigma_{21}^{(I)}(1;2) = 2^4A_2^*(1)A_2(2)\langle b^{(21)*}(1)b^{(21)}(2)\rangle A_1^*(1)A_1(2) . \tag{3.3.148}$$

Thus the procedure of deriving the cross sections therefrom becomes the same as in (3.2.89–95), with the result that, in terms of the boundary power spectrum function $\Phi(\boldsymbol{u})$ of (3.2.90), and with the constants ε_j, $j = 1,2$, defined by $k_j^2 = \varepsilon_j k_0^2$,

$$\begin{aligned}\eta_j^2\sigma_{j1}^{(I)}(\widehat{\boldsymbol{\Omega}}|\widehat{\boldsymbol{\Omega}}') &= k_0^4\Phi\left[k_0(\varepsilon_1^{1/2}\boldsymbol{\Omega}' - \varepsilon_j^{1/2}\boldsymbol{\Omega})\right] \\ &\times\left|\eta_jA_j(\widehat{\boldsymbol{\Omega}})\left[f_{j1}(\widehat{\boldsymbol{\Omega}}|\widehat{\boldsymbol{\Omega}}') + \varepsilon_1\eta_1^{-1} - \varepsilon_2\eta_2^{-1}\right]A_1(\widehat{\boldsymbol{\Omega}}')\right|^2 .\end{aligned} \tag{3.3.149}$$

Here,

$$\begin{aligned}A_1(\widehat{\boldsymbol{\Omega}}) &= \Omega_z\left[\Omega_z + (\eta_1/\eta_2)(\varepsilon_2/\varepsilon_1 - \boldsymbol{\Omega}^2)^{1/2}\right]^{-1} , \\ A_2(\widehat{\boldsymbol{\Omega}}) &= \Omega_z\left[\Omega_z + (\eta_2/\eta_1)(\varepsilon_1/\varepsilon_2 - \boldsymbol{\Omega}^2)^{1/2}\right]^{-1} ,\end{aligned} \tag{3.3.150}$$

and, from (3.3.44, 46b)

$$\begin{aligned}f_{11}(\widehat{\boldsymbol{\Omega}}|\widehat{\boldsymbol{\Omega}}') &= (\eta_1^{-1} - \eta_2^{-1})\varepsilon_1 \\ &\times\left[\eta_1\eta_2^{-1}(\varepsilon_2/\varepsilon_1 - \boldsymbol{\Omega}^2)^{1/2}(\varepsilon_2/\varepsilon_1 - \boldsymbol{\Omega}'^2)^{1/2} - \boldsymbol{\Omega}\cdot\boldsymbol{\Omega}'\right] ,\end{aligned} \tag{3.3.151}$$

$$\begin{aligned}f_{21}(\widehat{\boldsymbol{\Omega}}|\widehat{\boldsymbol{\Omega}}') &= (\eta_2^{-1} - \eta_1^{-1})(\varepsilon_1\varepsilon_2)^{1/2} \\ &\times\left[(\varepsilon_1/\varepsilon_2 - \boldsymbol{\Omega}^2)^{1/2}(\varepsilon_2/\varepsilon_1 - \boldsymbol{\Omega}'^2)^{1/2} + \boldsymbol{\Omega}\cdot\boldsymbol{\Omega}'\right] .\end{aligned} \tag{3.3.152}$$

In the case of electro-magnetic waves of vertical polarization in two-dimensional space (for example, scattered within the plane of incidence), $\eta = \varepsilon$ so that $\varepsilon_1\eta_1^{-1} - \varepsilon_2\eta_2^{-1} = 0$ in (3.3.149), while, in the case of horizontal polarization, $\eta = \mu = 1$ so that there is entirely no contribution from all the f_{j1} terms, $j = 1, 2$, in view of (3.3.151, 152). In either case, the cross sections for the reflected waves agree with those by previous authors [3.13–15].

4. System of Random Media and Rough Boundaries

In the previous chapters, we considered a system of either a purely random medium (Chap. 2) or a purely random boundary (Chap. 3). In the following Sect.'s, we consider a system of a random medium and random boundaries, typically in two cases of a semi-infinite random layer with one rough boundary, and a random layer with finite width with two rough boundaries. We begin this subject by constructing a solution of the BS equation for the entire system in such a way that the medium and the boundaries are involved on exactly the same footing [4.1]. This enables us to obtain several expressions of the mutual coherence function by interchanging the roles of the medium and the boundaries, including those expressions using an effective scattering matrix of each boundary as modified by the medium fluctuation, on the one hand, and using the scattering matrix of a purely random medium on the other. The method is applicable not only to the layer problems but also to fixed scatterers embedded in a random medium (Chap. 6).

4.1 Bethe-Salpeter Equation for the Entire System and Scattering Matrices

We first consider two random layers separated by a rough boundary, S, described by $z = -\zeta(\rho) < 0$, and illustrated in Fig. 4.1. As in Chap. 3, a scalar wave, $\psi(\hat{\boldsymbol{x}})$, is considered, and the wave function in each space, R_a, $a = 1, 2$, is denoted by $\psi_a(\hat{\boldsymbol{x}})$, whose wave equation is

$$[L_a - q_a(\hat{\boldsymbol{x}})]\,\psi_a(\hat{\boldsymbol{x}}) = j_a(\hat{\boldsymbol{x}}) \;, \tag{4.1.1a}$$

$$L_a = -\left(\frac{\partial}{\partial \hat{\boldsymbol{x}}}\right)^2 - k_a^2, \qquad \mathrm{Im}\,\{k_a\} < 0 \;. \tag{4.1.1b}$$

Here, $q_a(\hat{\boldsymbol{x}}) = q_a^*(\hat{\boldsymbol{x}})$ is the random part of the medium, and $j_a(\hat{\boldsymbol{x}})$ is a source term; k_a is the propagation constant when the medium is free from the random part, and the medium is assumed to be non-dissipative for the time being. The boundary condition is assumed to be the continuity of ψ_a, and

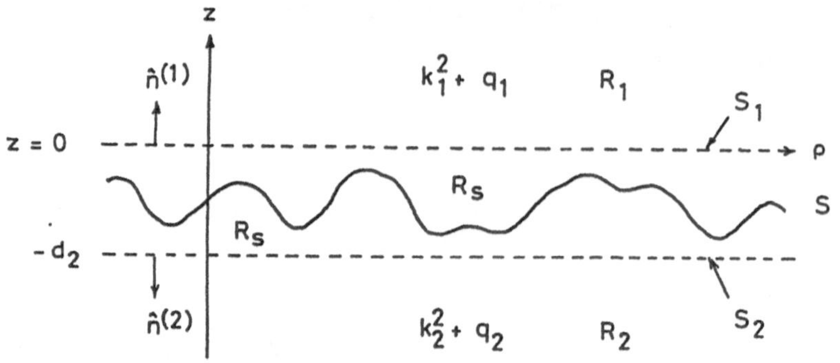

Fig. 4.1. Geometry and notations of the rough boundary for (4.1.1)

its gradient normal to the (real) boundary surface. Consistent with this, the power vector $\widehat{\boldsymbol{w}}_a$ in space R_a is defined by

$$\widehat{\boldsymbol{w}}_a = \psi_a^* \widehat{\boldsymbol{\alpha}} \psi_a \ , \tag{4.1.2a}$$

in terms of the operator, $\widehat{\boldsymbol{\alpha}}$, defined by (2.3.19), i.e.,

$$\widehat{\boldsymbol{\alpha}} = (2\mathrm{i})^{-1}\left(\frac{\overleftarrow{\partial}}{\partial \widehat{\boldsymbol{x}}} - \frac{\overrightarrow{\partial}}{\partial \widehat{\boldsymbol{x}}}\right) \ . \tag{4.1.2b}$$

Hence the power equation is

$$\frac{\partial}{\partial \widehat{\boldsymbol{x}}} \cdot \sum_a \widehat{\boldsymbol{w}}_a(\widehat{\boldsymbol{x}}) = \sum_a \frac{1}{2\mathrm{i}} [\psi_a^* j_a(\widehat{\boldsymbol{x}}) - j_a^* \psi_a(\widehat{\boldsymbol{x}})] \ . \tag{4.1.3}$$

The boundary condition can be transferred from the real boundary onto two reference boundary planes, S_1 and S_2 at $z = d_1 = 0$ and $z = -d_2$, respectively, chosen such that the change of the boundary height is ranged between S_1 and S_2 (Fig. 4.1). Hence, with the notation $\partial_n^{(a)} = \widehat{\boldsymbol{n}}^{(a)} \cdot \partial/\partial \widehat{\boldsymbol{x}}$ where $\widehat{\boldsymbol{n}}^{(a)}$ is the unit vector directed outward normal to S_a, the boundary equation can be written as

$$-\partial_n^{(a)} \psi_a(\boldsymbol{\rho}) = \sum_{b=1}^{2} \int d\boldsymbol{\rho}' \, B_{ab}^{(12)}(\boldsymbol{\rho}|\boldsymbol{\rho}') \psi_b(\boldsymbol{\rho}') \ , \tag{4.1.4}$$

which is the same as (3.3.4), except that $\eta_j = 1$ in the present case. Here, $\psi_a(\boldsymbol{\rho})$ denotes $\psi_a(\widehat{\boldsymbol{x}})$ on S_a (the same convention will hereafter be used for other quantities on S_a). $B_{ab}^{(12)}$ has been referred to as the surface impedance and obtained exactly for given $\zeta(\boldsymbol{\rho})$ (Sect. 3.3.2), including the case when S has a constant impedance (Sect. 3.1.5). In the boundary space R_S, where

$0 > z > -d_2$, the medium is assumed to be deterministic ($q = 0$). When the boundary is non-dissipative, condition (3.3.20) imposes the constraint that

$$[B_{ab}^{(12)}(\rho|\rho')]^\dagger \equiv [B_{ba}^{(12)}(\rho'|\rho)]^* = B_{ab}^{(12)}(\rho|\rho') = B_{ba}^{(12)}(\rho'|\rho) \ , \qquad (4.1.5)$$

i.e., that the matrix $B^{(12)}$ (defined by the elements $B_{ab}^{(12)}(\rho|\rho')$ and written by boldface letter in Sect. 3.3 for convenience) be Hermitian, having real and symmetrical elements with respect to both the coordinates and the subscripts. We have seen in Sect. 3.3.3 that with a Hermitian $B^{(12)}$, the power equation (4.1.3) (which is the same as when $q = 0$) is replaced by (3.3.74) when the boundary surface is included, showing that the continuity of power flux is guaranteed only with an additional power flux of the surface wave, given by the $\boldsymbol{\beta}^{(B)}$ term. Summarizing, the entire space is divided into three parts: R_1 on and above S_1, R_2 on and below S_2, and R_S enclosed by $S_{12} = S_1 + S_2$; and $q_a = j_a = \psi_a = 0$ in space R_b with $b \neq a$, while $q(\widehat{\boldsymbol{x}}) = j(\widehat{\boldsymbol{x}}) = 0$ is assumed within the boundary space, R_s.

The wave equation (4.1.1a) and the boundary equation (4.1.4) can be unified and written as one wave equation of the form

$$(L_a - q_a)\psi_a - \sum_{b=1}^{2} B_{ab}^{(12)}\psi_b = j_a \ . \qquad (4.1.6)$$

Here, both $B_{ab}^{(12)}$ and q_a are regarded as $\widehat{\boldsymbol{x}}$-coordinate matrices, defined by the elements

$$B_{ab}^{(12)}(\widehat{\boldsymbol{x}}|\widehat{\boldsymbol{x}}') = \delta(z + d_a) B_{ab}^{(12)}(\rho|\rho')\delta(z' + d_b) \qquad (4.1.7)$$

and $q_a(\widehat{\boldsymbol{x}}|\widehat{\boldsymbol{x}}') = q_a(\widehat{\boldsymbol{x}})\delta(\widehat{\boldsymbol{x}} - \widehat{\boldsymbol{x}}')$, and the solution is subject to the new boundary condition that $\partial_n^{(a)}\psi_a = 0$ ($a = 1, 2$) infinitesimally inside of S_a at $z = -0$ and $-d_2 + 0$. The proof can be given by integrating (4.1.6) with respect to z over two infinitesimal regions enclosing S_1 and S_2, separately; hence the boundary equation (4.1.4) is reproduced.

With a new matrix, v_{ab}, defined by

$$v_{ab} = q_a\delta_{ab} + B_{ab}^{(12)} \ , \qquad (4.1.8)$$

the equation of the deterministic Green's function of the new wave equation (4.1.6), say $g_{ab}(\widehat{\boldsymbol{x}}|\widehat{\boldsymbol{x}}')$ when the points $\widehat{\boldsymbol{x}}$ and $\widehat{\boldsymbol{x}}'$ are respectively in the spaces of $\boldsymbol{k}_a$ and $\boldsymbol{k}_b$, can be written as

$$\sum_c (L_a\delta_{ac} - v_{ac})g_{cb}(\widehat{\boldsymbol{x}}|\widehat{\boldsymbol{x}}') = \delta_{ab}\delta(\widehat{\boldsymbol{x}} - \widehat{\boldsymbol{x}}') \ , \qquad (4.1.9a)$$

or, in matrix form, as

$$(L - v)g = 1, \qquad v = g + B^{(12)} \ . \tag{4.1.9b}$$

Here, $v = v^\dagger$ may be regarded as an effective medium representing both the medium and the boundary on an equal basis, and ensuring the symmetry,

$$g^T = g, \qquad v^T = v = v^* \ . \tag{4.1.10}$$

The symmetry of the Green's function, $g_{ab}(\hat{x}|\hat{x}')$, can be directly shown by use of Green's theorem, as follows: With the boundary equation (4.1.4) and the outward propagating wave condition at infinity, we first prepare a Green's theorem applicable to the space $R_1 + R_2$ (excluding R_S) by writing

$$\sum_a \int d\hat{x}\, \psi_a'(\overleftarrow{L}_a - \overrightarrow{L}_a)\psi_a = \sum_a \int_{S_a} d\rho\, \psi_a'(\overleftarrow{\partial}_n^{(a)} - \overrightarrow{\partial}_n^{(a)})\psi_a \ ,$$

where $\rightarrow$ and $\leftarrow$ mean operation on the right- and left-hand sides, respectively. Hence, letting ψ_a' be the corresponding solution of a different wave equation,

$$(L_a - q_a')\psi_a' = j_a' \ ,$$

subject to the boundary condition,

$$-\partial_n^{(a)}\psi_a'(\rho) = \sum_b \int d\rho'\, B_{ab}'(\rho|\rho')\psi_b'(\rho') \ ,$$

and also the condition at infinity, we obtain

$$\sum_a \int d\hat{x} \Big[\psi_a'(q_a' - q_a)\psi_a + j_a'\psi_a - j_a\psi_a'\Big]$$
$$= \sum_{ab} \int d\rho\, d\rho' \psi_a'(\rho)\Big[-B_{ba}'(\rho'|\rho) + B_{ab}^{(12)}(\rho|\rho')\Big]\psi_b(\rho') \ .$$

A similar equation is obtained by applying Green's theorem in the boundary space R_S where $q_a' = j_a' = 0$, as long as ψ_a and ψ_a' are both subject to the same continuity condition on the boundary S so that $B' = B^{(12)}$. This shows that the right-hand side is zero because so is the left-hand side in the space R_S. Hence, $B_{ba}^{(12)}(\rho'|\rho) = B_{ab}^{(12)}(\rho|\rho')$, i.e.,

$$B^{(12)T} = B^{(12)} = B^{(12)*} \ ,$$

where the last equality holds only in the non-dissipative case. Thus setting $\psi_a(\hat{x}) = g_{ac}(\hat{x}|\hat{x}')$, $j_a(\hat{x}) = \delta_{ac}\delta(\hat{x} - \hat{x}')$, $\psi_a'(\hat{x}) = g_{ad}(\hat{x}|\hat{x}'')$, $j_a'(\hat{x}) = \delta_{ad}\delta(\hat{x} - \hat{x}'')$, and $q_a' = q_a$ in the above equation leads to

$$g_{dc}(\hat{x}''|\hat{x}') = g_{cd}(\hat{x}'|\hat{x}'') \ .$$

Proof of the symmetry (4.1.14) can also be shown in a similar fashion by setting $\psi_a = G_{ac}$ and $\psi_a' = G_{ad}$, and hence replacing $q \rightarrow M^{(q)}$ and $B^{(12)} \rightarrow M^{(12)}$, in view of (4.1.13b), as

$$M_{ab} \equiv M_{ab}^{(q)} + M_{ab}^{(12)} = M_{ab}^{(q)T} + M_{ab}^{(12)T} \ .$$

Note that the Green's theorem leads to a result involving the medium and the boundary on exactly the same footing.

4.1.1 Statistical Green's Functions

Equation (4.1.9b) enables us to obtain the statistical Green's functions in exactly the same form as those in a homogeneous random medium, v, according to the procedure of Chaps. 2 and 3, as follows [4.1]: The averaged version of (4.1.9b) is written by

$$(L - M)G = 1, \qquad G = \langle g \rangle \ , \tag{4.1.11}$$

in terms of an effective medium, M of v, defined by

$$MG = \langle vg \rangle, \qquad M = M^{(q)} + M^{(12)} \ . \tag{4.1.12}$$

Here, $M^{(q)}$ and $M^{(12)}$ are also defined in the same fashion, by

$$M^{(q)}G = \langle qg \rangle, \qquad M^{(12)}G = \langle B^{(12)}g \rangle \ , \tag{4.1.13a}$$

or, more precisely,

$$\sum_b M_{ab}^{(q)} G_{bc} = \langle q_a g_{ac} \rangle \ ,$$

$$\sum_b M_{ab}^{(12)} G_{bc} = \sum_b \langle B_{ab}^{(12)} g_{bc} \rangle \ , \tag{4.1.13b}$$

and are approximately equal to the independent contributions from the medium and the boundary, respectively, with the elements $M_a^{(q)}\delta_{ab}$ and $M_{ab}^{(12)}$. Here,

$$M^T = M, \qquad G^T = G \ , \tag{4.1.14}$$

although $M^{(q)T} \neq M^{(q)}$ and $M^{(12)T} \neq M^{(12)}$, strictly speaking [see Fig 4.2a].

For the statistical Green's function of second order, defined by

$$I_{ab;cd}(\hat{\boldsymbol{x}}_1, \hat{\boldsymbol{x}}_2 | \hat{\boldsymbol{x}}_1', \hat{\boldsymbol{x}}_2') = \langle g_{ac}^*(\hat{\boldsymbol{x}}_1 | \hat{\boldsymbol{x}}_1') g_{bd}(\hat{\boldsymbol{x}}_2 | \hat{\boldsymbol{x}}_2') \rangle \ , \tag{4.1.15a}$$

or, in matrix form,

$$I(1;2) = \langle g^*(1) g(2) \rangle \ , \tag{4.1.15b}$$

(here and also hereafter, the subscript 1 is attached to the coordinates of quantities of the complex–conjugate wave function, and the subscript 2 is attached to those of the original wave function) we first introduce a matrix, Δv, defined by

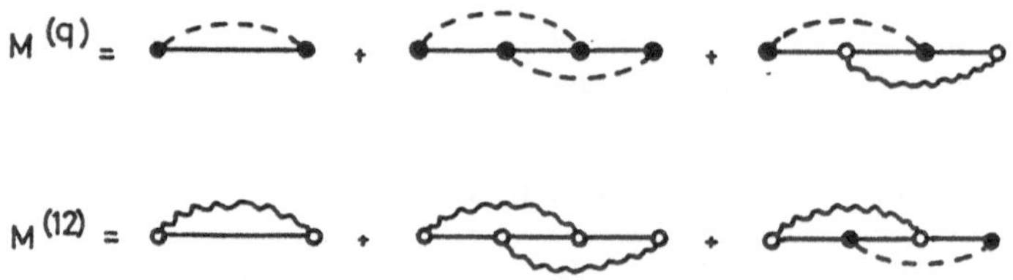

Fig. 4.2a. Schematic diagrams of $M^{(q)}$ and $M^{(12)}$ defined by (4.1.13a,b) are shown to the fourth order of q and $b^{(12)}$, assuming Gaussian statistics. Here G, q, and $b^{(12)}$ are represented, respectively, by a *solid line*, *filled circles*, and *open circles*, and are connected in the order of their matrix multiplication. $\langle g \ldots g \rangle$ is represented by *dashed lines* connecting the filled circles of the g's, and $\langle b^{(12)} \ldots b^{(12)} \rangle$ is represented by *wavy lines* connecting the open circles of the $b^{(12)}$'s (4.1.22a,b)

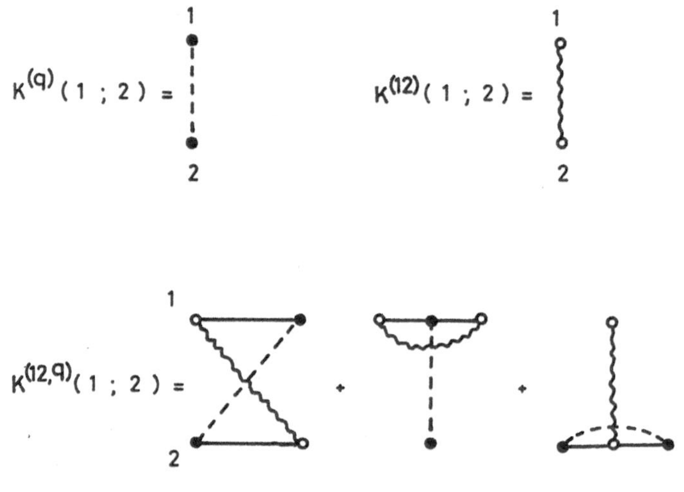

Fig. 4.2b. Non-vanishing elements of $K^{(q)}(1;2)$, $K^{(12)}(1;2)$, and $K^{(12,q)}(1;2)$ defined by (4.1.21) are shown to the lowest order of approximation, with the same notations as in Fig. 4.2a

$$\Delta v = v - M = \Delta q + \Delta B^{(12)} , \tag{4.1.16a}$$

where

$$\Delta q = q - M^{(q)}, \qquad \Delta B^{(12)} = B^{(12)} - M^{(12)} , \tag{4.1.16b}$$

and employ the expressions

$$g = G(1 + \Delta v g), \qquad \langle \Delta v g \rangle = 0 , \tag{4.1.17a}$$

$$g^* = G^*(1 + \Delta v^* g^*), \qquad \langle \Delta v^* g^* \rangle = 0 , \tag{4.1.17b}$$

for $g(2)$ and $g^*(1)$ in the right-hand side of (4.1.15b). We then obtain the following expression:

$$I(1;2) = G^*(1)G(2)\left[1 + K(1;2)I(1;2)\right] \tag{4.1.18}$$

which is an equation of the form of the Bethe–Salpeter equation, with a matrix $K(1;2)$, defined by

$$K(1;2)I(1;2) = \left\langle \Delta v^*(1)\Delta v(2) g^*(1) g(2)\right\rangle \ , \tag{4.1.19}$$

in the same fashion as M by (4.1.12). Here, $K(1;2)$ has matrix elements of the form $K_{ab;cd}(\widehat{\boldsymbol{x}}_1, \widehat{\boldsymbol{x}}_2 | \widehat{\boldsymbol{x}}_1', \widehat{\boldsymbol{x}}_2')$, and can be divided into two major parts of medium and boundary. That is,

$$\begin{aligned} K(1;2) = {} & K^{(q)}(1;2) + K^{(12)}(1;2) \\ & + K^{(q,12)}(1;2) + K^{(12,q)}(1;2) \ , \end{aligned} \tag{4.1.20}$$

where the $K^{(\ )}(1;2)$'s are defined according to

$$K^{(q)}(1;2)I(1;2) = \left\langle \Delta q^*(1)\Delta q(2) g^*(1) g(2)\right\rangle \ , \tag{4.1.21a}$$

$$K^{(12)}(1;2)I(1;2) = \left\langle \Delta B^{(12)*}(1)\Delta B^{(12)}(2) g^*(1) g(2)\right\rangle \ , \tag{4.1.21b}$$

$$K^{(q,12)}(1;2)I(1;2) = \left\langle \Delta q^*(1)\Delta B^{(12)}(2) g^*(1) g(2)\right\rangle \ , \tag{4.1.21c}$$

etc. Hence, to the ladder approximation,

$$K^{(q)}(1;2) = \left\langle q(1) q(2)\right\rangle, \qquad \langle q\rangle = 0 \ , \tag{4.1.22a}$$

$$K^{(12)}(1;2) = \left\langle b^{(12)}(1) b^{(12)}(2)\right\rangle, \qquad \langle b^{(12)}\rangle = 0 \ ; \tag{4.1.22b}$$

whereas $K^{(q,12)}$ and $K^{(12,q)}$ are of higher order and are negligible to the same approximation [see Fig. 4.2b]. On the other hand, when the medium and/or the boundary is composed of randomly distributed particles and/or bosses, respectively, M and K are obtained in terms of the scattering amplitudes of the particles and/or bosses [(2.2.14,19) and (3.3.87b,c)].

Thus, we can write

$$K(1;2) \simeq K^{(q)}(1;2) + K^{(12)}(1;2) \ , \tag{4.1.23}$$

as being an independent sum of $K^{(q)}$ from the medium, and $K^{(12)}$ from the boundary.

4.1.2 Optical Relations and a Dispersive Medium

The matrices M and K, as defined by (4.1.12, 19), respectively, are subject to a local (optical) relation

$$\begin{aligned} \frac{\partial}{\partial \widehat{\boldsymbol{x}}} \cdot \boldsymbol{\beta}(\widehat{\boldsymbol{x}}|1;2) = {} & (2\mathrm{i})^{-1}\delta(\widehat{\boldsymbol{x}}|1;2) \\ & \times\left\{M^*(1) - M(2) - \left[G^*(1) - G(2)\right]K(1;2)\right\} \ , \end{aligned} \tag{4.1.24}$$

with the same form as (3.3.92b) for a rough boundary. Here, the matrix $\delta(\hat{\boldsymbol{x}}|1;2)$ is the same as $\boldsymbol{\delta}(\hat{\boldsymbol{x}}|1;2)$, of boldface letter, by (3.3.72a–72c), being defined by the matrix elements $\delta_{ij}(\hat{\boldsymbol{x}}|\hat{\boldsymbol{x}}_1;\hat{\boldsymbol{x}}_2)$ of (3.3.72a). Hence (3.3.71), for example, can be rewritten as

$$\frac{\partial}{\partial\hat{\boldsymbol{x}}}\cdot\boldsymbol{\beta}^{(B)}(\hat{\boldsymbol{x}}|1;2)=(2\mathrm{i})^{-1}\delta(\hat{\boldsymbol{x}}|1;2)\left[v(1)-v(2)\right]\ , \tag{4.1.25}$$

in view of no contribution from the medium part q with diagonal matrix elements. Now the proof of relation (4.1.24) is straightforward if use is made of the relations

$$\langle v(1)g^*(1)g(2)\rangle=\left[M^*(1)+G(2)K(1;2)\right]I(1;2)\ , \tag{4.1.26a}$$
$$\langle v(2)g^*(1)g(2)\rangle=\left[M(2)+G^*(1)K(1;2)\right]I(1;2)\ , \tag{4.1.26b}$$

that follow from definitions (4.1.19, 16), and the relation given by averaging (3.3.70) with $\psi\to g,\quad \psi^*\to g^*$, and $B^{(12)}\to v$, i.e.,

$$-\sum_{a=1}^{2}\hat{\boldsymbol{n}}^{(a)}\cdot\langle\hat{\boldsymbol{w}}_a(\rho)\rangle\delta(z+d_a)$$
$$=-\sum_{a}\hat{\boldsymbol{n}}^{(a)}\cdot\langle\hat{\boldsymbol{w}}_a(\hat{\boldsymbol{x}})\rangle=\frac{\partial}{\partial\hat{\boldsymbol{x}}}\cdot\langle\boldsymbol{\beta}^{(B)}(\hat{\boldsymbol{x}}|1;2)g^*(1)g(2)\rangle \tag{4.1.27}$$
$$=\frac{\partial}{\partial\hat{\boldsymbol{x}}}\cdot\boldsymbol{\beta}(\hat{\boldsymbol{x}}|1;2)I(1;2) \tag{4.1.28}$$
$$=(2\mathrm{i})^{-1}\delta(\hat{\boldsymbol{x}}|1;2)\langle\left[v(1)-v(2)\right]g^*(1)g(2)\rangle\ . \tag{4.1.29}$$

Here, the matrix $\boldsymbol{\beta}(\hat{\boldsymbol{x}}|1;2)$ is defined by (4.1.28), as $K(1;2)$ has been defined by (4.1.19), and, in the last derivation, use has been made of (4.1.25); hence relation (4.1.24) is derived from (4.1.29) by substituting (4.1.26a,b).

To express power equations associated with $I(1;2)$, we introduce here, for any matrix $A(1;2)$, the convention (3.3.72b),

$$A(\hat{\boldsymbol{x}})\equiv A(\hat{\boldsymbol{x}}|1;2)\equiv\delta(\hat{\boldsymbol{x}}|1;2)A(1;2)\ , \tag{4.1.30a}$$
$$AB(\hat{\boldsymbol{x}})=\delta(\hat{\boldsymbol{x}}|1;2)AB(1;2)=A(\hat{\boldsymbol{x}}|1;2)B(1;2)=A(\hat{\boldsymbol{x}})B\ ; \tag{4.1.30b}$$

so that the entire power equation, obtained by averaging (3.3.74) with $\psi\to g$, $\psi^*\to g^*$, and using ΔG of (2.3.16b), becomes

$$\frac{\partial}{\partial\hat{\boldsymbol{x}}}\cdot(\hat{\boldsymbol{\alpha}}+\boldsymbol{\beta})I(\hat{\boldsymbol{x}})=\Delta G(\hat{\boldsymbol{x}})\ , \tag{4.1.31}$$

as a consequence of (4.1.27, 28). Here, the matrix elements of $\hat{\boldsymbol{\alpha}}(\hat{\boldsymbol{x}}|1;2)$ are given by (3.3.75b) with $\eta_j=1$, consistent with $\hat{\boldsymbol{\alpha}}$ by (4.1.2b, 30a), or, alternatively,

$$\widehat{\alpha}_{ab}(\widehat{x}|\widehat{x}_1;\widehat{x}_2) = \delta_{ab}\delta(\widehat{x}-\widehat{x}_1)\widehat{\alpha}\,\delta(\widehat{x}-\widehat{x}_2) \ . \tag{4.1.32}$$

Equation (4.1.31) has exactly the same form as (3.2.33b) for a rough surface, and can also be reproduced from the BS equation (4.1.18) with the aid of optical relation (4.1.24); i.e., by multiplying the difference of the two equations,

$$\begin{aligned}\left[L(2)-M(2)\right]I(1;2) &= G^*(1)\left[1+K(1;2)I(1;2)\right] \ ,\\ \left[L(1)-M^*(1)\right]I(1;2) &= G(2)\left[1+K(1;2)I(1;2)\right] \ ,\end{aligned} \tag{4.1.33}$$

to the left with $(2\mathrm{i})^{-1}\delta(\widehat{x}|1;2)$, and use of (4.1.24) leads directly to (4.1.31).

As in the previous chapters, we introduce here a coherent propagator, $U^{(C)}(1;2)$, and an attenuation matrix, $\Gamma(1;2)$, defined by

$$U^{(C)}(1;2) = G^*(1)G(2) \ , \tag{4.1.34}$$

$$\begin{aligned}\Gamma(1;2) &= (2\mathrm{i})^{-1}\left[M^*(1)-M(2)\right]\\ &= \Gamma^{(q)}(1;2)+\Gamma^{(12)}(1;2) \ ,\end{aligned} \tag{4.1.35}$$

where $\Gamma^{(\ )}$ is the contribution from $M^{(\ )}$. Hence the power equation for the coherent wave and optical relation (4.1.24) can be written, respectively, as

$$\left[\frac{\partial}{\partial\widehat{x}}\cdot\widehat{\alpha}+\Gamma\right]U^{(C)}(\widehat{x}) = \Delta G(\widehat{x}) \ , \tag{4.1.36}$$

$$\frac{\partial}{\partial\widehat{x}}\cdot\beta(\widehat{x}) = \Gamma(\widehat{x})-\Delta G(\widehat{x})K \ , \tag{4.1.37}$$

in exactly the same form as (3.2.52, 54b). And we again reproduce power equation (4.1.31), in the same fashion as (3.2.55) with the BS equation, which is now written as

$$I = U^{(C)}(1+KI) \ . \tag{4.1.38}$$

The optical relation (4.1.37) also holds true for each $M^{(\ )}$ and $K^{(\ )}$ of the medium and the boundary, approximately, such that

$$\frac{\partial}{\partial\widehat{x}}\cdot\beta^{(12)}(\widehat{x}) = \Gamma^{(12)}(\widehat{x})-\Delta G(\widehat{x})K^{(12)} \ , \tag{4.1.39a}$$

$$\frac{\partial}{\partial\widehat{x}}\cdot\widehat{\beta}^{(q)} = \Gamma^{(q)}(\widehat{x})-\Delta G^{(0)}(\widehat{x})K^{(q)} \ . \tag{4.1.39b}$$

Here, $\widehat{\beta}^{(q)}(\widehat{x}) = 0$ in the present case of a non-dispersive medium, and $G^{(0)}$ is the Green's function in a homogeneous medium, $M^{(q)}$; while, for $\Delta G(\widehat{x})$ in (4.1.39a), the effect of the medium fluctuation is negligible. On the other hand, when the medium q_a is dispersive, having off-diagonal matrix elements $q_a(\widehat{x}|\widehat{x}')$, an equation similar to (3.3.69) for the boundary holds true [4.2],

$$\frac{\partial}{\partial\hat{\boldsymbol{x}}}\cdot\hat{\boldsymbol{s}}^{(q)}(\hat{\boldsymbol{x}}|\hat{\boldsymbol{x}}_1;\hat{\boldsymbol{x}}_2)=\frac{1}{2\mathrm{i}}\left[q^*(\hat{\boldsymbol{x}}|\hat{\boldsymbol{x}}_1)\delta(\hat{\boldsymbol{x}}-\hat{\boldsymbol{x}}_2)-q(\hat{\boldsymbol{x}}|\hat{\boldsymbol{x}}_2)\delta(\hat{\boldsymbol{x}}-\hat{\boldsymbol{x}}_1)\right]\ , \tag{4.1.40a}$$

or, in matrix form,

$$\frac{\partial}{\partial\hat{\boldsymbol{x}}}\cdot\hat{\boldsymbol{s}}^{(q)}(\hat{\boldsymbol{x}}|1;2)=\frac{1}{2\mathrm{i}}\delta(\hat{\boldsymbol{x}}|1;2)[q^*(1)-q(2)]\ , \tag{4.1.40b}$$

whenever q is Hermitian. The matrix $\widehat{\boldsymbol{\beta}}^{(q)}(\hat{\boldsymbol{x}})$ in (4.1.39b) is defined by

$$\frac{\partial}{\partial\hat{\boldsymbol{x}}}\cdot\widehat{\boldsymbol{\beta}}^{(q)}(\hat{\boldsymbol{x}}|1;2)I(1;2)=\frac{\partial}{\partial\hat{\boldsymbol{x}}}\cdot\langle\hat{\boldsymbol{s}}^{(q)}(\hat{\boldsymbol{x}}|1;2)g^*(1)g(2)\rangle\ , \tag{4.1.40c}$$

in the same fashion as $\boldsymbol{\beta}(\hat{\boldsymbol{x}}|1;2)$ is defined by (4.1.28). Thus, the power equation (4.1.31) is changed by the replacement of $\boldsymbol{\beta}\to\boldsymbol{\beta}^{(12)}+\widehat{\boldsymbol{\beta}}^{(q)}$.

$K^{(q)}$ is a diagonal matrix with respect to the subscripts, and only has the elements $K_a^{(q)}\equiv K_{aa;aa}^{(q)}$, while the important elements of $K^{(12)}$ are $K_{ab}^{(12)}\equiv K_{aa;bb}^{(12)}$. Hence, in terms of the notations $I_{ab}^{(q+12)}=I_{aa;bb}$ and

$$U_{ab}^{(C)}(1;2)=G_{ab}^*(1)G_{ab}(2)\ , \tag{4.1.41}$$

the BS equation (4.1.38) can be written, in 2×2 matrix form, as

$$I^{(q+12)}=U^{(C)}\left[1+(K^{(q)}+K^{(12)})I^{(q+12)}\right]\ . \tag{4.1.42}$$

4.1.3 Case of Three Random Layers

The situation is also the same for the case of random layers, as illustrated in Fig. 4.3, and various equations formally remain unchanged with the setting

$$M=M^{(q)}+M^{(12)}+M^{(23)}\ , \tag{4.1.43a}$$

$$K\simeq K^{(q)}+K^{(12)}+K^{(23)}\ . \tag{4.1.43b}$$

Thus, using the notation $I_{ab}^{(q+12+23)}$, $a,b=1,2,3$, for the second order Green's function in this case, we obtain the BS equation, in 3×3 matrix form, as

$$I^{(q+12+23)}=U^{(C)}\left[1+(K^{(q)}+K^{(12)}+K^{(23)})I^{(q+12+23)}\right]\ . \tag{4.1.44}$$

Here, $K^{(q)}$ is a diagonal matrix with the elements $K_{ab}^{(q)}=K_a^{(q)}\delta_{ab}$, and $K^{(12)}$ and $K^{(23)}$ are the contributions purely from the boundaries, S_{12} and S_{23}, with the non-vanishing elements $K_{ab}^{(12)}$, $a,b=1,2$, and $K_{ab}^{(23)}$, $a,b=2,3$.

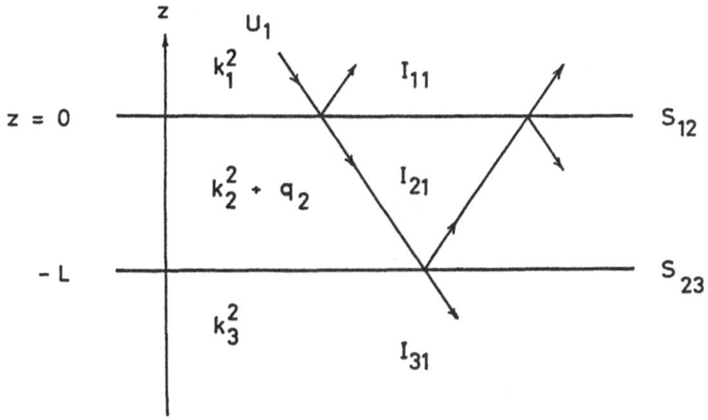

Fig. 4.3. Geometry and notations of a random layer for (4.1.44). The boundary space R_S (Fig. 4.1) is omitted because we regard it to be infinitesimally thin

4.1.4 Scattering Matrices and Solutions

To obtain the solution of the BS equation (4.1.42), we first introduce the solution in the special case, $K^{(q)} = 0$ (keeping $M^{(q)} \neq 0$, however), say, $I^{(12)}$, so that

$$I^{(12)} = U^{(C)}(1 + K^{(12)}I^{(12)}) \ . \tag{4.1.45}$$

Here, the solution can be written as

$$I^{(12)} = U^{(C)} + U^{(C)}S^{(12)}U^{(C)} \ , \tag{4.1.46}$$

in terms of an incoherent scattering matrix, $S^{(12)}$ of $K^{(12)}$, defined by

$$S^{(12)} = K^{(12)}(1 + U^{(C)}S^{(12)}) \tag{4.1.47a}$$

$$= (1 - K^{(12)}U^{(C)})^{-1}K^{(12)} \ . \tag{4.1.47b}$$

Hence, $S^{(12)}$ is equivalent to $\boldsymbol{S}^{(B)}$ of (3.3.89a,b) when $K^{(12)} = \boldsymbol{K}^{(B)}$, except for the redefinition of $\boldsymbol{U}^{(C)}$ by $U^{(C)}$ of (4.1.34), which is a dissipative propagator in view of $\mathrm{Im}\,\{\widetilde{M}^{(q)}\} < 0$. $I^{(12)}$ of (4.1.46) is therefore equivalent to $\boldsymbol{I}^{(B)}$ of (3.3.113), in the same sence, and, from (3.3.119a,b) can be written in terms of the resultant scattering matrix of the boundary, $\sigma^{(12)}$, as

$$I^{(12)} = U + U\sigma^{(12)}U \ , \tag{4.1.48a}$$

$$\sigma^{(12)} = V^{(12)} + (1 + V^{(12)}U)S^{(12)}(UV^{(12)} + 1) \ . \tag{4.1.48b}$$

Here, U is a diagonal 2×2 matrix with the elements $U_{ab} = U_a\delta_{ab}$, defined by

$$U_a(1;2) = G_a^{(0)*}(1)G_a^{(0)}(2) \ , \tag{4.1.49}$$

and, from (3.3.120a,b),

$$U^{(C)} = U + UV^{(12)}U \ , \tag{4.1.50a}$$

$$\begin{aligned} V^{(12)}(1;2) = &\langle T^{(12)*}(1)\rangle\langle T^{(12)}(2)\rangle \\ &+ \langle T^{(12)*}(1)\rangle G^{-1}(2) + \langle T^{(12)}(2)\rangle\left[G^*(1)\right]^{-1} \ , \end{aligned} \tag{4.1.50b}$$

where, from (3.3.107),

$$\langle T_{ab}^{(12)}\rangle = 2\mathrm{i}h_a\langle R_{ab}^{(12)}\rangle = \langle T_{ba}^{(12)}\rangle \ , \tag{4.1.51a}$$

and $\langle R_{ab}^{(12)}\rangle \neq \langle R_{ba}^{(12)}\rangle$ is the reflection–transmission coefficient of the boundary, given by (3.3.105). In (4.1.50b), the last two interference terms are negligible in a region separated far enough from the boundary, so that

$$V^{(12)}(1;2) \simeq \langle T^{(12)*}(1)\rangle\langle T^{(12)}(2)\rangle \ , \tag{4.1.51b}$$

and, from (3.3.117, 104),

$$\begin{aligned} F^{(C)} &\equiv 1 + UV^{(12)}(1;2) \\ &= \left(1 + \langle R^{(12)*}(1)\rangle\right)\left(1 + \langle R^{(12)}(2)\rangle\right) \ , \end{aligned} \tag{4.1.52a}$$

where

$$1 + \langle R^{(12)}\rangle = 2(h + \mathrm{i}M^{(12)})^{-1}h \ , \tag{4.1.52b}$$

and $\overline{F}^{(C)}$ is given by the transposition of $F^{(C)}$.

The introduction of $I^{(12)}$ by (4.1.45) enables the BS equation (4.1.42) to be rewritten as

$$I^{(q+12)} = I^{(12)}(1 + K^{(q)}I^{(q+12)}) \ , \tag{4.1.53}$$

which can be expressed as

$$I^{(q+12)} = I^{(12)} + I^{(12)}S^{(q/12)}I^{(12)} \ , \tag{4.1.54}$$

in terms of a scattering matrix, $S^{(q/12)}$ of $K^{(q)}$, defined by

$$S^{(q/12)}I^{(12)} = K^{(q)}I^{(q+12)} \ , \tag{4.1.55a}$$

$$I^{(12)}S^{(q/12)} = I^{(q+12)}K^{(q)} \ , \tag{4.1.55b}$$

and governed by

$$S^{(q/12)} = K^{(q)}(1 + I^{(12)}S^{(q/12)}) \ , \tag{4.1.55c}$$

with the superscript $(q/12)$ to mean the dependence on $\sigma^{(12)}$ through $I^{(12)}$. Here, the effect of $\sigma^{(12)}$ can be made explicit by introducing a solution of (4.1.55c) in the special case, $\sigma^{(12)} = 0$, say, $S^{(0q)}$, governed by

$$S^{(0q)} = K^{(q)}(1 + US^{(0q)}) \ , \tag{4.1.56}$$

so that (4.1.55c) can be written, using (4.1.48a), as

$$S^{(q/12)} = S^{(0q)}(1 + U\sigma^{(12)}US^{(q/12)}) \tag{4.1.57a}$$
$$= (1 - S^{(0q)}U\sigma^{(12)}U)^{-1}S^{(0q)} \ . \tag{4.1.57b}$$

Hence (4.1.54) is written in the final form,

$$I^{(q+12)} = I^{(12)} + (1 + U\sigma^{(12)})\mathcal{I}^{(q/12)}(\sigma^{(12)}U + 1) \ . \tag{4.1.58}$$

Here, the entire effect of the random medium appears only through a new matrix, $\mathcal{I}^{(q/12)}$, defined by

$$\mathcal{I}^{(q/12)} = US^{(q/12)}U \ , \tag{4.1.59}$$

and given as the solution of

$$\mathcal{I}^{(q/12)} = \mathcal{I}^{(0q)}(1 + \sigma^{(12)}\mathcal{I}^{(q/12)}) \ , \tag{4.1.60}$$

where

$$\mathcal{I}^{(0q)} = US^{(0q)}U = UK^{(q)}I^{(0q)} \tag{4.1.61a}$$
$$= UK^{(q)}(U + \mathcal{I}^{(0q)}) \ , \tag{4.1.61b}$$

which is a diagonal matrix with respect to the subscripts, and tends to zero as $K^{(q)} \to 0$.

Example: Case of a Semi-Infinite Random Medium ($q_1 = 0, \ q_2 \neq 0$)

We have $\mathcal{I}_1^{(0q)} = \mathcal{I}_{1b}^{(q/12)} = 0$, and the only non-vanishing element of $\mathcal{I}^{(q/12)}$ is $\mathcal{I}_{22}^{(q/12)}$, which is the solution of

$$\mathcal{I}_{22}^{(q/12)} = \mathcal{I}_2^{(0q)}(1 + \sigma_{22}^{(12)}\mathcal{I}_{22}^{(q/12)}) \ . \tag{4.1.62}$$

Here, $\mathcal{I}_2^{(0q)}$ is the solution of

$$\mathcal{I}_2^{(0q)} = U_2K_2^{(q)}(U_2 + \mathcal{I}_2^{(0q)}) \ , \tag{4.1.63}$$

and, therefore, is subject to the condition of no reflection at the boundary of $K_2^{(q)}$ which is distributed over $0 \geq z \geq -\infty$.

Hence, when the wave source is located in the space k_1 (Fig. 4.1), $I^{(q+12)}$ in the same space is given, according to (4.1.58), by

$$I_{11}^{(q+12)} = I_{11}^{(12)} + U_1\sigma_{12}^{(12)}\mathcal{I}_{22}^{(q/12)}\sigma_{21}^{(12)}U_1 \ , \tag{4.1.64}$$

and the wave transmitted into the space k_2 is given by

$$I_{21}^{(q+12)} = I_{21}^{(12)} + (1 + U_2\sigma_{22}^{(12)})\mathcal{I}_{22}^{(q/12)}\sigma_{21}^{(12)}U_1 \ . \qquad (4.1.65)$$

Also, for the case of three random layers, as illustrated in Fig. 4.3, the situation becomes exactly the same by introduction of a solution of when $K_a^{(q)} = 0$, $a = 1, 2, 3$, say, $I^{(12+23)}$, and letting $I^{(12+23)}$ do all the roles of $I^{(12)}$ in (4.1.58); that is, the basic Eqs. (4.1.58–61) remain unchanged with the replacement of the superscript (12) by (12+23) and use of the expression

$$I_{ab}^{(12+23)} = U_a\delta_{ab} + U_a\sigma_{ab}^{(12+23)}U_b \ . \qquad (4.1.66)$$

Here, when L, the distance between the two boundaries, is sufficiently large compared with the wave coherence distance, γ_2^{-1}, so that $\gamma_2 L \gg 1$, $\sigma_{ab}^{(12+23)}$ can be approximated by [see Sect. 4.2.2 for the detail]

$$\sigma_{ab}^{(12+23)} \simeq \sigma_{ab}^{(12)} + \sigma_{ab}^{(23)}, \qquad a, b = 1, 2, 3 \ , \qquad (4.1.67)$$

as the independent sum of the two boundary scattering matrices, $\sigma_{ab}^{(12)}$ of S_{12}, and $\sigma_{ab}^{(23)}$ of S_{23}. Hence (4.1.58) is replaced by a 3×3 matrix equation,

$$I^{(q+12+23)} = I^{(12+23)} + (1 + U\sigma^{(12+23)}) \times \mathcal{I}^{(q/12+23)}(\sigma^{(12+23)}U + 1) \ , \qquad (4.1.68)$$

and, using (4.1.67), (4.1.60) is changed to

$$\mathcal{I}^{(q/12+23)} = \mathcal{I}^{(0q)}[1 + (\sigma^{(12)} + \sigma^{(23)})\mathcal{I}^{(q/12+23)}] \ . \qquad (4.1.69)$$

$\mathcal{I}^{(0q)}$ is still governed by (4.1.61b).

It may be remarked that besides expressions (4.1.58, 68), several other expressions are possible, which are also useful, depending on the particular situations and information required (Sect. 4.2).

Example: Case of a Random Layer ($q_1 = q_3 = 0$, $q_2 \neq 0$)

In this case, the only non-vanishing element of $\mathcal{I}^{(q/12+23)}$ in (4.1.68) is $\mathcal{I}_{22}^{(q/12+23)}$. Hence, when the source is in the space k_1, and the layer width L is large enough such that $\gamma_2 L \gg 1$, $I^{(q+12+23)}$ within the same space is given by

$$I_{11}^{(q+12+23)} = I_{11}^{(12)} + U_1\sigma_{12}^{(12)}\mathcal{I}_{22}^{(q/12+23)}\sigma_{21}^{(12)}U_1 \ , \qquad (4.1.70)$$

on using (4.1.67); and the transmitted wave into the space k_3 is given by

$$I_{31}^{(q+12+23)} = U_3\sigma_{32}^{(23)}\mathcal{I}_{22}^{(q/12+23)}\sigma_{21}^{(12)}U_1 \ , \qquad (4.1.71)$$

where the contribution from $I_{31}^{(12+23)}$ is presently negligible. Here, the random medium is only involved through $\mathcal{I}_{22}^{(q/12+23)}$, which is the solution of

$$\mathcal{I}_{22}^{(q/12+23)} = \mathcal{I}_{2}^{(0q)}[1 + (\sigma_{22}^{(12)} + \sigma_{22}^{(23)})\mathcal{I}_{22}^{(q/12+23)}] \ , \tag{4.1.72}$$

where $\mathcal{I}_2^{(0q)}$ is the solution of (4.1.63), with the $K_2^{(q)}$ distributed over the range $0 \geq z \geq -L$. Here, as for $\sigma^{(12)}$ and $\sigma^{(23)}$, we may utilize experimental instead of theoretical values for the boundaries.

Thus the problem is reduced to finding the solution of (4.1.72), which will be obtained in Sect. 5.2 as a simple boundary–value solution of the diffusion equation, instead of solving the conventional transport equation by numerical methods.

4.2 Effective Boundary Scattering Matrices in a Random Medium and Construction of Solutions

In expression (4.1.54) or (4.1.58) for $I^{(q+12)}$, the random medium is involved only through $S^{(q/12)}$, which means an effective scattering matrix of $K^{(q)}$ under the effect of boundary scattering through $\sigma^{(12)}$; the situation is the same also for expression (4.1.68) of $I^{(q+12+23)}$ using $S^{(q/12+23)}$ which is dependent on $\sigma^{(12+23)}$. By interchanging the roles of the medium and the boundary, another expression is possible as follows. Equations (4.1.57a,b) can be rewritten in the form

$$S^{(q/12)} = S^{(0q)} + S^{(0q)}U\sigma^{(12/q)}US^{(0q)} \ , \tag{4.2.1}$$

with an effective scattering matrix, $\sigma^{(12/q)}$ of $\sigma^{(12)}$, defined by

$$\begin{aligned} \sigma^{(12/q)} &= \sigma^{(12)}[1 + US^{(0q)}U\sigma^{(12/q)}] \\ &= [1 - \sigma^{(12)}US^{(0q)}U]^{-1}\sigma^{(12)} \ . \end{aligned} \tag{4.2.2}$$

It may be remarked that (4.2.2) leads to another expression,

$$\sigma^{(12/q)} = \sigma^{(12)} + \sigma^{(12)}US^{(q/12)}U\sigma^{(12)} \ , \tag{4.2.3}$$

in terms of $\sigma^{(12)}$ and $S^{(q/12)}$. This expression is mathematically similar to (4.2.1) for $S^{(q/12)}$ which is given in terms of $S^{(0q)}$ and $\sigma^{(12/q)}$. There are existing relations,

$$S^{(q/12)}U\sigma^{(12)} = S^{(0q)}U\sigma^{(12/q)} \ , \tag{4.2.4a}$$

$$\sigma^{(12)}US^{(q/12)} = \sigma^{(12/q)}US^{(0q)} \ , \tag{4.2.4b}$$

similar also to (4.1.55a,b).

As we show below, (4.2.1) leads to another expression of $I^{(q+12)}$:

$$I^{(q+12)} = I^{(0q)} + I^{(0q)}\sigma^{(12/q)}I^{(0q)} \ . \tag{4.2.5}$$

Here, $I^{(0q)}$, with the diagonal elements $I_a^{(0q)}\delta_{ab}$, is the solution in an unbounded space of the random medium, q, and is hence defined by

$$I^{(0q)} = U[1 + K^{(q)}I^{(0q)}] \tag{4.2.6a}$$

$$= U + US^{(0q)}U = U + \mathcal{I}^{(0q)} \ . \tag{4.2.6b}$$

The proof of (4.2.5) is given by utilizing the relations

$$S^{(0q)}U = K^{(q)}I^{(0q)}, \qquad US^{(0q)} = I^{(0q)}K^{(q)} \ , \tag{4.2.7a}$$

$$S^{(0q)} = K^{(q)} + K^{(q)}I^{(0q)}K^{(q)} \ , \tag{4.2.7b}$$

that follow from (4.1.56) [more generally, see formulas (4.2.10, 11)] and also a similar relation from (4.1.55b,c),

$$S^{(q/12)} = K^{(q)} + K^{(q)}I^{(q+12)}K^{(q)} \ . \tag{4.2.8}$$

Hence, by substituting (4.2.7a,b, 8) into the right- and left-hand sides of (4.2.1), respectively, (4.2.5) is found immediately after dropping the two common factors, $K^{(q)}$, from both sides. In the expression (4.2.5) of $I^{(q+12)}$, the boundary is involved only through $\sigma^{(12/q)}$, an effective scattering matrix of $\sigma^{(12)}$ as affected by the medium fluctuation, whereas in the expression (4.1.54), the random medium is involved only through $S^{(q/12)}$, an effective scattering matrix of $K^{(q)}$ subject to the boundary condition.

More generally, let $I^{(A+B)}$ and $S^{(B/A)}$ be the solutions of two integral equations,

$$I^{(A+B)} = I^{(A)}[1 + K^{(B)}I^{(A+B)}] \ , \tag{4.2.9a}$$

$$S^{(B/A)} = K^{(B)}[1 + I^{(A)}S^{(B/A)}] \ , \tag{4.2.9b}$$

similar in form to the BS equation. Here, $I^{(A)}$ and $K^{(B)}$ are given matrices, and their roles are just interchanged in the two equations (therefore the notation $I^{(A/B)}$ may be more appropriate than $I^{(A+B)}$ but, in the previous equations, the latter has been used exclusively). Then the following relations hold:

$$I^{(A+B)} = I^{(A)} + I^{(A)}S^{(B/A)}I^{(A)} \ , \tag{4.2.10a}$$

$$S^{(B/A)} = K^{(B)} + K^{(B)}I^{(A+B)}K^{(B)} \ , \tag{4.2.10b}$$

together with

$$K^{(B)}I^{(A+B)} = S^{(B/A)}I^{(A)} \ , \tag{4.2.11a}$$

$$I^{(A+B)}K^{(B)} = I^{(A)}S^{(B/A)} \ . \tag{4.2.11b}$$

The proof is straightforward, writing the solutions of (4.2.9) in matrix form, as

$$I^{(A+B)} = (1 - I^{(A)}K^{(B)})^{-1}I^{(A)} \ ,$$

$$S^{(B/A)} = (1 - K^{(B)}I^{(A)})^{-1}K^{(B)} \ .$$

In terms of the incoherent waves, $\mathcal{I}^{(q/12)}$ and $\mathcal{I}^{(0q)}$, defined by (4.1.59, 61), respectively, the previous equations can be written in a more symmetrical form; i.e., from (4.1.57a) and (4.2.1),

$$\mathcal{I}^{(q/12)} = \mathcal{I}^{(0q)}(1 + \sigma^{(12)}\mathcal{I}^{(q/12)}) \tag{4.2.12}$$

$$= \mathcal{I}^{(0q)} + \mathcal{I}^{(0q)}\sigma^{(12/q)}\mathcal{I}^{(0q)} \ , \tag{4.2.13}$$

and, from (4.2.4),

$$\mathcal{I}^{(q/12)}\sigma^{(12)} = \mathcal{I}^{(0q)}\sigma^{(12/q)} \ , \tag{4.2.14a}$$

$$\sigma^{(12)}\mathcal{I}^{(q/12)} = \sigma^{(12/q)}\mathcal{I}^{(0q)} \ . \tag{4.2.14b}$$

Here from (4.2.2),

$$\sigma^{(12/q)} = \sigma^{(12)}[1 + \mathcal{I}^{(0q)}\sigma^{(12/q)}] \tag{4.2.15}$$

$$= [1 - \sigma^{(12)}\mathcal{I}^{(0q)}]^{-1}\sigma^{(12)} \ . \tag{4.2.16}$$

Hence, with (4.2.15, 14a), we obtain

$$\sigma^{(12/q)} = \sigma^{(12)} + \sigma^{(12)}\mathcal{I}^{(q/12)}\sigma^{(12)} \ , \tag{4.2.17}$$

in a form similar to (4.2.13).

It may be remarked here that (4.2.13, 15) for $\mathcal{I}^{(q/12)}$ and $\sigma^{(12/q)}$ are written in the same form as (4.1.54, 55c) for $I^{(q+12)}$ and $S^{(q/12)}$, respectively, wherein $K^{(q)}$ and $I^{(12)}$ play the roles of $\sigma^{(12)}$ and $\mathcal{I}^{(0q)}$, respectively; the same is also true between (4.2.17) and (4.2.8). Expression (4.2.13) for $\mathcal{I}^{(q/12)}$ is the same as (4.2.5) for $I^{(q+12)}$ except $\mathcal{I}^{(0q)} \to I^{(0q)}$, with the difference,

$$I^{(q+12)} - \mathcal{I}^{(q/12)} = U + U\sigma^{(12/q)}U$$

$$+ U\sigma^{(12/q)}\mathcal{I}^{(0q)} + \mathcal{I}^{(0q)}\sigma^{(12/q)}U \ . \tag{4.2.18}$$

Now, $I^{(q+12)}$ is written in the form

$$I^{(q+12)} = U + U\sigma^{(q+12)}U \ , \tag{4.2.19}$$

so that $\sigma^{(q+12)}$ denotes a resultant scattering matrix of the present system. Then, we obtain the two alternative expressions from (4.2.5) and (4.1.54),

$$\sigma^{(q+12)} = S^{(0q)} + (1 + S^{(0q)}U)\sigma^{(12/q)}(US^{(0q)} + 1) \tag{4.2.20a}$$

$$= \sigma^{(12)} + (1 + \sigma^{(12)}U)S^{(q/12)}(U\sigma^{(12)} + 1) \ ; \tag{4.2.20b}$$

each of which can be derived from the other by interchanging $S^{(0q)}$ and $\sigma^{(12)}$, in view of (4.2.2) and (4.1.57b), having shown that $\sigma^{(q+12)}$ is invariant against their interchange. Here, $\sigma^{(12/q)}$ has non-vanishing matrix elements only on the boundary, S_{12}, and provides an effective boundary scattering matrix when the medium is random and $\sigma^{(q+12)}$ of (4.2.20a) has a form similar to (4.1.48a) for $\sigma^{(12)}$, while $S^{(q/12)}$ has non-vanishing matrix elements over all the region of $K^{(q)} \neq 0$ and provides a random-volume scattering matrix including the boundary effect. In [4.3], a matrix similar to $\sigma^{(12/q)}$ of (4.2.16) was introduced based on the transport equation, with some numerical applications.

For given $I^{(\ )}$, it is often convenient to introduce two factors, $F^{(\ )}$ and $\overline{F}^{(\ )}$, defined by

$$I^{(\ )} = F^{(\ )}U = U\overline{F}^{(\ )} \ , \tag{4.2.21}$$

in the same fashion as (3.3.115), so that, from (4.1.48) and (4.2.6b),

$$F^{(12)} = 1 + U\sigma^{(12)}, \qquad \overline{F}^{(12)} = 1 + \sigma^{(12)}U \ , \tag{4.2.22a}$$

$$F^{(0q)} = 1 + US^{(0q)}, \qquad \overline{F}^{(0q)} = 1 + S^{(0q)}U \ ; \tag{4.2.22b}$$

which enables (4.2.20) to be written in a compact form.

Factors $Y^{(12/q)}$ and $\overline{Y}^{(12/q)}$ are also used hereafter, and are defined by

$$\mathcal{I}^{(q/12)} = Y^{(12/q)}\mathcal{I}^{(0q)} = \mathcal{I}^{(0q)}\overline{Y}^{(12/q)} \ , \tag{4.2.23}$$

so that, from (4.2.13),

$$Y^{(12/q)} = 1 + \mathcal{I}^{(0q)}\sigma^{(12/q)} = (1 - \mathcal{I}^{(0q)}\sigma^{(12)})^{-1} \ , \tag{4.2.24a}$$

$$\overline{Y}^{(12/q)} = 1 + \sigma^{(12/q)}\mathcal{I}^{(0q)} = (1 - \sigma^{(12)}\mathcal{I}^{(0q)})^{-1} \ . \tag{4.2.24b}$$

Substitution of (4.2.23) into (4.2.14) shows that

$$\sigma^{(12/q)} = \overline{Y}^{(12/q)}\sigma^{(12)} = \sigma^{(12)}Y^{(12/q)} \ , \tag{4.2.25}$$

being relations similar to (4.2.23) for $\mathcal{I}^{(q/12)}$. As an application, we observe that relation (4.1.55a) can be rewritten by multiplying both sides to the left with U and using (4.2.22a, 23), i.e.,

$$
\begin{aligned}
UK^{(q)}I^{(q+12)} &= \mathcal{I}^{(q/12)}\overline{F}^{(12)} \\
&= \mathcal{I}^{(0q)}\overline{Y}^{(12/q)}\overline{F}^{(12)} \ .
\end{aligned}
\tag{4.2.26}
$$

Now, using (4.1.61a) for $\mathcal{I}^{(0q)}$ on the right-hand side, leads to another expression of $I^{(q+12)}$,

$$I^{(q+12)} = I^{(0q)}\overline{Y}^{(12/q)}\overline{F}^{(12)} \tag{4.2.27a}$$

$$= F^{(12)}Y^{(12/q)}I^{(0q)} \ , \tag{4.2.27b}$$

where the last expression is obtained by the transposition. Hence, comparing with (4.2.5), we find the relations

$$F^{(12)}Y^{(12/q)} = 1 + I^{(0q)}\sigma^{(12/q)} \ , \tag{4.2.28a}$$

$$\overline{Y}^{(12/q)}\overline{F}^{(12)} = 1 + \sigma^{(12/q)}I^{(0q)} \ , \tag{4.2.28b}$$

which differ from $Y^{(12/q)}$ and $\overline{Y}^{(12/q)}$ of (4.2.24a,b) only by the replacement of $\mathcal{I}^{(0q)} \longrightarrow I^{(0q)}$. Equations (4.2.27) for $I^{(q+12)}$ are similar to (4.2.23) for $\mathcal{I}^{(q/12)}$ as (4.2.5) is similar to (4.2.13); and the expression (4.2.27b) provides $I^{(q+12)}$ in terms of a transfer matrix, $F^{(12)}Y^{(12/q)}$, of the boundary, S_{12}, instead of the resultant scattering matrix, $\sigma^{(q+12)}$ of (4.2.20).

Example: Case of a Semi-Infinite Random Layer ($q_1 = 0,\ q_2 \neq 0$)

In this case, the only non-vanishing element of $S^{(0q)}$ is $S_2^{(0q)}$; hence, $I_1^{(0q)} = U_1$, and, from (4.2.5),

$$I_{11}^{(q+12)} = U_1 + U_1\sigma_{11}^{(12/q)}U_1 \ , \tag{4.2.29a}$$

$$I_{21}^{(q+12)} = I_2^{(0q)}\sigma_{21}^{(12/q)}U_1 \ . \tag{4.2.29b}$$

Here, from (4.2.17, 14a),

$$\sigma_{11}^{(12/q)} = \sigma_{11}^{(12)} + \sigma_{12}^{(12)}\mathcal{I}_2^{(q/12)}\sigma_{21}^{(12)} \tag{4.2.30a}$$

$$= \sigma_{11}^{(12)} + \sigma_{12}^{(12)}\mathcal{I}_2^{(0q)}\sigma_{21}^{(12/q)} \tag{4.2.30b}$$

where, from (4.2.16),

$$\sigma_{21}^{(12/q)} = [1 - \sigma_{22}^{(12)}\mathcal{I}_2^{(0q)}]^{-1}\sigma_{21}^{(12)} \ , \tag{4.2.31a}$$

$$\mathcal{I}_2^{(0q)} = U_2K_2^{(q)}I_2^{(0q)} \ , \tag{4.2.31b}$$

and $\mathcal{I}_2^{(0q)}$ may be obtained to the optical approximation by solving the conventional transport equation, see (2.3.46).

4.2.1 Case of Three Random Layers with Two Rough Boundaries

The BS equation (4.1.44) for $I^{(q+12+23)}$ can be rewritten in terms of $I^{(12+23)}$, by

$$I^{(q+12+23)} = I^{(12+23)}[1 + K^{(q)} I^{(q+12+23)}] \ , \tag{4.2.32a}$$

and the solution is given by (4.1.68) or, in the present notation,

$$I^{(q+12+23)} = I^{(12+23)} + F^{(12+23)}\mathcal{I}^{(q/12+23)}\overline{F}^{(12+23)} \ . \tag{4.2.32b}$$

Here,

$$I^{(12+23)} = F^{(12+23)}U = U\overline{F}^{(12+23)} \tag{4.2.33a}$$

$$= U + U\sigma^{(12+23)}U \ , \tag{4.2.33b}$$

and is obtained as the solution of the BS equation (4.2.56), to be investigated in some detail. Here $\sigma^{(12+23)}$ may be divided into two parts,

$$\sigma^{(12+23)} = \sigma^{(12)} + {`\sigma}^{(23)} \tag{4.2.34a}$$

$$\simeq \sigma^{(12)} + \sigma^{(23)}, \quad \gamma_2 L \gg 1 \ , \tag{4.2.34b}$$

where $\sigma^{(12)}$ is the matrix when S_{23} is absent so that $V^{(23)} = K^{(23)} = 0$, and, therefore, is dependent purely on S_{12}; whereas ${`\sigma}^{(23)}$ is dependent on both $S_{23}$ and $S_{12}$. As the separation, $L$, between $S_{23}$ and $S_{12}$, increases, ${`\sigma}^{(23)} \to \sigma^{(23)}$, and (4.2.34a) reduces to (4.2.34b), whose explicit expression is obtained in Sect. 4.2.2.

The basic equation to be used hereafter is (4.2.12), which is presently replaced, on using (4.2.34b), by

$$\mathcal{I}^{(q/12+23)} = \mathcal{I}^{(0q)}[1 + (\sigma^{(12)} + \sigma^{(23)})\mathcal{I}^{(q/12+23)}] \tag{4.2.35}$$

$$= \mathcal{I}^{(q/12)}[1 + \sigma^{(23)}\mathcal{I}^{(q/12+23)}] \ . \tag{4.2.36}$$

Hence the equation can be written in the form

$$\mathcal{I}^{(q/12+23)} = \mathcal{I}^{(q/12)} + \mathcal{I}^{(q/12)}\sigma^{(23/q/12)}\mathcal{I}^{(q/12)} \ , \tag{4.2.37}$$

in terms of a new matrix, $\sigma^{(23/q/12)}$, defined by

$$\sigma^{(23/q/12)}\mathcal{I}^{(q/12)} = \sigma^{(23)}\mathcal{I}^{(q/12+23)} \ , \tag{4.2.38a}$$

$$\mathcal{I}^{(q/12)}\sigma^{(23/q/12)} = \mathcal{I}^{(q/12+23)}\sigma^{(23)} \ . \tag{4.2.38b}$$

Here, substitution of (4.3.37) into (4.2.38a) yields

$$\sigma^{(23/q/12)} = \sigma^{(23)}[1 + \mathcal{I}^{(q/12)}\sigma^{(23/q/12)}] \ , \tag{4.2.39}$$

which, upon substitution of (4.2.13) for $\mathcal{I}^{(q/12)}$, can be rewritten as

$$\sigma^{(23/q/12)} = \sigma^{(23/q)}[1 + \mathcal{I}^{(0q)}\sigma^{(12/q)}\mathcal{I}^{(0q)}\sigma^{(23/q/12)}] \,, \tag{4.2.40}$$

by applying (4.2.15) to $\sigma^{(23/q)}$. Thus, for the solution we obtain

$$\sigma^{(23/q/12)} = [1 - \sigma^{(23/q)}\mathcal{I}^{(0q)}\sigma^{(12/q)}\mathcal{I}^{(0q)}]^{-1}\sigma^{(23/q)} \,, \tag{4.2.41}$$

in terms of the effective scattering matrices, $\sigma^{(23/q)}$ and $\sigma^{(12/q)}$, of the involved boundaries.

A relation similar to (4.2.17) is obtained from (4.2.39) by using (4.2.38b),

$$\sigma^{(23/q/12)} = \sigma^{(23)} + \sigma^{(23)}\mathcal{I}^{(q/12+23)}\sigma^{(23)} \,. \tag{4.2.42}$$

And, from (4.2.37), $\mathcal{I}^{(q/12+23)}$ can be written in the form of (4.2.13) for $\mathcal{I}^{(q/12)}$, as

$$\mathcal{I}^{(q/12+23)} = \mathcal{I}^{(0q)} + \mathcal{I}^{(0q)}\sigma^{(12+23/q)}\mathcal{I}^{(0q)} \,, \tag{4.2.43}$$

where, as for the factor $\sigma^{(12+23/q)}$, use of relation (4.2.23) leads to the expression

$$\sigma^{(12+23/q)} = \sigma^{(12/q)} + \overline{Y}^{(12/q)}\sigma^{(23/q/12)}Y^{(12/q)} \tag{4.2.44a}$$

$$= \sigma^{(12/q)} + [1 + \sigma^{(12/q)}\mathcal{I}^{(0q)}] \times \sigma^{(23/q/12)}[\mathcal{I}^{(0q)}\sigma^{(12/q)} + 1] \,, \tag{4.2.44b}$$

with $\sigma^{(23/q/12)}$ of (4.2.41), in terms of $\sigma^{(12/q)}$ and $\sigma^{(23/q)}$.

Thus

$$I^{(q+12+23)} = U + U\sigma^{(q+12+23)}U \tag{4.2.45}$$

can be obtained from either of two expressions,

$$I^{(q+12+23)} = I^{(0q)} + I^{(0q)}\sigma^{(12+23/q)}I^{(0q)} \tag{4.2.46a}$$

$$= I^{(12+23)} + I^{(12+23)}S^{(q/12+23)}I^{(12+23)} \,, \tag{4.2.46b}$$

according to (4.2.5) and (4.1.54), respectively. From the former,

$$\sigma^{(q+12+23)} = S^{(0q)} + \overline{F}^{(0q)}\sigma^{(12+23/q)}F^{(0q)} \,, \tag{4.2.47}$$

while another expression from the latter is written with two terms dependent and independent of the medium fluctuation, respectively.

Alternatively, $I^{(q+12+23)}$ can be written in the form

$$I^{(q+12+23)} = F^{(12+23)}Y^{(12+23/q)}I^{(0q)} \,, \tag{4.2.48}$$

which is similar to (4.2.27b) for $I^{(q+12)}$. Here, the factor, $Y^{(12+23/q)}$, can be found from (4.2.38a) by substituting expression (4.2.23) for $\mathcal{I}^{(q/12)}$, and the corresponding

$$\mathcal{I}^{(q/12+23)} = Y^{(12+23/q)}\mathcal{I}^{(0q)} \ , \tag{4.2.49}$$

to obtain

$$\sigma^{(23/q/12)}Y^{(12/q)} = \sigma^{(23)}Y^{(12+23/q)} \ . \tag{4.2.50}$$

Now if we substitute (4.2.41) for $\sigma^{(23/q/12)}$ in the left-hand side of (4.2.50) and apply a formula similar to (4.2.25) for $\sigma^{(23/q)}$, then the following expression for $Y^{(12+23/q)}$ is obtained

$$Y^{(12+23/q)} = Y^{(23/q)}[1 - \mathcal{I}^{(0q)}\sigma^{(12/q)}\mathcal{I}^{(0q)}\sigma^{(23/q)}]^{-1}Y^{(12/q)} \ , \tag{4.2.51}$$

which is in terms of $\sigma^{(12/q)}$, $\sigma^{(23/q)}$, and $\mathcal{I}^{(0q)}$ (independent of U). While, from relation (4.2.28a), we find the relation

$$F^{(12+23)}Y^{(12+23/q)} = 1 + I^{(0q)}\sigma^{(12+23/q)} \ . \tag{4.2.52}$$

Example: Case of a Random Layer ($q_1 = q_3 = 0, \quad q_2 \neq 0$)
In this case the only non-vanishing element of $\mathcal{I}^{(0q)}$ is $\mathcal{I}_2^{(0q)}$. Hence, the matrix element $\sigma_{11}^{(q+12+23)}$ for reflected waves in the space, k_1, is given, from (4.2.47, 44b), by

$$\begin{aligned}\sigma_{11}^{(q+12+23)} &= \sigma_{11}^{(12+23/q)}\\ &= \sigma_{11}^{(12/q)} + \sigma_{12}^{(12/q)}\mathcal{I}_2^{(0q)}\\ &\quad \times [1 - \sigma_{22}^{(23/q)}\mathcal{I}_2^{(0q)}\sigma_{22}^{(12/q)}\mathcal{I}_2^{(0q)}]^{-1}\sigma_{22}^{(23/q)}\mathcal{I}_2^{(0q)}\sigma_{21}^{(12/q)} \ ,\end{aligned} \tag{4.2.53}$$

similar to (4.2.30) for $\sigma_{11}^{(12/q)}$. Here, the second term gives the entire effect of the boundary, S_{23}, exclusively in terms of $\sigma_{22}^{(23/q)}$.

For the transmitted wave, $I_{31}^{(q+12+23)}$, expression (4.2.48) with (4.2.51) is also convenient to use. Therein the factor $F^{(12+23)}$ can be replaced by $F^{(23)}$, in view of $\sigma^{(12)}$ having no matrix element with subscript 3, and, from (4.2.28a),

$$F^{(23)}Y^{(23/q)} = 1 + I^{(0q)}\sigma^{(23/q)} \ .$$

Hence, with $I_3^{(0q)} = U_3$ and $I_1^{(0q)} = U_1$,

$$\begin{aligned}I_{31}^{(q+12+23)} &= U_3\sigma_{32}^{(23/q)}[1 - \mathcal{I}_2^{(0q)}\sigma_{22}^{(12/q)}\mathcal{I}_2^{(0q)}\sigma_{22}^{(23/q)}]^{-1}\\ &\quad \times \mathcal{I}_2^{(0q)}\sigma_{21}^{(12/q)}U_1 \ .\end{aligned} \tag{4.2.54}$$

Finally, by replacing $\sigma^{(23)} \to {}^{`}\sigma^{(23)}$ of (4.2.34a), the results (4.2.53) become exact for an arbitrary separation of the two boundaries [for a specific expression of (4.2.53) in the case of $K^{(12)} = K^{(23)} = 0$, refer to (6.1.32–43)].

4.2.2 $I^{(12+23)}$ and Boundary Scattering Matrices

In the case of two boundaries, S_{12} and S_{23}, $U^{(C)}(1;2)$ can also be written in the same form as (3.3.120a), i.e.,

$$U^{(C)} = U + UV^{(C)}U \ , \tag{4.2.55a}$$

$$V^{(C)} = V^{(12+23)} = V^{(12)} + {}^{`}V^{(23)} \ . \tag{4.2.55b}$$

Here, $V^{(12)}$ is the scattering matrix when the other boundary, S_{23}, is perfectly random so that no coherent scattering is made [${}^{`}V^{(23)} = 0$], and ${}^{`}V^{(23)}$ designates an entire coherent effect of S_{23}, including the interference terms. Therefore, the element ${}^{`}V_{ab}^{(23)}$, or, more precisely, ${}^{`}V_{ij;kl}^{(23)}$, is generally nonzero, even when the indices refer to numbers 1 and/or 2 of S_{12}. But, as the separation of the boundaries increases ($\gamma_2 L \to \infty$), ${}^{`}V^{(23)} \to V^{(23)}$, i.e., the scattering matrix of $S_{23}$ when $S_{12}$ is absent [for ${}^{`}V^{(23)}$, refer to (6.1.35–39)]. On the other hand, the incoherent scattering by S_{12} and S_{23} is caused, through the BS equation,

$$I^{(12+23)} = U^{(C)}[1 + (K^{(12)} + K^{(23)})I^{(12+23)}] \ , \tag{4.2.56}$$

by the matrices, $K^{(12)}$ and $K^{(23)}$, with non-vanishing elements, $K_{ab}^{(12)}$, $a, b = 1, 2$, and $K_{ab}^{(23)}$, a, $b = 2$, 3, only on the respective boundaries.

The solution $I^{(12+23)}$ is obtained in terms of $\sigma^{(12+23)}$ which can be written, from (4.2.20) for $\sigma^{(q+12)}$ with $K^{(q)} \to K^{(23)}$, by two alternative expressions, as

$$\sigma^{(12+23)} = S^{(023)} + (1 + S^{(023)}U)\sigma_C^{(12/23)}(US^{(023)} + 1) \tag{4.2.57a}$$

$$= \sigma_C^{(12)} + (1 + \sigma_C^{(12)}U)S_C^{(23/12)}(U\sigma_C^{(12)} + 1) \ . \tag{4.2.57b}$$

Here, $S^{(023)}$ is the same operator of $K^{(23)}$ as $S^{(0q)}$ is of $K^{(q)}$; and the same is also true of $\sigma_C^{(12/23)}$ and $S_C^{(23/12)}$, i.e.,

$$\sigma_C^{(12/23)} = [1 - \sigma_C^{(12)}\mathcal{I}^{(023)}]^{-1}\sigma_C^{(12)} \ , \tag{4.2.58a}$$

$$S_C^{(23/12)} = [1 - S^{(023)}U\sigma_C^{(12)}U]^{-1}S^{(023)}$$

$$\simeq [1 - S^{(23)}U\sigma^{(12)}U]^{-1}S^{(23)} \ , \tag{4.2.58b}$$

$$\mathcal{I}^{(023)} = US^{(023)}U = UK^{(23)}I^{(023)} \ . \tag{4.2.58c}$$

Here, $\mathcal{I}^{(0B)}$, $B = 12, 23$, is a short-range function appreciable only in the neighborhood of the boundary within a distance of the order of the wave coherence distance, γ_2^{-1}, in contrast to a long-range function, $\mathcal{I}^{(0q)}$. $\sigma_C^{(12)}$ is the same as $\sigma^{(12)}$, except that the $V^{(12)}$ is replaced by $V^{(C)}$, and can be obtained from (4.2.20) for $\sigma^{(q+12)}$ with $K^{(q)} \to K^{(12)}$ and $\sigma^{(12)} \to V^{(C)}$, as

$$\sigma_C^{(12)} = S^{(012)} + (1 + S^{(012)}U)V^{(C/12)}(US^{(012)} + 1) \quad (4.2.59a)$$

$$= V^{(C)} + (1 + V^{(C)}U)S_C^{(12)}(UV^{(C)} + 1) \; . \quad (4.2.59b)$$

Hence we write it in the form

$$\sigma_C^{(12)} = \sigma^{(12)} + \sigma_{(23)}^{(12)} \; . \quad (4.2.60)$$

An explicit expression of the last term can be obtained from (4.2.59a), using (4.2.98a,b) for $V^{(C1+C2/q)}$ with $V^{(C1)} \to V^{(12)}$, and $V^{(C2)} \to {}^{`}V^{(23)}$, and $K^{(q)} \to K^{(12)}$, as

$$\sigma_{(23)}^{(12)} = \overline{F}^{(012)}\overline{Y}_{(12)}^{(12/12)}\,{}^{`}V^{(23/12/12)}Y_{(12)}^{(12/12)}F^{(012)} \; , \quad (4.2.61a)$$

where

$$F^{(012)} = 1 + US^{(012)}, \qquad Y_{(12)}^{(12/12)} = 1 + \mathcal{I}^{(012)}V^{(12/12)} \; ,$$
$$V^{(12/12)} = (1 - V^{(12)}\mathcal{I}^{(012)})^{-1}V^{(12)} \; ; \quad (4.2.61b)$$

$\overline{F}^{(012)}$ and $\overline{Y}_{(12)}^{(12/12)}$ are obtained by the transposition. Here, ${}^{`}V^{(23/12/12)} \sim {}^{`}V^{(23)}$ when S_{12} and S_{23} are separated far enough; and, on S_{23}, $\sigma_{(23)}^{(12)} \sim V^{(23)}$.

Thus, upon substitution of (4.2.34a, 60) into the left- and right-hand sides of (4.2.57b), respectively, ${}^{`}\sigma^{(23)}$ is found to be

$${}^{`}\sigma^{(23)} = \sigma_{(23)}^{(12)} + (1 + \sigma_C^{(12)}U)S_C^{(23/12)}(U\sigma_C^{(12)} + 1) \; , \quad (4.2.62)$$

with $\sigma_{(23)}^{(12)}$ and $S_C^{(23/12)}$ given by (4.2.61a, 58b), respectively.

4.2.3 Another Expression for $I^{(q+12+23)}$

By interchanging the roles of $K^{(23)}$ and $K^{(q)}$ in (4.2.46b), we obtain another expression for $I^{(q+12+23)}$, of the form

$$I^{(q+12+23)} = I_C^{(q+12)} + I_C^{(q+12)}S^{(23/q+12)}I_C^{(q+12)} \; . \quad (4.2.63)$$

Here (and also hereafter) the subscript C is attached to any quantity to indicate the dependence on $V^{(C)}$ (which may or may not be $V^{(12)}$) whenever there is confusion; and, from (4.2.5) for $I^{(q+12)}$,

$$I_C^{(q+12)} = I^{(0q)} + I^{(0q)}\sigma_C^{(12/q)}I^{(0q)} \; , \quad (4.2.64)$$

where $\sigma_C^{(12/q)}$ is the same as $\sigma^{(12/q)}$ except $\sigma^{(12)} \to \sigma_C^{(12)}$ which is given by (4.2.60). The matrix $S^{(23/q+12)}$ is defined by

$$S^{(23/q+12)} I_C^{(q+12)} = K^{(23)} I^{(q+12+23)} \ , \tag{4.2.65a}$$

and is hence given by

$$\begin{aligned} S^{(23/q+12)} &= K^{(23)}[1 + I_C^{(q+12)} S^{(23/q+12)}] \\ &= \sigma_0^{(23/q)}[1 + I^{(0q)} \sigma_C^{(12/q)} I^{(0q)} S^{(23/q+12)}] \ . \end{aligned} \tag{4.2.65b}$$

Here, the last equation is a result of using (4.2.64), with $\sigma_0^{(23/q)}$ defined by

$$\sigma_0^{(23/q)} = K^{(23)}[1 + I^{(0q)} \sigma_0^{(23/q)}] \tag{4.2.66a}$$

$$\begin{aligned} &= S^{(023)}[1 + \mathcal{I}^{(0q)} \sigma_0^{(23/q)}] \\ &= [1 - S^{(023)} \mathcal{I}^{(0q)}]^{-1} S^{(023)} \ , \end{aligned} \tag{4.2.66b}$$

which is the same as that obtained from $\sigma^{(23/q)}$ when $\sigma^{(23)} = S^{(023)}$ or $V^{(C)} = 0$, see (4.2.59a). The solution of (4.2.65b) is

$$S^{(23/q+12)} = [1 - \sigma_0^{(23/q)} I^{(0q)} \sigma_C^{(12/q)} I^{(0q)}]^{-1} \sigma_0^{(23/q)} \ . \tag{4.2.67}$$

Thus, from (4.2.63, 64), expression (4.2.46a) for $I^{(q+12+23)}$ is reproduced, with a different expression of $\sigma^{(12+23/q)}$ [cf.(4.2.44b)]

$$\begin{aligned} \sigma^{(12+23/q)} &= \sigma_C^{(12/q)} + [1 + \sigma_C^{(12/q)} I^{(0q)}] \\ &\quad \times S^{(23/q+12)}[I^{(0q)} \sigma_C^{(12/q)} + 1] \ . \end{aligned} \tag{4.2.68a}$$

Another expression can be obtained based on the invariance of $\sigma^{(12+23)}$ against the interchange of $\sigma_C^{(12)}$ and $S^{(023)}$ from (4.2.57a,b); hence, with (4.2.67), (4.2.68a) can be replaced by

$$\begin{aligned} \sigma^{(12+23/q)} &= \sigma_0^{(23/q)} + [1 + \sigma_0^{(23/q)} I^{(0q)}] \\ \times [1 - \sigma_C^{(12/q)} I^{(0q)} \sigma_0^{(23/q)} I^{(0q)}]^{-1} & \sigma_C^{(12/q)} [I^{(0q)} \sigma_0^{(23/q)} + 1] \ . \end{aligned} \tag{4.2.68b}$$

Two more expressions are obtained by interchanging the roles of S_{12} and S_{23}, or the superscripts 12 and 23.

In the special case of $V^{(23)} = 0$ when the boundary S_{23} is perfectly random, we obtain $\sigma_C^{(12/q)} = \sigma^{(12/q)}$ and $\sigma_0^{(23/q)} = \sigma^{(23/q)}$, showing that even then, $\sigma^{(23)} \neq \sigma^{(23)} = S^{(023)}$, in view of (4.2.62) (unless $\gamma_2 L \gg 1$), although $\sigma_{(23)}^{(12)} = 0$, as a consequence of multiple scattering between S_{12} and S_{23} through the propagator U. Also, comparing expressions (4.2.68) for $\sigma^{(12+23/q)}$

with the previous (4.2.44b), as corrected by $\sigma^{(23)} \to {}^{`}\sigma^{(23)}$, we observe that all the $I^{(0q)}$'s involved in (4.2.68) are replaced by the incoherent parts $\mathcal{I}^{(0q)}$. This is because in the latter (4.2.44), the entire effect of the coherent scattering by the medium is already included in $\sigma^{(12+23)}$, and the only addition is the effect of the incoherent scattering caused by $K^{(q)}$. A great advantage of using expressions (4.2.68a,b) in the present case is that they are available without any constraint of the boundary separation, being applicable to the case $\gamma_2 L \lesssim 1$, as well as $\gamma_2 L \gg 1$. Expression (4.2.44) is available more simply in the case $\gamma_2 L \gg 1$, where ${}^{`}\sigma^{(23)} \simeq \sigma^{(23)}$ is realized, whereas, in the case $\gamma_2 L \lesssim 1$, ${}^{`}\sigma^{(23)}$ has to be obtained from (4.2.62, 58b), which, when ${}^{`}V^{(23)} = 0$, becomes

$$
{}^{`}\sigma^{(23)} = (1 + \sigma^{(12)}U)(1 - \sigma^{(23)}U\sigma^{(12)}U)^{-1}\sigma^{(23)}(U\sigma^{(12)} + 1) \ , \qquad (4.2.69)
$$

where $\sigma^{(23)} = S^{(023)}$. Note that ${}^{`}\sigma^{(23)}$ has the matrix elements on both $S_{12}$ and $S_{23}$ at $z = 0$ and $-L$, respectively, including cross matrix elements, ${}^{`}\sigma^{(23)}_{12,23}$, with row and column indices referring to different boundaries.

In the case of a random layer where $q_1 = q_3 = 0$, $q_2 \neq 0$, and ${}^{`}V^{(23)} = 0$, (4.2.68a) gives the scattering matrix for reflected waves as

$$
\begin{aligned}
\sigma_{11}^{(12+23/q)} &= \sigma_{11}^{(12/q)} + \sigma_{12}^{(12/q)} I_2^{(0q)} \\
&\times [1 - \sigma_{22}^{(23/q)} I_2^{(0q)} \sigma_{22}^{(12/q)} I_2^{(0q)}]^{-1} \sigma_{21}^{(23/q)} I_2^{(0q)} \sigma_{21}^{(12/q)} \ ,
\end{aligned} \qquad (4.2.70)
$$

wherein $\sigma^{(23)} = S^{(023)}$. The corresponding $\sigma_{33}^{(12+23/q)}$ from (4.2.68b) gives the expression when the waves are reflected by a perfectly random boundary S_{23} which also can be obtained from (4.2.70) by interchanging the roles of $\sigma^{(12)}$ and $\sigma_0^{(23)}$, and replacing all the subscripts $1 \to 3$. Expression (4.2.70) is similar to that intuitively obtained based on the transport equation in [4.3], with some numerical applications. Note that the expression is correct as long as one of the boundaries is perfectly random, independent of whether the other boundary is smooth or not, and that, even to the optical approximation, the result is not equivalent to that which would be obtained by solving the transport equation subjected to the conventional boundary condition. The latter method is valid only when the layer width is sufficiently large compared with the wave coherence distance, and the two results tend to agree with each other as $\gamma_2 L \to \infty$.

Even when neither $V^{(12)}$ nor ${}^{`}V^{(23)}$ is zero, expressions (4.2.44) and (4.2.68a,b) are all correct, but they are not of much use because $\sigma_C^{(12)}$ is no longer an observable parameter of S_{12}, depending also on S_{23}; expressions can be useful only when they are written in terms of good observable parameters. And, in the case of expressions (4.2.68a,b), the situation remains unimproved even when the layer width is large enough that $\gamma_2 L \gg 1$; whereas expression (4.2.44b) provides a useful expression when $\gamma_2 L \gg 1$, in view of

(4.2.34b), showing that $\sigma^{(12+23)}$ is given by the independent sum of $\sigma^{(12)}$ and $\sigma^{(23)}$, in this case, and, to the optical approximation, the result is equivalent to the boundary-value solution of the transport equation.

Alternatively, as is shown below, $I^{(q+12+23)}$ can be written in the form (4.2.48), in terms of a transfer matrix, $Y^{(12+23/q)}$, as

$$I^{(q+12+23)} = [1 + I^{(0q)}\sigma_0^{(23/q)}] \times [1 - I^{(0q)}\sigma_C^{(12/q)}I^{(0q)}\sigma_0^{(23/q)}]^{-1}I_C^{(q+12)} \ , \tag{4.2.71}$$

with $\sigma_0^{(23/q)}$ of (4.2.66a,b) and

$$I_C^{(q+12)} = [1 + I^{(0q)}\sigma_C^{(12/q)}]I^{(0q)} \ , \tag{4.2.72}$$

from (4.2.64). The expression is derived from the definition (4.2.65a)

$$S^{(23/q+12)}I_C^{(q+12)} = K^{(23)}I^{(q+12+23)} \tag{4.2.73}$$

for $S^{(23/q+12)}$; i.e., (4.2.71) is derived from (4.2.73) by substituting $S^{(23/q+12)}$ of (4.2.67), and followed by dropping the common factor, $K^{(23)}$, from both sides with the aid of (4.2.66a). In the case of a random layer where $q_1 = q_3 = 0$, $q_2 \neq 0$, and the boundary, S_{23}, is perfectly random ($`V^{(23)} = 0$), (4.2.71) for the transmitted wave becomes

$$I_{31}^{(q+12+23)} = U_3\sigma_{32}^{(23/q)} \times [1 - I_2^{(0q)}\sigma_{22}^{(12/q)}I_2^{(0q)}\sigma_{22}^{(23/q)}]^{-1}I_2^{(0q)}\sigma_{21}^{(12/q)}U_1 \ , \tag{4.2.74}$$

where $\sigma^{(23)} = S^{(023)}$, although $\sigma^{(12)} = S^{(012)}$ is also possible by interchanging the roles of S_{12} and S_{23}. Together with the corresponding expression (4.2.70) for $\sigma_{11}^{(12+23/q)}$, (4.2.74) is available in the range $\gamma_2 L \lesssim 1$, as well as $\gamma_2 L \gg 1$, beyond the range of applicability of the corresponding boundary-value solution of the transport equation, in contrast to expression (4.2.54).

4.2.4 Addition Formulas of Scattering Matrices Utilized

It may be worthwhile to summarize here the addition formulas utilized in the previous sections. Let $I^{(A)}$ be the solution of the BS equation:

$$I^{(A)} = U^{(C)}(1 + K^{(A)}I^{(A)}) \tag{4.2.75a}$$

$$= U + U\sigma^{(A)}U \ , \tag{4.2.75b}$$

$$U^{(C)} = U + UV^{(C)}U \ . \tag{4.2.76}$$

Here, $\sigma^{(A)} = \sigma_C^{(A)}$ is the resultant scattering matrix depending on both $V^{(C)}$ and $K^{(A)}$, and can be written as two alternative expressions from (4.2.20a,b), with $K^{(q)} \to K^{(A)}$ and $K^{(12)} = 0$,

$$\sigma^{(A)} = S^{(0A)} + (1 + S^{(0A)}U)V^{(C/A)}(US^{(0A)} + 1) \tag{4.2.77a}$$
$$= V^{(C)} + (1 + V^{(C)}U)S^{(A)}(UV^{(C)} + 1) \ . \tag{4.2.77b}$$

Here,

$$S^{(0A)} = K^{(A)}(1 + US^{(0A)}) = (1 - K^{(A)}U)^{-1}K^{(A)} \ , \tag{4.2.78a}$$
$$S^{(A)} = K^{(A)}(1 + U^{(C)}S^{(A)}) = S^{(0A)}(1 + UV^{(C)}US^{(A)})$$
$$= (1 - S^{(0A)}UV^{(C)}U)^{-1}S^{(0A)} \ , \tag{4.2.78b}$$
$$V^{(C/A)} = (1 - V^{(C)}\mathcal{I}^{(0A)})^{-1}V^{(C)} \ , \tag{4.2.79}$$

where

$$\mathcal{I}^{(0A)} = US^{(0A)}U = UK^{(A)}I^{(0A)} \ , \tag{4.2.80a}$$

being the incoherent part of

$$I^{(0A)} \equiv I^{(A)} \,|_{C=0} = U + \mathcal{I}^{(0A)} \ , \tag{4.2.80b}$$

with the mark $|_{C=0}$ denoting set $V^{(C)} = 0$. Expressions (4.2.77a,b) show that $\sigma^{(A)}$ is invariant against the interchange of $V^{(C)}$ and $S^{(0A)}$.

Similarly, $I^{(B)}$ and $\sigma^{(B)}$ are defined by

$$I^{(B)} = U^{(C)}(1 + K^{(B)}I^{(B)}) \tag{4.2.81a}$$
$$= U + U\sigma^{(B)}U \ ; \tag{4.2.81b}$$

and $I^{(A+B)}$ and $\sigma^{(A+B)}$ are also defined by

$$I^{(A+B)} = U^{(C)}[1 + (K^{(A)} + K^{(B)})I^{(A+B)}] \tag{4.2.82a}$$
$$= U + U\sigma^{(A+B)}U \ , \tag{4.2.82b}$$

with the same $U^{(C)}$. Then, from (4.2.5) and (4.1.54) with $K^{(q)} \to K^{(A)}$ and $K^{(12)} \to K^{(B)}$, we obtain two alternative expressions of $I^{(A+B)}$:

$$I^{(A+B)} = I^{(0A)} + I^{(0A)}\sigma^{(B/A)}I^{(0A)} \tag{4.2.83a}$$
$$= I^{(B)} + I^{(B)}S^{(A/B)}I^{(B)} \ . \tag{4.2.83b}$$

Here,

$$\sigma^{(B/A)} = (1 - \sigma^{(B)}\mathcal{I}^{(0A)})^{-1}\sigma^{(B)} \ , \tag{4.2.84}$$

being the same as $V^{(C/A)}$ of (4.2.79), with $V^{(C)} \to \sigma^{(B)}$, and

$$S^{(A/B)} = (1 - S^{(0A)} U \sigma^{(B)} U)^{-1} S^{(0A)} \ , \tag{4.2.85}$$

which is converted to $\sigma^{(B/A)}$ by the interchange of $S^{(0A)}$ and $\sigma^{(B)}$. Equations (4.2.83) indicate that

$$\sigma^{(A+B)} = S^{(0A)} + (1 + S^{(0A)} U) \sigma^{(B/A)} (U S^{(0A)} + 1) \tag{4.2.86a}$$

$$= \sigma^{(B)} + (1 + \sigma^{(B)} U) S^{(A/B)} (U \sigma^{(B)} + 1) \ , \tag{4.2.86b}$$

and is invariant against the interchange of $S^{(0A)}$ and $\sigma^{(B)}$, as well as the interchange of the superscripts A (or $K^{(A)}$) and B (or $K^{(B)}$). Also, $\sigma^{(A+B)}$ can be obtained from $\sigma^{(A)}$ of (4.2.77) by the replacement of $V^{(C)} \to \sigma^{(B)}$, in view of (4.2.75a,76) for $I^{(A)}$ with $U^{(C)} \to I^{(B)}$.

In the same fashion as for the BS equation (4.2.32a) for $I^{(q+12+23)}$, we obtain

$$I^{(q+A+B)} = I^{(A+B)} [1 + K^{(q)} I^{(q+A+B)}] \tag{4.2.87a}$$

$$= I^{(0q)} + I^{(0q)} \sigma^{(A+B/q)} I^{(0q)} \ , \tag{4.2.87b}$$

where the last expression is from (4.2.83a) with $K^{(A)} \to K^{(q)}$, $K^{(B)} \to K^{(A)} + K^{(B)}$, and $\sigma^{(A+B/q)}$ to be given by (4.2.92, 93). An alternative expression is

$$I^{(q+A+B)} = I_C^{(q+B)} [1 + K^{(A)} I^{(q+A+B)}] \tag{4.2.88a}$$

$$= I_C^{(q+B)} + I_C^{(q+B)} S^{(A/q+B)} I_C^{(q+B)} \ . \tag{4.2.88b}$$

Here, using formula (4.2.64),

$$I_C^{(q+B)} = I^{(0q)} + I^{(0q)} \sigma_C^{(B/q)} I^{(0q)} \ , \tag{4.2.89}$$

where the subscript C is attached to make sure the dependence on $V^{(C)}$ (which can be neither $V^{(A)}$ nor $V^{(B)}$), and $\sigma_C^{(B/q)}$ is given by (4.2.84) for $\sigma^{(B/A)}$ with $\sigma^{(B)} \to \sigma_C^{(B)}$ and $\mathcal{I}^{(0A)} \to \mathcal{I}^{(0q)}$. And, in (4.2.88b),

$$S^{(A/q+B)} = [1 - \sigma_0^{(A/q)} I^{(0q)} \sigma_C^{(B/q)} I^{(0q)}]^{-1} \sigma_0^{(A/q)} \ , \tag{4.2.90}$$

from (4.2.67) with $K^{(12)} \to K^{(B)}$ and $K^{(23)} \to K^{(A)}$; wherein

$$\sigma_0^{(A/q)} \equiv \sigma_C^{(A/q)} \,|_{C=0} = S^{(0A/q)} \ , \tag{4.2.91}$$

having been given by $\sigma_C^{(A/q)}$ of when $\sigma_C^{(A)} = S^{(0A)}$.

On substituting (4.2.89) into (4.2.88b), the matrix $\sigma^{(A+B/q)}$ in expression (4.2.87b) can be written as

$$\sigma^{(A+B/q)} = \sigma_C^{(B/q)} + [1 + \sigma_C^{(B/q)} I^{(0q)}] \times S^{(A/q+B)} [I^{(0q)} \sigma_C^{(B/q)} + 1] \ , \tag{4.2.92}$$

which is invariant against the interchange of $\sigma_C^{(B)}$ and $S^{(0A)}$, in view of (4.2.86a,b) for $\sigma^{(A+B)}$. Hence, using (4.2.90), we obtain another expression with this interchange:

$$\sigma^{(A+B/q)} = \sigma_0^{(A/q)} + [1 + \sigma_0^{(A/q)} I^{(0q)}] \times [1 - \sigma_C^{(B/q)} I^{(0q)} \sigma_0^{(A/q)} I^{(0q)}]^{-1} \sigma_C^{(B/q)} [I^{(0q)} \sigma_0^{(A/q)} + 1] \ . \tag{4.2.93}$$

On the other hand, from (4.2.71, 72) we obtain

$$I^{(q+12+23)} = [1 + I^{(0q)} \sigma_0^{(A/q)}][1 - I^{(0q)} \sigma_C^{(B/q)} I^{(0q)} \sigma_0^{(A/q)}]^{-1} \times [1 + I^{(0q)} \sigma_C^{(B/q)}] I^{(0q)} \ , \tag{4.2.94}$$

and also another expression which is given by the interchange of $S^{(0A)}$ and $\sigma_C^{(B)}$, or just by the transposition.

So far, $V^{(C)}$ has been held fixed in $U^{(C)}$ of (4.2.76) when adding the incoherent factors, $K^{(A)}$'s, in the BS equation. Another sort of addition also exists, resulting from the factor $U^{(C)}$, which is the propagator of the coherent wave, and which, therefore, may be divided into several parts owing to the possible boundaries and fixed scatterers. We set [cf. (6.1.35–39)]

$$V^{(C)} = V^{(C1+C2)} \equiv V^{(C1)} + V^{(C2)} \ , \tag{4.2.95}$$

with $K = K^{(q)}$, held fixed. Then, by comparing $U^{(C)}$ with $I^{(12+23)}$ of (4.2.33b, 34b), we observe that, in (4.2.35) we can make the replacements of $\sigma^{(12)} \to V^{(C1)}$ and $\sigma^{(23)} \to V^{(C2)}$, to obtain

$$\mathcal{I}_{C1+C2}^{(q)} = \mathcal{I}^{(0q)} [1 + (V^{(C1)} + V^{(C2)}) \mathcal{I}_{C1+C2}^{(q)}] \ , \tag{4.2.96}$$

with $\mathcal{I}^{(q/12+23)} \to \mathcal{I}_{C1+C2}^{(q)}$. Equation (4.2.96) has the same form as the BS equation (2.4.3) when $K = K^{(A)} + K^{(B)}$, and hence provides us with a basic equation for the present addition by letting $\mathcal{I}^{(0q)}$ play the role of U (more generally, see (4.2.9a,b) written in an original form).

The solution of (4.2.96) can be written in the form of (4.2.43), so that

$$\mathcal{I}_{C1+C2}^{(q)} = \mathcal{I}^{(0q)} + \mathcal{I}^{(0q)} V^{(C1+C2/q)} \mathcal{I}^{(0q)} \ . \tag{4.2.97}$$

Here, $V^{(C1+C2/q)}$ is given by the same equation as (4.2.44b) for $\sigma^{(12+23/q)}$, with the replacements $\sigma^{(12)} \to V^{(C1)}$ and $\sigma^{(23)} \to V^{(C2)}$, i.e.,

$$V^{(C1+C2/q)} = V^{(C1/q)} + [1 + V^{(C1/q)}\mathcal{I}^{(0q)}] \times V^{(C2/q/C1)}[\mathcal{I}^{(0q)}V^{(C1/q)} + 1] , \quad (4.2.98a)$$

$$V^{(C2/q/C1)} = [1 - V^{(C2/q)}\mathcal{I}^{(0q)}V^{(C1/q)}\mathcal{I}^{(0q)}]^{-1}V^{(C2/q)} . \quad (4.2.98b)$$

We observe that both $V^{(C1+C2/q)}$ and $\sigma^{(A+B/q)}$ have the same structure, as the expressions (4.2.92, 93) indicate, except that all the $\mathcal{I}^{(0q)}$ are replaced with $I^{(0q)} = U + \mathcal{I}^{(0q)}$ in the latter (except in $\sigma_0^{(A/q)}$ and $\sigma_C^{(B/q)}$), reflecting the fact that when $V^{(C)}$ is divided into two parts in the form (4.2.95), their interaction term due to the coherent multiple scattering is necessarily included in either or both of the two parts, in contrast to the case where K is divided into two independent parts.

4.2.5 Optical Relations of Random Layers

From (4.2.19), $\sigma_{ab}^{(q+12)}$ is defined by

$$I_{ab}^{(q+12)} = U_a\delta_{ab} + U_a\sigma_{ab}^{(q+12)}U_b . \quad (4.2.99)$$

and, in the case of a semi-infinite random layer where $q_1 = 0$ and $q_2 \neq 0$, $I_{11}^{(q+12)}$ is given by (4.1.64). Hence,

$$\sigma_{11}^{(q+12)} = \sigma_{11}^{(12)} + \sigma_{12}^{(12)}\mathcal{I}_{22}^{(q/12)}\sigma_{21}^{(12)} , \quad (4.2.100)$$

and means a resultant scattering matrix for the entire random volume (taking into account both both boundary and medium scatterings) when the source and the observer are both in the space k_1. To find the optical condition for $\sigma_{11}^{(q+12)}$, we observe that the integrated power of $I_{11}^{(q+12)}$, away from the boundary, S_1, is necessarily zero because all the waves propagated into the (non-dissipative) space, k_2, are finally scattered back to the space, k_1, as will be investigated in more detail later. Hence,

$$\int_{S_1} d\rho\, \hat{\boldsymbol{n}}^{(1)} \cdot \hat{\boldsymbol{\alpha}}\, I_{11}^{(q+12)}(\hat{\boldsymbol{x}}) = 0 ,$$

or, for short,

$$\langle S_1 | I_{11}^{(q+12)} = 0 , \quad (4.2.101)$$

which is equivalent to

$$\langle S_2 | I_{21}^{(q+12)} = 0 , \quad (4.2.102)$$

in view of no dissipation by S_{12}. Using (4.2.99),

$$\langle S_1 | U_1(1 + \sigma_{11}^{(q+12)}U_1) = 0 ,$$

$$\langle S_2 | U_2\sigma_{21}^{(q+12)}U_1 = 0 , \quad (4.2.103)$$

and provide basic equations to derive optical relations for $\sigma_{ab}^{(q+12)}$.

More basically, the corresponding equation for $I^{(12)}$ can be written as

$$\langle S_1 | I_{1a}^{(12)} + \langle S_2 | I_{2a}^{(12)} = 0, \qquad a = 1, 2 \ , \tag{4.2.104}$$

which is a surface integral over $S_{12} = S_1 + S_2$, enclosing the whole boundary space, R_S (Fig. 4.1), and $a = 1$ or 2, depending on whether the source is in the space k_1 or k_2, respectively. Hence, using (4.1.48a), we obtain

$$\langle S_1 | U_1 (1 + \sigma_{11}^{(12)} U_1) + \langle S_2 | U_2 \sigma_{21}^{(12)} U_1 = 0 \ , \tag{4.2.105a}$$

$$\langle S_2 | U_2 (1 + \sigma_{22}^{(12)} U_2) + \langle S_1 | U_1 \sigma_{12}^{(12)} U_2 = 0 \ , \tag{4.2.105b}$$

which can be shown to lead directly to an optical relation (5.1.3) for $\sigma_{ab}^{(12)}$ to be obtained [and equivalent to (3.3.140)].

A similar equation for the medium counterpart, $S_{22}^{(q/12)}$, can also be found by substituting (4.2.99, 100) into (4.2.101), and using relation (4.2.105a) to eliminate the $\sigma_{11}^{(12)}$ term; hence,

$$\langle S_1 | U_1 \sigma_{12}^{(12)} \mathcal{I}_{22}^{(q/12)} \sigma_{21}^{(12)} U_1 = \langle S_2 | U_2 \sigma_{21}^{(12)} U_1 \ , \tag{4.2.106}$$

where only the boundary value of $\mathcal{I}_{22}^{(q/12)}$ on S_2 is involved in the left-hand side. It simply says that the entire incident power through the boundary, as given by the right-hand side, is the same as the integrated power of the waves scattered back to the space, k_1, by the random medium, q_2. A direct proof of relation (4.2.106) using the basic equation (4.2.12) for $\mathcal{I}^{(q/12)}$ is shown, first by observing that the power equation for $I_2^{(0q)}$ is given from (4.1.31a), by

$$\frac{\partial}{\partial \hat{\boldsymbol{x}}} \cdot \hat{\boldsymbol{\alpha}} \, I_2^{(0q)}(\hat{\boldsymbol{x}}) = \Delta G_2^{(0)}(\hat{\boldsymbol{x}}) \ , \tag{4.2.107}$$

where the right-hand side is zero in the present case when the source is located in the outside space, k_1. Hence, integrating both sides over the space, $z \leq 0$, we obtain, in terms of notation (4.2.101),

$$\langle S_2 | \mathcal{I}_2^{(0q)} = -\langle S_1 | U_2 \ . \tag{4.2.108}$$

Therefore, from (4.1.62),

$$\begin{aligned} \langle S_2 | \mathcal{I}_{22}^{(q/12)} &= \langle S_2 | \mathcal{I}_2^{(0q)} (1 + \sigma_{22}^{(12)} \mathcal{I}_{22}^{(q/12)}) \\ &= -\langle S_2 | U_2 (1 + \sigma_{22}^{(12)} \mathcal{I}_{22}^{(q/12)}) \ , \end{aligned} \tag{4.2.109}$$

which can be rewritten using (4.2.105b), as

$$\langle S_1|U_1\sigma_{12}^{(12)}\mathcal{I}_{22}^{(q/12)} = \langle S_2|U_2 \ , \tag{4.2.110}$$

which is equivalent to (4.2.106) (except for the common factor $\sigma_{21}^{(12)}U_1$ on both sides), and hence also to the optical relations (4.2.103) for $\sigma_{11}^{(q+12)}$ and $\sigma_{21}^{(q+12)}$, as can be shown by using (4.2.105a).

In the case of a random layer where $q_1 = q_3 = 0$, $q_2 \neq 0$, and $\gamma_2 L \gg 1$ so that (4.1.67) is valid (Fig. 4.3), relation (4.2.110) is replaced with

$$[\langle S_1|U_1\sigma_{12}^{(12)} + \langle S_3|U_3\sigma_{32}^{(23)}]\mathcal{I}_{22}^{(q/12+23)} = \langle S_2|U_2 \ . \tag{4.2.111}$$

Here, the boundary S_2 on the right-hand side refers to either S_{+2} at $z = 0$, or S_{-2} at $z = -L$. As to the scattering matrix for the entire system, $\sigma_{ab}^{(q+12+23)}$, defined by

$$\begin{aligned} I_{11}^{(q+12+23)} &= U_1 + U_1\sigma_{11}^{(q+12+23)}U_1 \ , \\ I_{31}^{(q+12+23)} &= U_3\sigma_{31}^{(q+12+23)}U_1 \ , \end{aligned} \tag{4.2.112}$$

the optical relations are given by the same equations as (4.2.105) with the replacement of $\sigma^{(12)} \to \sigma^{(q+12+23)}$ and $S_2 \to S_3$, e.g.,

$$\langle S_1|U_1(1 + \sigma_{11}^{(q+12+23)}U_1) + \langle S_3|U_3\sigma_{31}^{(q+12+23)}U_1 = 0 \ , \tag{4.2.113}$$

which can also be shown directly, based on (4.1.70, 71), by using (4.2.111, 105).

It may be noticed that optical relations of a random volume are described exclusively in terms of the boundary values, as given by (4.2.110, 111), independent of details of the inside [see also the cross sections (5.1.27)].

5. Optical Cross Sections of a Random Layer

In this chapter, cross sections of a random layer and their optical relations are obtained, in terms of cross sections of the two boundaries, from Sect. 3.3.6, and the medium counterpart, $S_2^{(0q)}$, which can be obtained as a boundary-value solution of the transport equation subject to the conventional condition of no reflection at the boundaries. The contribution from the entire random medium, $S_{22}^{(q/12+23)}$, is thereby constructed based on the equations of Sect. 4.2.1, first, and the resultant cross section of the layer is then obtained. We then apply the diffusion approximation to the medium part of the contribution by introducing an appropriate boundary condition for the diffusion equation. This enables us to obtain specific expressions of the cross sections in terms of a simple boundary-value solution of the diffusion equation. Finally, the boundary conditions are generalized to the case in which media are random on both sides of a rough boundary.

5.1 Construction of the Cross Sections

In the same manner that (3.3.119a) for $\boldsymbol{I}^{(B)}$ has been optically written as (3.3.128), the optical expression of (4.1.48a) for $I^{(12)}$ can be written in the form

$$I_{ab}^{(12)}(\widehat{\boldsymbol{\Omega}}, z|\widehat{\boldsymbol{\Omega}}', z') = U_a(\widehat{\boldsymbol{\Omega}}, z - z')\delta_{ab}\delta^2(\widehat{\boldsymbol{\Omega}} - \widehat{\boldsymbol{\Omega}}') + U_a(\widehat{\boldsymbol{\Omega}}, z)\sigma_{ab}^{(12)}(\widehat{\boldsymbol{\Omega}}|\widehat{\boldsymbol{\Omega}}')U_b(\widehat{\boldsymbol{\Omega}}', -z') \ . \tag{5.1.1}$$

Here, $U_a(\widehat{\boldsymbol{\Omega}}, z)$ is the $\boldsymbol{\lambda}$ Fourier transforms of $U_a(\widehat{\boldsymbol{\Omega}}, \widehat{\boldsymbol{\rho}})$ (2.3.38). Hence, using (2.3.34) for $\widetilde{U}_a(\widehat{\boldsymbol{\Omega}}, \widehat{\boldsymbol{\lambda}})$, we can write

$$\begin{aligned} U_a(\widehat{\boldsymbol{\Omega}}, z) &= (2\pi)^{-1} \int d\lambda_z \exp(-\mathrm{i}\lambda_z z)\widetilde{U}_a(\widehat{\boldsymbol{\Omega}}, \widehat{\boldsymbol{\lambda}}) \\ &= \begin{cases} |\Omega_z|^{-1} \exp\left[-\Omega_z^{-1}(\gamma_a - \mathrm{i}\boldsymbol{\Omega}\cdot\boldsymbol{\lambda})z\right], & \Omega_z z > 0 \\ 0, & \Omega_z z < 0 \ ; \end{cases} \end{aligned} \tag{5.1.2}$$

while $\sigma_{ab}^{(12)}(\widehat{\boldsymbol{\Omega}}|\widehat{\boldsymbol{\Omega}}')$ is given by (3.3.129) with $\eta_a = 1$, meaning the cross section per unit area of S_{12}. The component $\boldsymbol{\lambda}$ is hereafter suppressed unless otherwise noted. By the $\boldsymbol{\lambda}$ Fourier inversion of (5.1.1), the three-dimensional optical expression $I_{ab}^{(12)}(\widehat{\boldsymbol{\Omega}},\widehat{\boldsymbol{\rho}}|\widehat{\boldsymbol{\Omega}}',\widehat{\boldsymbol{\rho}}')$ is obtained, which is formally the same as the original (5.1.1), except for the replacement of $U_a(\widehat{\boldsymbol{\Omega}},z)$ with $U_a(\widehat{\boldsymbol{\Omega}},\widehat{\boldsymbol{\rho}})$, as long as the $\boldsymbol{\lambda}$ dependence of $\sigma_{ab}^{(12)}$ is neglected. Expression (3.3.128) is thereby reproduced by the $\widehat{\boldsymbol{\Omega}}$ integration when $\gamma_a = 0$. Expression (5.1.1) should be regarded as the matrix element of $I^{(12)}$, with respect to $\widehat{\boldsymbol{\Omega}}$ and z, to be multiplied by a $\widehat{\boldsymbol{\Omega}}-z$ vector $f(\widehat{\boldsymbol{\Omega}},z)$. The cross section, $\sigma_{ab}^{(12)}$, is subject to the optical relation (3.3.140), which is presently written as

$$\sum_{a=1}^{2}\int_{2\pi} d\widehat{\boldsymbol{\Omega}}'\, k_a\sigma_{ab}^{(12)}(\widehat{\boldsymbol{\Omega}}'|\widehat{\boldsymbol{\Omega}}) = -k_b\Omega_n^{(b)} > 0 \ , \tag{5.1.3}$$

where $\Omega_n^{(b)} = \widehat{\boldsymbol{n}}^{(b)}\cdot\widehat{\boldsymbol{\Omega}}$.

The $\widehat{\boldsymbol{\Omega}}-z$ expression of integral equation (4.1.72) for $\mathcal{I}_{22}^{(q/12+23)}$ becomes, according to (2.3.33–35),

$$\begin{aligned}\mathcal{I}_{22}^{(q/12+23)}(\widehat{\boldsymbol{\Omega}},z|\widehat{\boldsymbol{\Omega}}',z') &= \mathcal{I}_2^{(0q)}(\widehat{\boldsymbol{\Omega}},z|\widehat{\boldsymbol{\Omega}}',z')\\ &+\int d\widehat{\boldsymbol{\Omega}}''\int d\widehat{\boldsymbol{\Omega}}'''\,\mathcal{I}_2^{(0q)}(\widehat{\boldsymbol{\Omega}},z|\widehat{\boldsymbol{\Omega}}'',0)\\ &\times\sigma_{22}^{(12)}(\widehat{\boldsymbol{\Omega}}''|\widehat{\boldsymbol{\Omega}}''')\mathcal{I}_{22}^{(q/12+23)}(\widehat{\boldsymbol{\Omega}}''',0|\widehat{\boldsymbol{\Omega}}',z')\\ &+\int d\widehat{\boldsymbol{\Omega}}''\int d\widehat{\boldsymbol{\Omega}}'''\,\mathcal{I}_2^{(0q)}(\widehat{\boldsymbol{\Omega}},z|\widehat{\boldsymbol{\Omega}}'',-L)\\ &\times\sigma_{22}^{(23)}(\widehat{\boldsymbol{\Omega}}''|\widehat{\boldsymbol{\Omega}}''')\mathcal{I}_{22}^{(q/12+23)}(\widehat{\boldsymbol{\Omega}}''',-L|\widehat{\boldsymbol{\Omega}}',z')\ .\end{aligned} \tag{5.1.4}$$

Here, $\mathcal{I}_2^{(0q)}$ is equivalent to the solution of the transport equation (2.3.46), which is presently written as

$$\begin{aligned}(\gamma_2+\Omega_z\partial_z)\mathcal{I}_2^{(0q)}(\widehat{\boldsymbol{\Omega}},z|\widehat{\boldsymbol{\Omega}}',z') &= K_2^{(q)}(\widehat{\boldsymbol{\Omega}}|\widehat{\boldsymbol{\Omega}}')U_2(\widehat{\boldsymbol{\Omega}}',z-z')\\ &+\int d\widehat{\boldsymbol{\Omega}}''\,K_2^{(q)}(\widehat{\boldsymbol{\Omega}}|\widehat{\boldsymbol{\Omega}}'')\mathcal{I}_2^{(0q)}(\widehat{\boldsymbol{\Omega}}'',z|\widehat{\boldsymbol{\Omega}}',z')\ ,\end{aligned} \tag{5.1.5}$$

wherein $\partial_z = \partial/\partial z$, and the additional term $-i\boldsymbol{\Omega}\cdot\boldsymbol{\lambda}$ in ( ) on the left-hand side has been suppressed. Here the cross section $K_2^{(q)}(\widehat{\boldsymbol{\Omega}}|\widehat{\boldsymbol{\Omega}}')$ is subject to the optical relation (2.3.50),

$$\gamma_2(\widehat{\boldsymbol{\Omega}}) = \int_{4\pi} d\widehat{\boldsymbol{\Omega}}'\, K_2^{(q)}(\widehat{\boldsymbol{\Omega}}'|\widehat{\boldsymbol{\Omega}})\ , \tag{5.1.6}$$

while the solution is subject to the boundary conditions,

$$\mathcal{I}_2^{(0q)}(\widehat{\boldsymbol{\Omega}},z=0|\widehat{\boldsymbol{\Omega}}',z') = 0,\quad \Omega_z < 0\ , \tag{5.1.7a}$$

$$\mathcal{I}_2^{(0q)}(\widehat{\boldsymbol{\Omega}},z=-L|\widehat{\boldsymbol{\Omega}}',z') = 0,\quad \Omega_z > 0\ , \tag{5.1.7b}$$

as a consequence of (5.1.2), and satisfies the reciprocity (Sect. 2.5.1)

$$\mathcal{I}_2^{(0q)}(-\widehat{\Omega}', z'|-\widehat{\Omega}, z) = \mathcal{I}_2^{(0q)}(\widehat{\Omega}, z|\widehat{\Omega}', z') , \tag{5.1.8}$$

$$K_2^{(q)}(-\widehat{\Omega}'|-\widehat{\Omega}) = K_2^{(q)}(\widehat{\Omega}|\widehat{\Omega}') . \tag{5.1.9}$$

With known $\mathcal{I}_2^{(0q)}$, $\mathcal{I}_{22}^{(q/12+23)}$ is obtained as the solution of integral equation (5.1.4), and thereby $I_{11}^{(q+12+23)}$ is obtained from (4.1.70) ($\gamma_2 L \gg 1$), as

$$\begin{aligned} I_{11}^{(q+12+23)}(\widehat{\Omega}, z|\widehat{\Omega}', z') &= I_{11}^{(12)}(\widehat{\Omega}, z|\widehat{\Omega}', z') \\ &+ \int d\widehat{\Omega}'' \int d\widehat{\Omega}''' \, U_1(\widehat{\Omega}, z)\sigma_{12}^{(12)}(\widehat{\Omega}|\widehat{\Omega}'')\mathcal{I}_{22}^{(q/12+23)}(\widehat{\Omega}'', 0|\widehat{\Omega}''', 0) \\ &\times \sigma_{21}^{(12)}(\widehat{\Omega}'''|\widehat{\Omega}')U_1(\widehat{\Omega}', -z') , \end{aligned} \tag{5.1.10}$$

in the same form as the original matrix equation. The same is also true for $I_{31}^{(q+12+23)}$ of (4.1.71), leading to the expression

$$\begin{aligned} I_{31}^{(q+12+23)}(\widehat{\Omega}, z|\widehat{\Omega}', z') &= \int d\widehat{\Omega}'' \int d\widehat{\Omega}''' U_3(\widehat{\Omega}, z+L)\sigma_{32}^{(23)}(\widehat{\Omega}, |\widehat{\Omega}'') \\ &\times \mathcal{I}_{22}^{(q/12+23)}(\widehat{\Omega}'', -L|\widehat{\Omega}''', 0)\sigma_{21}^{(12)}(\widehat{\Omega}'''|\widehat{\Omega}')U_1(\widehat{\Omega}', -z') . \end{aligned} \tag{5.1.11}$$

Thus we find that various expressions of $I^{(q+12+23)}$ which were obtained in terms of the effective boundary scattering matrices, $\sigma^{(12/q)}$ and $\sigma^{(23/q)}$ in Sect. 4.2, also remain unchanged when writing their optical expressions in matrix form. That is, once $\sigma_{21}^{(12/q)}(\widehat{\Omega}|\widehat{\Omega}')$ is obtained in terms of known $\widehat{\Omega}$ matrices, $\mathcal{I}_2^{(0q)}(\widehat{\Omega}, 0|\widehat{\Omega}', 0)$ and $\sigma_{22}^{(12)}(\widehat{\Omega}|\widehat{\Omega}')$, according to the matrix equation (4.2.31a), and $\sigma_{22}^{(12/q)}(\widehat{\Omega}|\widehat{\Omega}')$ and $\sigma_{ab}^{(23/q)}(\widehat{\Omega}|\widehat{\Omega}')$, $a, b = 2, 3$, are also obtained in the same fashion, $\sigma_{11}^{(q+12+23)}$ can be constructed according to the $\widehat{\Omega}$ matrix equation with the same form as (4.2.53), wherein $\sigma_{11}^{(12/q)}$ is given by (4.2.30b).

The three-dimensional expression of (5.1.10) is

$$\begin{aligned} I_{11}^{(q+12+23)}(\widehat{\Omega}, \widehat{\rho}|\widehat{\Omega}', \widehat{\rho}') &= I_{11}^{(12)}(\widehat{\Omega}, \widehat{\rho}|\widehat{\Omega}', \widehat{\rho}') \\ &+ \int d\rho'' \, d\rho''' \int d\widehat{\Omega}'' \, d\widehat{\Omega}''' \, U_1(\widehat{\Omega}, \widehat{\rho} - \rho'')\sigma_{12}^{(12)}(\widehat{\Omega}|\widehat{\Omega}'') \\ &\times \mathcal{I}_{22}^{(q/12+23)}(\widehat{\Omega}'', 0|\rho'' - \rho'''|\widehat{\Omega}''', 0) \\ &\times \sigma_{21}^{(12)}(\widehat{\Omega}'''|\widehat{\Omega}')U_1(\widehat{\Omega}', \rho''' - \widehat{\rho}') . \end{aligned} \tag{5.1.12}$$

Here, $U_a(\widehat{\Omega}, \widehat{\rho})$ is given by (2.3.38),

$$U_a(\widehat{\Omega}, \widehat{\rho}) = |\widehat{\rho}|^{-2} \exp(-\gamma_a|\widehat{\rho}|)\delta^2(\widehat{\Omega} - \widehat{\rho}/|\widehat{\rho}|) . \tag{5.1.13}$$

$\mathcal{I}^{(\;)}(\widehat{\Omega}, z|\rho|\widehat{\Omega}', z')$ is the $\boldsymbol{\lambda}$ Fourier inversion of $\mathcal{I}^{(\;)}(\widehat{\Omega}, z|\boldsymbol{\lambda}|\widehat{\Omega}', z')$, with respect to the Fourier variable $\boldsymbol{\lambda}$ that has been suppressed so far, and $I^{(12)}(\widehat{\Omega}, \widehat{\rho}|\widehat{\Omega}', \widehat{\rho}')$ is reduced to (3.3.128) by the $\widehat{\Omega}$ integration when $\gamma_a = 0$.

The averaged power at a point $\widehat{\rho}$ in space k_a for the wave from a point source at $\widehat{\rho}'$ in space k_b, say $\langle \widehat{w}_{ab}(\widehat{\rho}|\widehat{\rho}')\rangle$, is given, according to the definition (4.1.2a,b) and the optical transformation (2.3.33–35), by

$$\begin{aligned}\langle \widehat{w}_{ab}(\widehat{\rho}|\widehat{\rho}')\rangle &= \mathrm{i}\frac{\partial}{\partial \widehat{\boldsymbol{r}}} I_{ab}(\widehat{\boldsymbol{r}}, \widehat{\rho}|\widehat{\boldsymbol{r}}', \widehat{\rho}')\bigg|_{\widehat{\boldsymbol{r}}=\widehat{\boldsymbol{r}}'=0} \\ &= (4\pi)^{-2}\int d\widehat{\Omega}\, d\widehat{\Omega}'\, k_a \widehat{\Omega}^{(a)} I_{ab}(\widehat{\Omega}, \widehat{\rho}|\widehat{\Omega}', \widehat{\rho}') \ . \qquad (5.1.14)\end{aligned}$$

This equation is expressed in terms of the relative coordinates, $\widehat{\boldsymbol{r}}$ and $\widehat{\boldsymbol{r}}'$ of (2.3.25), and the unit vector, $\widehat{\Omega}^{(a)}$ of $\widehat{\Omega}$, in space k_a, as introduced after (3.3.125).

5.1.1 Case Involving No Boundary Scattering

When the boundaries make no scattering so that $\sigma_{11}^{(12)} = \sigma_{22}^{(12)} = 0$,

$$\sigma_{12}^{(12)}(\widehat{\Omega}|\widehat{\Omega}') = |\Omega_z|\delta_s^2(\widehat{\Omega} - \widehat{\Omega}') \ , \qquad (5.1.15)$$

etc., $\mathcal{I}_{22}^{(q/12+23)}$ is reduced to $\mathcal{I}_2^{(0q)}$, and hence (5.1.12) is reduced to

$$\begin{aligned}I_{11}^{(q+12+23)}(\widehat{\Omega}, \widehat{\rho}|\widehat{\Omega}', \widehat{\rho}') = U_1(\widehat{\Omega}, \widehat{\rho} - \widehat{\rho}')\delta^2(\widehat{\Omega} - \widehat{\Omega}') \\ + \mathcal{I}_2^{(0q)}(\widehat{\Omega}, \widehat{\rho}|\widehat{\Omega}', \widehat{\rho}') \ . \qquad (5.1.16)\end{aligned}$$

Here, the last term is a space continuation of $\mathcal{I}_2^{(0q)}$ (originally defined in the space of $q_2 \neq 0$) to the outside space, k_1, and is obtained, in view of (5.1.15), by replacing $U_2(\widehat{\Omega}, \widehat{\rho} - \widehat{\rho}')$ with

$$U_{12}(\widehat{\Omega}, \widehat{\rho}|\widehat{\rho}') \equiv \int_{-\infty}^{\infty} d\rho''\, |\Omega_z| U_1(\widehat{\Omega}, \widehat{\rho} - \rho'') U_2(\widehat{\Omega}, \rho'' - \widehat{\rho}') \ . \qquad (5.1.17)$$

By using (5.1.2), (5.1.17) can be shown to be the same as that given by (5.1.13), except for the exponential factor which is replaced by $\exp(\gamma_2|\Omega_z|^{-1}z')$, $z' < 0$, yielding a dissipation only in the range of $q_2 \neq 0$. Hence, from (4.1.61a),

$$\begin{aligned}\mathcal{I}_2^{(0q)}(\widehat{\Omega}, \widehat{\rho}|\widehat{\Omega}', \widehat{\rho}') = \int d\widehat{\rho}''\, d\widehat{\rho}'''\, U_{12}(\widehat{\Omega}, \widehat{\rho}|\widehat{\rho}'') \\ \times S_2^{(0q)}(\widehat{\Omega}, \widehat{\rho}''|\widehat{\Omega}', \widehat{\rho}''')U_{21}(\widehat{\Omega}', \widehat{\rho}'''|\widehat{\rho}') \ , \qquad (5.1.18)\end{aligned}$$

where the integration is made over the entire space of the medium $q_2 \neq 0$.

Here, we suppose that in (5.1.18), the points $\hat{\rho}$ and $\hat{\rho}'$ are both separated far enough from the boundary, S_{12}, so that the total incoherent intensity, $\mathcal{I}_2^{(0q)}(\hat{\rho}|\hat{\rho}')$, given by

$$\mathcal{I}_2^{(0q)}(\hat{\rho}|\hat{\rho}') = \int d\hat{\Omega}\, d\hat{\Omega}'\, \mathcal{I}_2^{(0q)}(\hat{\Omega}, \hat{\rho}|\hat{\Omega}', \hat{\rho}') , \tag{5.1.19}$$

can be written in an asymptotic form, using the relative coordinates, $\Delta\rho$ and $\overline{\rho}$, defined by

$$\Delta\rho = \rho'' - \rho''', \qquad \overline{\rho} = \tfrac{1}{2}(\rho'' + \rho''') , \tag{5.1.20a}$$

$$\hat{\rho}'' = (\overline{\rho} + \tfrac{1}{2}\Delta\rho, z''), \qquad \rho''' = (\overline{\rho} - \tfrac{1}{2}\Delta\rho, z''') , \tag{5.1.20b}$$

$d\rho''\, d\rho''' = d(\Delta\rho)\, d\overline{\rho}$; and also using (5.1.13) for $U_2(\hat{\Omega}, \hat{\rho})$, as

$$\mathcal{I}_2^{(0q)}(\hat{\rho}|\hat{\rho}') = \int_{-\infty}^{\infty} d\overline{\rho}\, |\hat{\rho} - \overline{\rho}|^{-2} S_{+2,+2}^{(0q)}(\hat{\Omega}|\overline{\rho}|\hat{\Omega}')|\overline{\rho} - \hat{\rho}'|^{-2} . \tag{5.1.21}$$

Here, $\hat{\Omega}$ and $\hat{\Omega}'$ are in the directions of $\hat{\rho} - \overline{\rho}$ and $\overline{\rho} - \hat{\rho}'$, respectively, and $S_{+2,+2}^{(0q)}(\hat{\Omega}|\overline{\rho}|\hat{\Omega}')$ means the scattering cross section per unit area of the layer boundary at $\overline{\rho}$, for scattering of the wave in direction $\hat{\Omega}'$ to $\hat{\Omega}$, and, using

$$\int_{-\infty}^{\infty} d(\Delta\rho)\, S_2^{(0q)}(\hat{\Omega}, \hat{\rho}''|\hat{\Omega}', \hat{\rho}''') = S_2^{(0q)}(\hat{\Omega}, z''|\boldsymbol{\lambda} = 0|\hat{\Omega}', z''') ,$$

is given by

$$S_{+2,+2}^{(0q)}(\hat{\Omega}|\overline{\rho}|\hat{\Omega}') = \widetilde{S}_{+2,+2}^{(0q)}(\hat{\Omega}, \lambda_z|\boldsymbol{\lambda} = 0|\hat{\Omega}', \lambda_z'). . \tag{5.1.22}$$

The right-hand side is a Fourier transform, defined by

$$\widetilde{S}_{+2,+2}^{(0q)}(\hat{\Omega}, \lambda_z|\boldsymbol{\lambda}|\hat{\Omega}', \lambda_z')$$
$$= \int_{-L}^{0} dz\, dz'\, \exp[\mathrm{i}(\lambda_z z - \lambda_z' z')]\, S_2^{(0q)}(\hat{\Omega}, z|\boldsymbol{\lambda}|\hat{\Omega}', z') , \tag{5.1.23}$$

with

$$\lambda_z = -\mathrm{i}(\gamma_2 - \mathrm{i}\boldsymbol{\Omega}\cdot\boldsymbol{\lambda})\Omega_z^{-1}, \qquad \Omega_z > 0 , \tag{5.1.24a}$$

$$\lambda_z' = -\mathrm{i}(\gamma_2 - i\boldsymbol{\Omega}'\cdot\boldsymbol{\lambda})(\Omega_z')^{-1}, \quad \Omega_z' < 0 ; \tag{5.1.24b}$$

and is function of $\hat{\Omega}$, $\hat{\Omega}'$, and $\boldsymbol{\lambda}$.

Equations (5.1.24a,b) mean that when $\boldsymbol{\lambda} = 0$, $\widehat{\boldsymbol{\lambda}}_j$, $j = 1, 2, 3, 4$, of (2.3.26b) are assigned by

$$\widehat{\boldsymbol{\lambda}}_1 = k_2\widehat{\boldsymbol{\Omega}} + 2^{-1}\mathrm{i}\gamma_2(\Omega_z)^{-1}\widehat{\boldsymbol{n}}_z \ , \qquad (5.1.25a)$$

$$\widehat{\boldsymbol{\lambda}}_2 = k_2\widehat{\boldsymbol{\Omega}} - 2^{-1}\mathrm{i}\gamma_2(\Omega_z)^{-1}\widehat{\boldsymbol{n}}_z \ , \qquad (5.1.25b)$$

$$\widehat{\boldsymbol{\lambda}}_1' \equiv \widehat{\boldsymbol{\lambda}}_3 = k_2\widehat{\boldsymbol{\Omega}}' + 2^{-1}\mathrm{i}\gamma_2(\Omega_z')^{-1}\widehat{\boldsymbol{n}}_z \ , \qquad (5.1.25c)$$

$$\widehat{\boldsymbol{\lambda}}_2' \equiv \widehat{\boldsymbol{\lambda}}_4 = k_2\widehat{\boldsymbol{\Omega}}' - 2^{-1}\mathrm{i}\gamma_2(\Omega_z')^{-1}\widehat{\boldsymbol{n}}_z \ , \qquad (5.1.25d)$$

and with these $\widehat{\boldsymbol{\lambda}}_j$'s, (5.1.22,23) state that the cross section, $S_{+2,+2}^{(0q)}(\widehat{\boldsymbol{\Omega}}|\widehat{\boldsymbol{\Omega}}') \equiv S_{+2,+2}^{(0q)}(\widehat{\boldsymbol{\Omega}}|\overline{\boldsymbol{\rho}}|\widehat{\boldsymbol{\Omega}}')$, is given by the full Fourier transform $\widetilde{S}_2^{(0q)}(\widehat{\boldsymbol{\lambda}}_1;\widehat{\boldsymbol{\lambda}}_2|\widehat{\boldsymbol{\lambda}}_1';\widehat{\boldsymbol{\lambda}}_2')$, except for the factor $(2\pi)^2\delta(\boldsymbol{\lambda}-\boldsymbol{\lambda}')$.

5.1.2 Case Involving Boundary Scattering

From (5.1.12), the optical expression of $I_{11}^{(q+12+23)}$ can be written in terms of the cross section, $\sigma_{11}^{(q+12+23)}$, as

$$\begin{aligned} I_{11}^{(q+12+23)}(\widehat{\boldsymbol{\Omega}},\widehat{\boldsymbol{\rho}}|\widehat{\boldsymbol{\Omega}}',\widehat{\boldsymbol{\rho}}') &= U_1(\widehat{\boldsymbol{\Omega}},\widehat{\boldsymbol{\rho}}-\widehat{\boldsymbol{\rho}}')\delta^2(\widehat{\boldsymbol{\Omega}}-\widehat{\boldsymbol{\Omega}}') \\ &+ \int d\boldsymbol{\rho}''\,d\boldsymbol{\rho}'''\,U_1(\widehat{\boldsymbol{\Omega}},\widehat{\boldsymbol{\rho}}-\boldsymbol{\rho}'')\sigma_{11}^{(q+12+23)}(\widehat{\boldsymbol{\Omega}}|\boldsymbol{\rho}''-\boldsymbol{\rho}'''|\widehat{\boldsymbol{\Omega}}') \\ &\times U_1(\widehat{\boldsymbol{\Omega}}',\boldsymbol{\rho}'''-\widehat{\boldsymbol{\rho}}') \ . \end{aligned} \qquad (5.1.26)$$

Here, the second term is a surface integral on S_1, with

$$\begin{aligned} \sigma_{11}^{(q+12+23)}(\widehat{\boldsymbol{\Omega}}|\boldsymbol{\rho}''-\boldsymbol{\rho}'''|\widehat{\boldsymbol{\Omega}}') &= \sigma_{11}^{(12)}(\widehat{\boldsymbol{\Omega}}|\widehat{\boldsymbol{\Omega}}')\delta(\boldsymbol{\rho}''-\boldsymbol{\rho}''') \\ &+ \int d\widehat{\boldsymbol{\Omega}}''\,d\widehat{\boldsymbol{\Omega}}'''\,\sigma_{12}^{(12)}(\widehat{\boldsymbol{\Omega}}|\widehat{\boldsymbol{\Omega}}'') \\ &\times \mathcal{I}_{22}^{(q/12+23)}(\widehat{\boldsymbol{\Omega}}'',z''=0|\boldsymbol{\rho}''-\boldsymbol{\rho}'''|\widehat{\boldsymbol{\Omega}}''',z'''=0) \\ &\times \sigma_{21}^{(12)}(\widehat{\boldsymbol{\Omega}}'''|\widehat{\boldsymbol{\Omega}}') \ . \end{aligned} \qquad (5.1.27)$$

Hence, when the points $\widehat{\boldsymbol{\rho}}$ and $\widehat{\boldsymbol{\rho}}'$ are both separated far enough from the boundary, the total intensity from (5.1.26), say $I_{11}^{(q+12+23)}(\widehat{\boldsymbol{\rho}}|\widehat{\boldsymbol{\rho}}')$, is obtained in an asymptotic form similar to (5.1.21):

$$\begin{aligned} I_{11}^{(q+12+23)}(\widehat{\boldsymbol{\rho}}|\widehat{\boldsymbol{\rho}}') &= |\widehat{\boldsymbol{\rho}}-\widehat{\boldsymbol{\rho}}'|^{-2} \\ &+ \int d\overline{\boldsymbol{\rho}}\,|\widehat{\boldsymbol{\rho}}-\overline{\boldsymbol{\rho}}|^{-2}\sigma_{11}^{(q+12+23)}(\widehat{\boldsymbol{\Omega}}|\widehat{\boldsymbol{\Omega}}')|\overline{\boldsymbol{\rho}}-\widehat{\boldsymbol{\rho}}'|^{-2} \ . \end{aligned} \qquad (5.1.28)$$

Here, $\sigma_{11}^{(q+12+23)}(\widehat{\boldsymbol{\Omega}}|\widehat{\boldsymbol{\Omega}}')$ means the resultant cross section per unit area of the boundary at $\overline{\boldsymbol{\rho}}$ due to the scattering by both the medium and the boundaries, and is given by

$$\sigma_{11}^{(q+12+23)}(\widehat{\Omega}|\widehat{\Omega}') \equiv \int_{-\infty}^{\infty} d(\Delta\rho)\, \sigma_{11}^{(q+12+23)}(\widehat{\Omega}|\Delta\rho|\widehat{\Omega}') \tag{5.1.29a}$$

$$= \sigma_{11}^{(12)}(\widehat{\Omega}|\widehat{\Omega}') + \int d\widehat{\Omega}''\, d\widehat{\Omega}'''\, \sigma_{12}^{(12)}(\widehat{\Omega}|\widehat{\Omega}'')|\Omega_z''|^{-1} \times S_{+2,+2}^{(q/12+23)}(\widehat{\Omega}''|\widehat{\Omega}''')|\Omega_z'''|^{-1}\sigma_{21}^{(12)}(\widehat{\Omega}'''|\widehat{\Omega}') \ , \tag{5.1.29b}$$

where, in the last term, use has been made of

$$\mathcal{I}_{22}^{(q/12+23)}(\widehat{\Omega}'', z''=0|\widehat{\Omega}''', z'''=0) = |\Omega_z''\Omega_z'''|^{-1} S_{+2,+2}^{(q/12+23)}(\widehat{\Omega}''|\widehat{\Omega}''') \ , \tag{5.1.30a}$$

with

$$S_{+2,+2}^{(q/12+23)}(\widehat{\Omega}''|\widehat{\Omega}''') \equiv \widetilde{S}_{+2,+2}^{(q/12+23)}(\widehat{\Omega}''\lambda_z''|\boldsymbol{\lambda}=0|\widehat{\Omega}''', \lambda_z''') \ , \tag{5.1.30b}$$

which is a Fourier transform defined by

$$\widetilde{S}_{+2,+2}^{(q/12+23)}(\widehat{\Omega}, \lambda_z|\boldsymbol{\lambda}|\widehat{\Omega}', \lambda_z') = \int_{-L}^{0} dz\, dz'\, \exp\left[i(\lambda_z z - \lambda_z' z')\right] S_{22}^{(q/12+23)}(\widehat{\Omega}, z|\boldsymbol{\lambda}|\widehat{\Omega}', z') \ , \tag{5.1.30c}$$

similar to $\widetilde{S}_{+2,+2}^{(0q)}$ of (5.1.23). These equations are obtained from an expression of $\mathcal{I}^{(q/12+23)}$, similar to (4.1.59), using (5.1.2), with the same procedure that led to (5.1.22–24).

Also, for the transmitted wave, we similarly obtain the asymptotic expression,

$$I_{31}^{(q+12+23)}(\widehat{\rho}|\widehat{\rho}') = \int d\overline{\rho}\, |\widehat{\rho} - \overline{\rho}|^{-2} \sigma_{31}^{(q+12+23)}(\widehat{\Omega}|\widehat{\Omega}')|\overline{\rho} - \widehat{\rho}'|^{-2} \ , \tag{5.1.31a}$$

with

$$\sigma_{31}^{(q+12+23)}(\widehat{\Omega}|\widehat{\Omega}') = \int d\widehat{\Omega}''\, d\widehat{\Omega}'''\, \sigma_{32}^{(23)}(\widehat{\Omega}|\widehat{\Omega}'')|\Omega_z''|^{-1} \times S_{-2,+2}^{(q/12+23)}(\widehat{\Omega}''|\widehat{\Omega}''')|\Omega_z'''|^{-1}\sigma_{21}^{(12)}(\widehat{\Omega}'''|\widehat{\Omega}') \ . \tag{5.1.31b}$$

Here, Ω_z'', $\Omega_z''' < 0$, and

$$S_{-2,+2}^{(q/12+23)}(\widehat{\Omega}|\widehat{\Omega}') = \int_{-L}^{0} dz\, dz' \exp\{i[\lambda_z(z+L) - \lambda_z' z']\} \times S_{22}^{(q/12+23)}(\widehat{\Omega}, z|\boldsymbol{\lambda}|\widehat{\Omega}', z') \ , \tag{5.1.32}$$

where $\boldsymbol{\lambda} = 0$, and λ_z and λ_z' are defined by (5.1.24), except that $\Omega_z < 0$.

Summarizing, the scattered waves are described by the asymptotic expressions (5.1.28,29) and (5.1.31a,b) for the reflected and transmitted waves, respectively, in terms of the cross sections per unit area of the boundary surface, $\sigma_{ab}^{(q+12+23)}$, $a, b = 1, 3$, which are composed of the entire contributions from both the medium and the boundaries. Here the medium scattering is manifested only through the $\widehat{\Omega}$ matrices, $S_{ab}^{(q/12+23)}$, $a, b = \pm 2$, and the resulting cross sections can be constructed according to (5.1.29b,31b), by the successive $\widehat{\Omega}$ matrix multiplication of the boundary and medium cross section matrices, following the order of the scatterings, and with a weighting function, $|\Omega_z|^{-1}$, when making the $\widehat{\Omega}$ integration.

5.1.3 Optical Relations and Reciprocity

The optical relation (5.1.3) for $\sigma^{(12)}$ is reproduced by writing (4.2.105) in optical form; i.e. by using (5.1.2), $\langle s_1|U_1$ is expressed by

$$\int_{\Omega_z \gtrless 0} d\widehat{\Omega}\, k_1 \Omega_z U_1(\widehat{\Omega}, z = \pm 0)\delta^2(\widehat{\Omega} - \widehat{\Omega}')\bigg|_{\lambda=0} = \pm k_1 \ , \qquad (5.1.33)$$

and (4.2.105a) leads directly to

$$\int_{\Omega_z > 0} d\widehat{\Omega}\, k_1 \sigma_{11}^{(12)}(\widehat{\Omega}|\widehat{\Omega}') + \int_{\Omega_z < 0} d\widehat{\Omega}\, k_2 \sigma_{21}^{(12)}(\widehat{\Omega}|\widehat{\Omega}') = k_1 |\Omega_z'| \ . \qquad (5.1.34)$$

In the same way, (4.2.113) leads to the optical relation for $\sigma_{ab}^{(q+12+23)}$:

$$\int d\widehat{\Omega}\, k_1 \sigma_{11}^{(q+12+23)}(\widehat{\Omega}|\widehat{\Omega}') + \int d\widehat{\Omega}\, k_3 \sigma_{31}^{(q+12+23)}(\widehat{\Omega}|\widehat{\Omega}') = k_1 |\Omega_z'| \ . \qquad (5.1.35)$$

A more detailed version of the relation is obtained from (4.2.111), which is optically expressed as

$$\int d\widehat{\Omega}\, d\widehat{\Omega}''\, k_1 \sigma_{12}^{(12)}(\widehat{\Omega}|\widehat{\Omega}'')|\Omega_z''|^{-1} S_{+2,+2}^{(q/12+23)}(\widehat{\Omega}''|\widehat{\Omega}') + \int d\widehat{\Omega}\, d\widehat{\Omega}''\, k_3 \sigma_{32}^{(23)}(\widehat{\Omega}|\widehat{\Omega}'')|\Omega_z''|^{-1} \times S_{-2,+2}^{(q/12+23)}(\widehat{\Omega}''|\widehat{\Omega}') = k_2 |\Omega_z^{(2)\prime}|^{-1} \ , \qquad (5.1.36)$$

in terms of the medium cross sections, $S_{\pm 2,+2}^{(q/12+23)}(\widehat{\Omega}|\widehat{\Omega}')$, defined by (5.1.30b,c, 32); hence relation (5.1.35) is reproduced as a result of (5.1.29b, 31b), and optical relation (5.1.34). As for $S_{\pm 2,+2}^{(0q)}(\widehat{\Omega}|\widehat{\Omega}')$, where no boundary scattering is involved, we obtain

$$\int d\widehat{\Omega}\, S^{(0q)}_{+2,+2}(\widehat{\Omega}|\widehat{\Omega}') + \int d\widehat{\Omega}\, S^{(0q)}_{-2,+2}(\widehat{\Omega}|\widehat{\Omega}') = |\Omega'_z| \qquad (5.1.37)$$

as a special case of (5.1.36).

Here the reciprocity relations,

$$S^{(q/12+23)}_{\pm 2,+2}(\widehat{\Omega}|\widehat{\Omega}') = S^{(q/12+23)}_{+2,\pm 2}(-\widehat{\Omega}'|-\widehat{\Omega}) \ , \qquad (5.1.38)$$

are valid as a consequence of the reciprocities,

$$\begin{aligned} \sigma^{(\ )}_{ab}(\widehat{\Omega}|\widehat{\Omega}') &= \sigma^{(\ )}_{ba}(-\widehat{\Omega}'|-\widehat{\Omega}) \ , \\ K^{(q)}_{2}(\widehat{\Omega}|\widehat{\Omega}') &= K^{(q)}_{2}(-\widehat{\Omega}'|-\widehat{\Omega}) \ , \end{aligned} \qquad (5.1.39a)$$

$$S^{(0q)}_{2}(\widehat{\Omega}, z|\boldsymbol{\lambda}|\widehat{\Omega}', z') = S^{(0q)}_{2}(-\widehat{\Omega}', z'|-\boldsymbol{\lambda}|-\widehat{\Omega}, z) \ , \qquad (5.1.39b)$$

and the definitions (5.1.22–24); the reciprocity of the cross section, $\sigma^{q+12+23}_{ab}(\widehat{\Omega}|\widehat{\Omega}')$, for the entire system is thereby ensured.

5.2 Application of the Diffusion Approximation

When the source and the observer are both separated far enough from the layer boundaries, $I^{(q+12+23)}_{11}$ and $I^{(q+12+23)}_{31}$ are given by the asymptotic expressions (5.1.28,29) and (5.1.31), respectively, in terms of the cross section per unit area of the layer surfaces, $\sigma^{(q+12+23)}_{11}$ and $\sigma^{(q+12+23)}_{31}$, of the form ($\gamma_2 L \gg 1$)

$$\sigma^{(q+12+23)}_{11}(\widehat{\Omega}|\widehat{\Omega}') = \sigma^{(12)}_{11}(\widehat{\Omega}|\widehat{\Omega}') + \sigma^{(q/12+23)}_{11}(\widehat{\Omega}|\widehat{\Omega}') \ , \qquad (5.2.1a)$$

$$\sigma^{(q+12+23)}_{31}(\widehat{\Omega}|\widehat{\Omega}') = \sigma^{(q/12+23)}_{31}(\widehat{\Omega}|\widehat{\Omega}') \ , \qquad (5.2.1b)$$

Here, with (5.1.30a),

$$\begin{aligned} \sigma^{(q/12+23)}_{11}(\widehat{\Omega}|\widehat{\Omega}') = \int d\widehat{\Omega}'' \int d\widehat{\Omega}'''\, \sigma^{(12)}_{12}(\widehat{\Omega}|\widehat{\Omega}'') \\ \times\, \mathcal{I}^{(q/12+23)}_{22}(\widehat{\Omega}'', z''=0|\widehat{\Omega}''', z'''=0)\sigma^{(12)}_{21}(\widehat{\Omega}'''|\widehat{\Omega}') \ , \end{aligned} \qquad (5.2.2a)$$

being equivalent to the second term of (5.1.10) when $\boldsymbol{\lambda} = 0$, and, similarly,

$$\begin{aligned} \sigma^{(q+12+23)}_{31}(\widehat{\Omega}|\widehat{\Omega}') = \int d\widehat{\Omega}'' \int d\widehat{\Omega}'''\, \sigma^{(23)}_{32}(\widehat{\Omega}|\widehat{\Omega}'') \\ \times\, \mathcal{I}^{(q/12+23)}_{22}(\widehat{\Omega}'', z''=-L|\widehat{\Omega}''', z'''=0)\sigma^{(12)}_{21}(\widehat{\Omega}'''|\widehat{\Omega}') \ ; \end{aligned} \qquad (5.2.2b)$$

both can be regarded as additional boundary cross sections caused by the medium fluctuation, tending to zero as $q_2 \to 0$. Here, $\mathcal{I}_{22}^{(q/12+23)}$ is the solution of (4.1.72) whose optical expression is

$$
\begin{aligned}
\mathcal{I}_{22}^{(q/12+23)}(\widehat{\Omega}, z|\widehat{\Omega}', z') &= \mathcal{I}_{2}^{(0q)}(\widehat{\Omega}, z|\widehat{\Omega}', z') \\
&+ \int d\widehat{\Omega}''\, d\widehat{\Omega}'''\, \mathcal{I}_{2}^{(0q)}(\widehat{\Omega}, z|\widehat{\Omega}'', z''=0) \\
&\times \sigma_{22}^{(12)}(\widehat{\Omega}''|\widehat{\Omega}''')\mathcal{I}_{22}^{(q/12+23)}(\widehat{\Omega}''', 0|\widehat{\Omega}', z') \\
&+ \int d\widehat{\Omega}''\, d\widehat{\Omega}'''\, \mathcal{I}_{2}^{(0q)}(\widehat{\Omega}, z|\widehat{\Omega}'', -L) \\
&\times \sigma_{22}^{(23)}(\widehat{\Omega}''|\widehat{\Omega}''')\mathcal{I}_{22}^{(q/12+23)}(\widehat{\Omega}''', -L|\widehat{\Omega}', z') ,
\end{aligned}
\tag{5.2.3}
$$

and can also be obtained as a boundary-value solution of the diffusion equation when the boundaries are separated far enough so that $\gamma_2 L \gg 1$ [5.1].

In (5.2.3), $\mathcal{I}_{2}^{(0q)}$ can be expanded, according to (2.5.9,10), in an eigenfunction series of the form

$$
\mathcal{I}_{2}^{(0q)}(\widehat{\Omega}, z|\widehat{\Omega}', z') = \sum_{A} \phi_A(\widehat{\Omega}, \mathrm{i}\overrightarrow{\partial}_z) S_A^{(0q)}(z|z') \overline{\phi}_A(\widehat{\Omega}', -\mathrm{i}\overleftarrow{\partial}_z') \ . \tag{5.2.4}
$$

Here, $S_A^{(0q)}(z|z')$ is the solution of (2.5.6). Particularly for the diffusion term,

$$
\gamma_2^{-1}(\gamma^{(ab)} - D_2\partial_z^2) S_A^{(0q)}(z|z') = \delta(z - z') \tag{5.2.5}
$$

from (2.5.52) with $D \to D_2$, with the eigenfunctions defined by

$$
\begin{aligned}
\gamma_2 \phi_A(\widehat{\Omega}, \mathrm{i}\partial_z) &= 4\pi \overline{\phi}_A(\widehat{\Omega}, i\partial_z) \\
&= 1 - 3D_2 \Omega_z \partial_z
\end{aligned}
\tag{5.2.6}
$$

from (2.5.51) when $K(\widehat{\Omega}|\widehat{\Omega}') = K(\widehat{\Omega} \cdot \widehat{\Omega}')$.

The corresponding eigenfunction series for $\mathcal{I}_{22}^{(q/12+23)}$ is generally written, in view of (5.2.3), in the form

$$
\begin{aligned}
\mathcal{I}_{22}^{(q/12+23)}(\widehat{\Omega}, z|\widehat{\Omega}', z') &= \sum_{A,B} \phi_A(\widehat{\Omega}, \mathrm{i}\partial_z) \\
&\times S_{AB}^{(q/12+23)}(z|z') \overline{\phi}_B(\widehat{\Omega}', -\mathrm{i}\overleftarrow{\partial}_z') && (5.2.7) \\
&\sim \phi_A(\widehat{\Omega}, \mathrm{i}\partial_z) S_A^{(q/12+23)}(z|z') \overline{\phi}_A(\widehat{\Omega}', -\mathrm{i}\overleftarrow{\partial}_z') , && (5.2.8)
\end{aligned}
$$

where the last expression is an approximation obtained by using only the diffusion term, and is therefore valid only when the points z and z' are both separated far enough from the boundaries. To the same approximation, (5.2.3) is reduced to an integral equation for $S_A^{(q/12+23)}$, as

$$S_A^{(q/12+23)}(z|z') = S_A^{(0q)}(z|z')$$
$$+ S_A^{(0q)}(z|z'(=0))\langle\sigma_{22}^{(12)}(-\overleftarrow{\partial}'_z|\overrightarrow{\partial}_z)\rangle$$
$$\times S_A^{(q/12+23)}(z(=0)|z')$$
$$+ S_A^{(0q)}(z|z'(=-L))\langle\sigma_{22}^{(23)}(-\overleftarrow{\partial}'_z|\overrightarrow{\partial}_z)\rangle$$
$$\times S_A^{(q/12+23)}(z(=-L)|z') \ , \qquad (5.2.9)$$

in terms of the notation

$$\langle\sigma_{22}^{(12)}(-\overleftarrow{\partial}_z|\overrightarrow{\partial}_z)\rangle = \int d\widehat{\Omega} \int d\widehat{\Omega}' \, \phi_A(\widehat{\Omega}, -\mathrm{i}\overleftarrow{\partial}_z)$$
$$\times \sigma_{22}^{(12)}(\widehat{\Omega}|\widehat{\Omega}')\phi_A(\widehat{\Omega}', \mathrm{i}\overrightarrow{\partial}_z) \ . \qquad (5.2.10)$$

5.2.1 Boundary Condition of the Diffusion Equation

Equation (5.2.9) indicates that $S_A^{(q/12+23)}$ is a solution of the diffusion equation (5.2.5) except on the boundaries. Here, to find the boundary conditions, we investigate the (power) equation of continuity for $\mathcal{I}_{22}^{(q/12+23)}$ by first studying the equation for $\mathcal{I}_2^{(0q)}$ which upon the $\widehat{\Omega}$ integration of the transport equation (5.1.5), is given by

$$\partial_z \int_{4\pi} d\widehat{\Omega}\, \Omega_z \mathcal{I}_2^{(0q)}(\widehat{\Omega}, z|\widehat{\Omega}', z') = \gamma_2 U_2(\widehat{\Omega}', z - z') \ , \qquad (5.2.11)$$

as a consequence of the optical relation (5.1.6). Here, the power flux of $\mathcal{I}_2^{(0q)}$ in the z direction, say $\langle w_z^{(0q)}\rangle$, can be written as a sum of the two components $\langle w_z^{(0q)}\rangle^{\pm}$ propagating in the positive and negative directions, respectively, and given by

$$\langle w_z^{(0q)}(z|\widehat{\Omega}', z')\rangle^{\pm} \equiv \int_{\Omega_z \gtrless 0} d\widehat{\Omega}\, \Omega_z \mathcal{I}_2^{(0q)}(\widehat{\Omega}, z|\widehat{\Omega}', z') \ , \qquad (5.2.12)$$

except for a numerical factor. Hence,

$$\langle w_z^{(0q)}\rangle = \langle w_z^{(0q)}\rangle^{+} + \langle w_z^{(0q)}\rangle^{-} \ , \qquad (5.2.13)$$

and (5.2.11) is written as

$$\partial_z \langle w_z^{(0q)}(z|\widehat{\Omega}', z')\rangle = \gamma_2 U_2(\widehat{\Omega}', z - z') \ . \qquad (5.2.14)$$

Here we introduce a distance l subject to $\gamma_2 l \gg 1$, and integrate both sides of (5.2.14) over the range $0 \geq z \geq -l$ to obtain

$$\langle w_z^{(0q)}(z|\widehat{\Omega}', z')\rangle \Big|_{z=-l}^{0} = \int_{-l}^{0} dz\, \gamma_2 U_2(\widehat{\Omega}', z - z')$$
$$\simeq \begin{cases} 1, & \Omega_z' < 0, \quad z' = 0 \\ 0, & \Omega_z' \gtrless 0, \quad \gamma_2(l + z') \ll -1 \end{cases}, \tag{5.2.15}$$

where use been made of (5.1.2). Here, on the left-hand side, the part $\langle w_z^{(0q)}\rangle^-$ is zero at $z = 0$ in view of (5.1.7a), while, for the part $\langle w_z^{(0q)}\rangle^+$, we assume that the length $l \gg \gamma_2^{-1}$ can be minimized so that within the region $0 \geq z \geq -l$, $\langle w_z^{(0q)}\rangle^+$ (propagating toward the boundary S_{12}) remains almost unchanged; this is a severe condition, not quite realized, though. Thus, on the left-hand side of (5.2.15), $\langle w_z^{(0q)}\rangle^+$ makes no contribution, resulting in

$$-\int_{\Omega_z<0} d\widehat{\Omega}\, \Omega_z \mathcal{I}_2^{(0q)}(\widehat{\Omega}, z = -l|\widehat{\Omega}', z')$$
$$\simeq \begin{cases} 1, & \Omega_z' < 0, \quad z' = 0 & \text{(5.2.16a)} \\ 0, & \Omega_z' \gtrless 0, \quad \gamma_2(l + z') \ll -1\ . & \text{(5.2.16b)} \end{cases}$$

It is now straightforward to find the corresponding equation for the boundary dependent $\mathcal{I}_{22}^{(q/12+23)}$, by applying the above relations to the governing equation (5.2.3), hence,

$$-\int_{\Omega_z<0} d\widehat{\Omega}\, \Omega_z \mathcal{I}_{22}^{(q/12+23)}(\widehat{\Omega}, z = -l|\widehat{\Omega}', z')$$
$$\simeq \int_{\Omega_z<0} d\widehat{\Omega} \int_{\Omega_z'>0} d\widehat{\Omega}''\, \sigma_{22}^{(12)}(\widehat{\Omega}|\widehat{\Omega}'')$$
$$\times\, \mathcal{I}_{22}^{(q/12+23)}(\widehat{\Omega}'', z'' = 0|\widehat{\Omega}', z')\ , \tag{5.2.17}$$

where $\gamma_2(l + z') \ll -1$, $\Omega_z' \gtrless 0$, and the $\sigma_{22}^{(23)}$ term does not make a contribution, in view of (5.2.16b). Equation (5.2.17) simply says that the total backscattered power by the boundary S_{12} is transported without getting any change to an imaginary plane at $z = -l$, assumed in the diffusion region. Here we observe that the left-hand side of (5.2.17) can therefore be approximated by the diffusion term (5.2.8), and also that

$$S_A^{(q/12+23)}(z = -l|z') \simeq S_A^{(q/12+23)}(z = 0|z')\ , \tag{5.2.18}$$

being a slowly changing function of z; while, for the right-hand side, the $\widehat{\Omega}''$ integration makes the contribution from the diffusion term dominant, in view of that the $\widehat{\Omega}$ integrated $\sigma_{22}^{(12)}(\widehat{\Omega}|\widehat{\Omega}'')$ provides a slowly changing factor with respect to $\widehat{\Omega}''$ [refer to (5.1.34)].

Thus upon substitution of the expression (5.2.8) for $\mathcal{I}_{22}^{(q/12+23)}$, (5.2.17) is reduced to a boundary equation of the form

$$(2\gamma_2)^{-1}(\tfrac{1}{2}+D_2\partial_z)S_A^{(q/12+23)}(z(=0)|z') \\ = \langle\sigma_{22}^{(12)}(\partial_z)\rangle S_A^{(q/12+23)}(z(=0)|z') \ . \quad (5.2.19)$$

Here, with (5.2.6), use has been made of

$$(4\pi)^{-1}\int_{\Omega_z \gtrless 0} d\hat{\Omega}\,\Omega_z\phi_A(\hat{\Omega},\mathrm{i}\partial_z) = (2\gamma_2)^{-1}(\pm\tfrac{1}{2}-D_2\partial_z) \ , \quad (5.2.20)$$

and

$$\langle\sigma_{ab}^{(12)}(\vec{\partial}_z)\rangle = \langle\sigma_{ab}^{(12)}(-\overleftarrow{\partial}_z(=0)|\vec{\partial}_z)\rangle \quad (5.2.21a)$$

$$= (4\pi)^{-1}\int d\hat{\Omega}\int d\hat{\Omega}'\,\sigma_{ab}^{(12)}(\hat{\Omega}|\hat{\Omega}') \\ \times\phi_A(\hat{\Omega}',\mathrm{i}\vec{\partial}_z) \quad (5.2.21b)$$

$$= (2\gamma_2)^{-1}\left[\langle\sigma_{ab}^{(12)}\rangle_0 - \langle\sigma_{ab}^{(12)}\rangle_1 D_2\vec{\partial}_z\right] \ , \quad (5.2.21c)$$

in terms of the notations

$$\langle\sigma_{ab}^{(12)}\rangle_0 = \frac{1}{2\pi}\int d\hat{\Omega}\,d\hat{\Omega}'\,\sigma_{ab}^{(12)}(\hat{\Omega}|\hat{\Omega}') \ , \quad (5.2.22a)$$

$$\langle\sigma_{ab}^{(12)}\rangle_1 = \frac{3}{2\pi}\int d\hat{\Omega}\,d\hat{\Omega}'\,\sigma_{ab}^{(12)}(\hat{\Omega}|\hat{\Omega}')|\Omega_z'| \ . \quad (5.2.22b)$$

The boundary equation (5.2.19) can finally be written, on substituting (5.2.21c), in the form

$$-D_2\partial_z S_A^{(q/12+23)}\big|_{z=0} = Z^{(12)} S_A^{(q/12+23)}\big|_{z=0} \ , \quad (5.2.23)$$

with

$$Z^{(12)} = \left(\tfrac{1}{2}-\langle\sigma_{22}^{(12)}\rangle_0\right)/\left(1+\langle\sigma_{22}^{(12)}\rangle_1\right) \ . \quad (5.2.24)$$

Here, with (5.2.21b,20), optical relation (5.1.3,3.3.140) for $\sigma_{ab}^{(12)}$ leads to the relation

$$\sum_{a=1}^{2} k_a\langle\sigma_{a2}^{(12)}(\partial_z)\rangle = k_2(2\gamma_2)^{-1}(\tfrac{1}{2}-D_2\partial_z) \ . \quad (5.2.25)$$

Hence the boundary equation (5.2.19) can also be written as

$$\left[D_2\partial_z + (k_1/k_2)\gamma_2\langle\sigma_{12}^{(12)}(\partial_z)\rangle\right]S_A^{(q/12+23)}(z(=0)|z') = 0 \ , \quad (5.2.26)$$

in terms of $\langle\sigma_{12}^{(12)}(\partial_z)\rangle$, leading to the boundary equation (5.2.23) with $Z^{(12)}$ given as the root of

$$Z^{(12)} = (k_1/k_2)\gamma_2\langle\sigma_{12}^{(12)}(-D_2^{-1}Z^{(12)})\rangle \ , \tag{5.2.27}$$

which, by using (5.2.21c), gives

$$Z^{(12)} = \left[2(k_2/k_1) - \langle\sigma_{12}^{(12)}\rangle_1\right]^{-1}\langle\sigma_{12}^{(12)}\rangle_0 \ , \tag{5.2.28}$$

in terms of $\sigma_{12}^{(12)}$. Here, when the boundary is free from reflection, $\langle\sigma_{12}^{(12)}\rangle_1 = 1$ and $\langle\sigma_{12}^{(12)}\rangle_0 = \frac{1}{2}$, by (5.1.15), giving $Z^{(12)} = \frac{1}{2}$, which is consistent with the result from the original (5.2.24).

Another boundary equation at S_{23} is given in a similar way by

$$+D_2\partial_z S_A^{(q/12+23)}\Big|_{z=-L} = Z^{(23)} S_A^{(q/12+23)}\Big|_{z=-L} \ , \tag{5.2.29}$$

where $Z^{(23)}$ is determined by $\sigma^{(23)}$ as $Z^{(12)}$ is determined by $\sigma^{(12)}$, according to (5.2.24) or (5.2.28).

For later convenience, we note here that (5.2.27) and a similar equation for $Z^{(23)}$ can be written, using (5.2.6), as

$$Z^{(12)} = (4\pi)^{-1}(k_1/k_2)\int d\hat{\Omega}\,\sigma_{12}^{(12)}(\hat{\Omega}|Z^{(12)}) \ , \tag{5.2.30a}$$

$$Z^{(23)} = (4\pi)^{-1}(k_1/k_3)\int d\hat{\Omega}\,\sigma_{32}^{(23)}(\hat{\Omega}|Z^{(23)}) \ , \tag{5.2.30b}$$

in terms of the notations

$$\sigma_{12}^{(12)}(\hat{\Omega}|Z^{(12)}) = \int d\hat{\Omega}'\,\sigma_{12}^{(12)}(\hat{\Omega}|\hat{\Omega}')(1+3|\Omega_z'|Z^{(12)}) \ , \tag{5.2.31a}$$

$$\sigma_{21}^{(12)}(Z^{(12)}|\hat{\Omega}) = \int d\hat{\Omega}'\,(1+3Z^{(12)}|\Omega_z'|)\sigma_{21}^{(12)}(\hat{\Omega}'|\hat{\Omega}) \ , \tag{5.2.31b}$$

(upon suppression of the additional term $3D_2\mathrm{i}\boldsymbol{\Omega}'\cdot\boldsymbol{\lambda}$ in ( ) on each right-hand side), with the reciprocity

$$\sigma_{12}^{(12)}(\hat{\Omega}|Z^{(12)}) = \sigma_{21}^{(12)*}(Z^{(12)}|-\hat{\Omega})\,\Big|_{\boldsymbol{\lambda}\to-\boldsymbol{\lambda}} \ . \tag{5.2.32}$$

5.2.2 Boundary-Value Solution of the Diffusion Equation

In the diffusion expression (5.2.8), $S_A^{(q/12+23)}$ is the solution of the diffusion equation (5.2.5) subject to boundary equations (5.2.23,29), and is given by

$$S_A^{(q/12+23)}(z|z') = C_A\varphi^{(12)}(z_>)\varphi^{(23)}(z_<) \ . \tag{5.2.33}$$

Here $z_>$ and $z_<$ designate the larger and the smaller of z and z', respectively, and $\varphi^{(12)}(z)$ and $\varphi^{(23)}(z)$ are solutions of the homogeneous diffusion equation subject to the boundary conditions at $z = 0$ and $-L$ (Fig.5.1), respectively; C_A is a constant given by

$$C_A = (\gamma_2/D_2)\left[\varphi^{(12)}\partial_z\varphi^{(23)} - \varphi^{(23)}\partial_z\varphi^{(12)}\right]^{-1} . \tag{5.2.34}$$

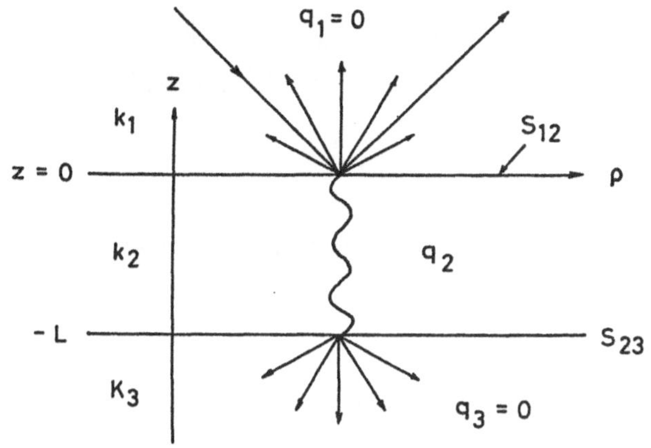

Fig. 5.1. Geometry of a random layer for (5.2.33)

Hence, with $\varkappa = \left(\gamma^{(ab)}D_2^{-1} + \boldsymbol{\lambda}^2\right)^{1/2}$, we can set

$$\varphi^{(12)}(z) = \cosh(\varkappa z) - Z^{(12)}(\varkappa D_2)^{-1}\sinh(\varkappa z) , \tag{5.2.35a}$$

$$\varphi^{(23)}(z) = \cosh\left[\varkappa(z+L)\right] + Z^{(23)}(\varkappa D_2)^{-1}\sinh\left[\varkappa(z+L)\right] , \tag{5.2.35b}$$

and determine C_A at $z = 0$, so that

$$C_A = \gamma_2\Big\{Z^{(12)}\left[\cosh(\varkappa L) + Z^{(23)}(\varkappa D_2)^{-1}\sinh(\varkappa L)\right] + \varkappa D_2\sinh(\varkappa L) + Z^{(23)}\cosh(\varkappa L)\Big\}^{-1} . \tag{5.2.36}$$

In the special case of $\varkappa = 0$ or $\gamma^{(ab)} = \boldsymbol{\lambda} = 0$, we obtain

$$S_A^{(q/12+23)}(-L|0) = \gamma_2\left[Z^{(12)}(1 + Z^{(23)}L/D_2) + Z^{(23)}\right]^{-1} , \tag{5.2.37}$$

$$Z^{(12)}S_A^{(q/12+23)}(0|0) + Z^{(23)}S_A^{(q/12+23)}(-L|0) = \gamma_2 . \tag{5.2.38}$$

Here, the last relation can also be shown directly by integrating the diffusion equation (5.2.5) over the range $0 \geq z \geq -L$ for $z' = -0$, followed by use of the boundary conditions.

In terms of a new variable, $\mathcal{T}$, defined by

$$\mathcal{T} = \varkappa D_2 \left[\frac{\varkappa D_2 \tanh(\varkappa L) + Z^{(23)}}{\varkappa D_2 + Z^{(23)} \tanh(\varkappa L)} \right] , \tag{5.2.39}$$

the boundary-value solutions can be written as

$$S_A^{(q/12+23)}(0|0) = \gamma_2 \left(Z^{(12)} + \mathcal{T} \right)^{-1} , \tag{5.2.40a}$$

$$S_A^{(q/12+23)}(-L|0) = S_A^{(q/12+23)}(0|0) \times \left[\cosh(\varkappa L) + (\varkappa D_2)^{-1} Z^{(23)} \sinh(\varkappa L) \right]^{-1} . \tag{5.2.40b}$$

Here, when $\varkappa = 0$,

$$\mathcal{T} \,\Big|_{\varkappa=0} \equiv \mathcal{T}_0 = Z^{(23)} \left[1 + (L/D_2) Z^{(23)} \right]^{-1} . \tag{5.2.41}$$

which tends to zero as $L \to \infty$, so that (5.2.37,38) are reproduced by (5.2.40a,b) with $\mathcal{T} \to \mathcal{T}_0$.

To find $\sigma_{11}^{(q/12+23)}$ from (5.2.2a), we only need

$$\mathcal{I}_{22}^{(q/12+23)}(\widehat{\Omega}, z|\widehat{\Omega}', z') = (4\pi\gamma_2)^{-1} (1 - 3D_2 \Omega_z \partial_z) \times (1 + 3D_2 \Omega_z' \partial_z') S_A^{(q/12+23)}(z|z') , \tag{5.2.42}$$

from (5.2.8), upon substitution of (5.2.6). Here, using (5.2.23),

$$\begin{aligned} D_2 \partial_z' S_A^{(q/12+23)}&(z = 0|z'(= -0)) \\ &= \left[D_2 (\partial_z' - \partial_z) + D_2 \partial_z \right] S_A^{(q/12+23)}(z(= 0)|z'(= -0)) \\ &= \gamma_2 - Z^{(12)} S_A^{(q/12+23)}(0|0) , \end{aligned} \tag{5.2.43}$$

as a consequence of (5.2.33,34). Hence (5.2.42), together with (5.2.40a), yields

$$\mathcal{I}_{22}^{(q/12+23)}(\widehat{\Omega}, z = 0|\widehat{\Omega}', z' = -0) = (4\pi)^{-1} \left(Z^{(12)} + \mathcal{T} \right)^{-1} \times (1 + 3\Omega_z Z^{(12)}) (1 + 3\Omega_z' \mathcal{T}) \tag{5.2.44}$$

$$= (4\pi Z^{(12)})^{-1} (1 + 3\Omega_z Z^{(12)}) \times \left[1 - (1 - 3\Omega_z' Z^{(12)}) \mathcal{T} \left(Z^{(12)} + \mathcal{T} \right)^{-1} \right] . \tag{5.2.45}$$

Hence, by the substitution in (5.2.2a,b), we obtain

$$\sigma_{11}^{(q/12+23)}(\widehat{\Omega}|\widehat{\Omega}') = (4\pi Z^{(12)})^{-1} \sigma_{12}^{(12)}(\widehat{\Omega}|Z^{(12)}) \times \left[\sigma_{21}^{(12)}(Z = 0|\widehat{\Omega}') - \sigma_{21}^{(12)}(Z^{(12)}|\widehat{\Omega}') \mathcal{T} \left(Z^{(12)} + \mathcal{T} \right)^{-1} \right] , \tag{5.2.46}$$

$$\sigma_{31}^{(q/12+23)}(\hat{\Omega}|\hat{\Omega}') = (4\pi\gamma_2)^{-1}\sigma_{32}^{(23)}(\hat{\Omega}|Z^{(23)}) \times S_A^{(q/12+23)}(-L|0)\sigma_{21}^{(12)}(Z^{(12)}|\hat{\Omega}') \ , \quad (5.2.47)$$

in terms of the notations of (5.2.31a,b); the variable $\mathcal{T}$ is given by (5.2.39), and $S_A^{(q/12+23)}$ by (5.2.40a,b). Here, when $\varkappa L \gg 1$,

$$\mathcal{T} \sim \varkappa D_2, \qquad \varkappa L \gg 1 \ , \quad (5.2.48)$$

and when $\varkappa = 0$, a relation equivalent to (5.1.36) holds:

$$\int d\hat{\Omega}\, k_1 \sigma_{11}^{(q/12+23)}(\hat{\Omega}|\hat{\Omega}') + \int d\hat{\Omega}\, k_3 \sigma_{31}^{(q/12+23)}(\hat{\Omega}|\hat{\Omega}') = k_2 \sigma_{21}^{(12)}(z=0|\hat{\Omega}') = k_2 \int d\hat{\Omega}\, \sigma_{21}^{(12)}(\hat{\Omega}|\hat{\Omega}') \ . \quad (5.2.49)$$

This is a direct result of relations (5.2.30a,b) and

$$S_A^{(q/12+23)}(-L|0) = (\gamma_2/Z^{(23)})\mathcal{T}_0\left(Z^{(12)} + \mathcal{T}_0\right)^{-1} \quad (5.2.50)$$

from (5.2.40b,41); and, together with optical relation (5.1.3) for $\sigma^{(12)}$, ensures the optical relation for the entire system, (5.1.35), as we will show.

In the case of a semi-infinite random layer where $L = \infty$ and when $\varkappa = 0$, (5.2.46) is reduced to

$$\sigma_{11}^{(q/12+\infty)}(\hat{\Omega}|\hat{\Omega}') \equiv (4\pi Z^{(12)})^{-1}\sigma_{12}^{(12)}(\hat{\Omega}|Z^{(12)})\sigma_{21}^{(12)}(Z=0|\hat{\Omega}') \ , \quad (5.2.51)$$

which depends on the medium characteristics only through $\sigma^{(12)}$ and $Z^{(12)}$. In terms of $\sigma_{11}^{(q/12+23)}$, (5.2.46) can be written as

$$\sigma_{11}^{(q/12+23)}(\hat{\Omega}|\hat{\Omega}') - \sigma_{11}^{(q/12+\infty)}(\hat{\Omega}|\hat{\Omega}') = -(4\pi Z^{(12)})^{-1}\sigma_{12}^{(12)}(\hat{\Omega}|Z^{(12)})\mathcal{T}\left(Z^{(12)} + \mathcal{T}\right)^{-1} \times \sigma_{21}^{(12)}(Z^{(12)}|\hat{\Omega}') \quad (5.2.52a)$$

$$= -(4\pi\gamma_2)^{-1}\left(Z^{(23)}/Z^{(12)}\right)\sigma_{12}^{(12)}(\hat{\Omega}|Z^{(12)}) \times S_A^{(q/12+23)}(-L|0)\sigma_{21}^{(12)}(Z^{(12)}|\hat{\Omega}') \ . \quad (5.2.52b)$$

Here, the second expression is valid only when $\gamma^{(ab)} = \boldsymbol{\lambda} = 0$, in view of (5.2.50), and, together with (5.2.47) for $\sigma_{31}^{(q/12+23)}$, leads directly to the basic optical relation (5.2.49) by perfect cancellation of the latter contribution, in virtue of relations (5.2.30a,b). This means that the right-hand side term of (5.2.52a) has the same accuracy as $\sigma_{31}^{(q/12+23)}$ for the transmitted wave has when $\gamma_2 L \gg 1$, i.e., a sufficient accuracy that can be expected by the diffusion approximation.

5.2.3 Case of a Random Layer with Smooth Boundaries

From (3.3.129), $\sigma_{ab}^{(12)}$, in this case, is given by

$$\sigma_{ab}^{(12)}(\widehat{\Omega}|\widehat{\Omega}') = |\Omega_z^{(a)}\langle R_{ab}^{(12)}(\widehat{\Omega})\rangle^2|\delta_s^2(\widehat{\Omega}^{(a)} - \widehat{\Omega}^{(a)\prime}) \tag{5.2.53a}$$

$$= |\Omega_z^{(b)}\langle R_{ba}^{(12)}(\widehat{\Omega})\rangle^2|\delta_s^2(\widehat{\Omega}^{(b)} - \widehat{\Omega}^{(b)\prime}) \,. \tag{5.2.53b}$$

Here, $\langle R_{ab}^{(12)}\rangle$ is the reflection-transmission coefficient, and the second expression is a consequence of [see (3.3.107)]

$$\delta_s^2(\widehat{\Omega}^{(a)} - \widehat{\Omega}^{(a)\prime}) = |\partial\widehat{\Omega}^{(b)}/\partial\widehat{\Omega}^{(a)}|\delta_s^2(\widehat{\Omega}^{(b)} - \widehat{\Omega}^{(b)\prime}) \,, \tag{5.2.54a}$$

$$|\partial\widehat{\Omega}^{(b)}/\partial\widehat{\Omega}^{(a)}| = (k_a/k_b)^2|\Omega_z^{(a)}/\Omega_z^{(b)}| \,. \tag{5.2.54b}$$

Hence,

$$\int d\widehat{\Omega}'\,\sigma_{12}^{(12)}(\widehat{\Omega}'|\widehat{\Omega}) = |\Omega_z^{(1)}\langle R_{12}^{(12)}(\widehat{\Omega})\rangle^2| \,, \tag{5.2.55}$$

and $Z^{(12)}$ is given by (5.2.28) with

$$\langle\sigma_{12}^{(12)}\rangle_0 = (2\pi)^{-1}\int d\widehat{\Omega}^{(2)}\,|\Omega_z^{(1)}\langle R_{12}^{(12)}(\widehat{\Omega})\rangle^2| \,, \tag{5.2.56a}$$

$$\langle\sigma_{12}^{(12)}\rangle_1 = 3(2\pi)^{-1}\int d\widehat{\Omega}^{(2)}\,|\Omega_z^{(1)}\Omega_z^{(2)}\langle R_{12}^{(12)}(\widehat{\Omega})\rangle^2| \tag{5.2.56b}$$

from (5.2.22a,b). Here, using (3.3.105,104),

$$\langle R_{12}^{(12)}(\widehat{\Omega})\rangle = 2\Omega_z^{(2)}\,/\,[\Omega_z^{(2)} + (k_1/k_2)\Omega_z^{(1)}] \,, \tag{5.2.57}$$

where, with $k_1/k_2 < 1$,

$$\begin{aligned} (k_1/k_2)\Omega_z^{(1)} &= \left[(\Omega_z^{(2)})^2 - C^2\right]^{1/2} \,, \\ C &= \left[1 - (k_1/k_2)^2\right]^{1/2} \,. \end{aligned} \tag{5.2.58}$$

Hence, changing the variable of integration by

$$\Omega_z^{(2)} = C\cosh x, \qquad (k_1/k_2)\Omega_z^{(1)} = C\sinh x \,, \tag{5.2.59}$$

(5.2.56a) leads to

$$\begin{aligned} \frac{k_1}{k_2}\langle\sigma_{12}^{(12)}\rangle_0 &= \frac{k_1}{k_2}\int_0^1 d\Omega_z^{(2)}\,\Omega_z^{(1)}|\langle R_{12}^{(12)}(\widehat{\Omega})\rangle^2| \\ &= 4C^2\int_0^{x_C} dx\,\frac{\cosh^2 x\sinh^2 x}{(\sinh x + \cosh x)^2} \,, \end{aligned} \tag{5.2.60}$$

where

$$x_C = \operatorname{arctanh}(k_1/k_2) \ . \tag{5.2.61}$$

In the same way, from (5.2.56b),

$$\frac{k_1}{k_2}\langle\sigma_{12}^{(12)}\rangle_1 = 12\,C^3 \int_0^{x_C} dx \, \frac{\cosh^3 x \sinh^2 x}{(\sinh x + \cosh x)^2} \ . \tag{5.2.62}$$

Hence $Z^{(12)}$ as given by (5.2.28) with (5.2.60,62), and when $x_C \sim k_1/k_2 \ll 1$, is reduced to

$$Z^{(12)} \sim \tfrac{2}{3}(k_1/k_2)^3, \qquad |k_1/k_2| \ll 1 \ , \tag{5.2.63}$$

while, when $k_1/k_2 \sim 1$ so that $x_C \sim +\infty$, $Z^{(12)} = 0.5$, consistent with (5.2.24) when $\sigma_{22}^{(12)} = 0$.

Thus the cross section per unit area of the layer surface, from (5.2.1a,52a), is given by

$$\begin{aligned}\sigma_{11}^{(q+12+23)}(\widehat{\Omega}|\widehat{\Omega}') &= \sigma_{11}^{(12)}(\widehat{\Omega}|\widehat{\Omega}') + \sigma^{(q/12+\infty)}(\widehat{\Omega}|\widehat{\Omega}') \\ &- \left(4\pi Z^{(12)}\right)^{-1}\sigma_{12}^{(12)}(\widehat{\Omega}|Z^{(12)})\mathcal{T}\left(Z^{(12)}+\mathcal{T}\right)^{-1} \\ &\times \sigma_{21}^{(12)}(Z^{(12)}|\widehat{\Omega}') \ ,\end{aligned} \tag{5.2.64}$$

where, using (5.2.53b) in (5.2.31a),

$$\begin{aligned}\sigma_{12}^{(12)}(\widehat{\Omega}|Z^{(12)}) =& \sigma_{21}^{(12)}(Z^{(12)}|-\widehat{\Omega}) \\ &= \left(1+3|\Omega_z^{(2)}|Z^{(12)}\right)|\Omega_z^{(2)}\langle R_{21}^{(12)}(\widehat{\Omega})\rangle^2| \ ,\end{aligned} \tag{5.2.65}$$

with

$$\langle R_{21}^{(12)}(\widehat{\Omega})\rangle = 2(k_1/k_2)\Omega_z^{(1)} \ / \ [\Omega_z^{(2)} + (k_1/k_2)\Omega_z^{(1)}] \ . \tag{5.2.66}$$

Now we can do the same thing for the cross section, $\sigma_{31}^{(q+12+23)}(\widehat{\Omega}|\widehat{\Omega}')$, as given by (5.2.1b) with (5.2.47, 50) and $\sigma_{32}^{(23)}(\widehat{\Omega}|Z^{(23)})$ given by an expression similar to (5.2.65).

Shown in Fig.5.2 is a curve of $Z^{(12)}$ versus k_1/k_2, given by (5.2.28) when the boundary is perfectly smooth. As k_1/k_2 changes from 0 to 1, $Z^{(12)}$ changes from 0 (perfectly conducting surface) to 0.5 for a boundary of no reflection, as is expected. Shown in Fig.5.3 are the angle distributions of both backscattered (solid curves) and transmitted waves (dash-dot curves) through a random layer with diffusion constant D_2 (2.5.49), width L, averaged propagation constant k_2, and separated by two smooth boundaries from the outside free spaces with the equal propagation constants, $k_1 = k_3$ (Fig.5.1). The curves are shown for a normally incident wave, and (a) $k_1/k_2 = 0.99$, $Z^{(12)} = 0.4863$; (b) $k_1/k_2 = 0.2$, $Z^{(12)} = 0.004122$, by using the numerical distance L/D_2 as

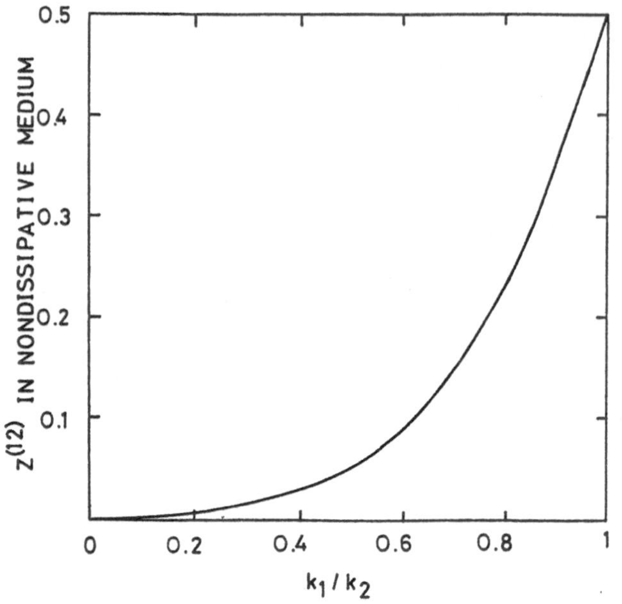

Fig. 5.2. $Z^{(12)}$ of (5.2.28) is shown as a function of k_1/k_2 for a perfectly smooth boundary in a non-dissipative medium, with the eigenfunction of (5.2.6)

a parameter. The sum of the backscattered and transmitted wave intensities are independent of L/D_2 in each case, reflecting the power conservation; here the sum in case (b) is smaller than that in case (a) by the different amount of specular reflection by the respective upper boundaries. The backscattered waves are generally stronger than the transmitted waves when $L/D_2 = 100$, whereas the situation is inverse when $L/D_2 = 5$, although the two waves are of nearly the same intensity in case (b); the smaller k_1/k_2 is, the more the waves are trapped inside the layer as a consequence of an enhanced multiple reflection between the two boundaries.

5.2.4 Reciprocity

The reciprocity,

$$\sigma_{31}^{(q/12+23)}(\hat{\Omega}|\hat{\Omega}') = \sigma_{13}^{(q/12+23)}(-\hat{\Omega}'|-\hat{\Omega}) \ , \qquad (5.2.67)$$

follows directly from (5.2.47,32), and the same holds true also for the right-hand side terms of (5.2.52a,b) for $\sigma_{11}^{(q/12+23)}$. However in view of (5.2.51), this is not the case of the term $\sigma_{11}^{(q/12+\infty)}$ on the left-hand side, unless $Z^{(12)} \ll 1$, as realized when $k_1/k_2 \ll 1$, in view of (5.2.63), in spite of satisfying the correct optical relation [as can be shown by using (5.2.30a)]

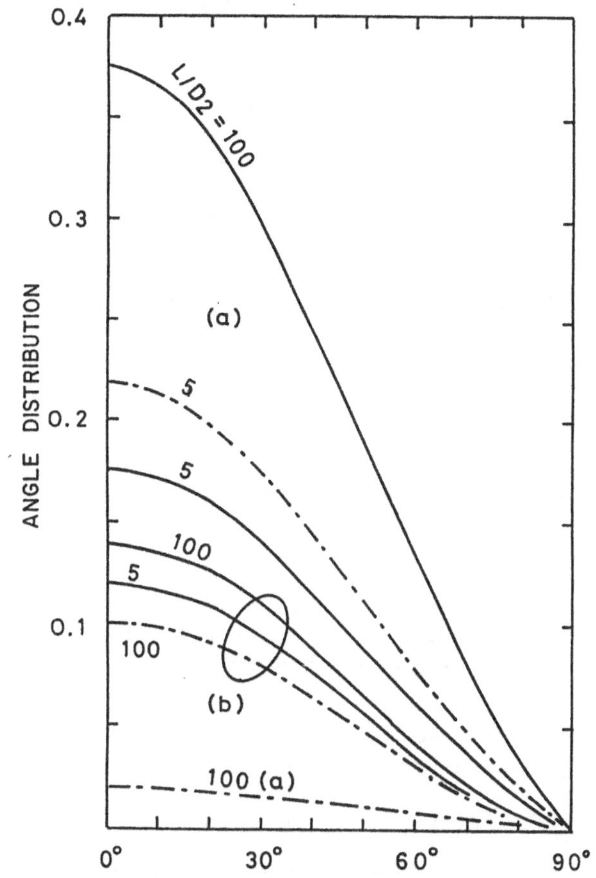

Fig. 5.3. Angle distributions of both backscattered and transmitted waves through a random layer with perfectly smooth boundaries and numerical width L/D_2. The wave is normally incident and the medium is non-dissipative with (a) $k_1/k_2 = 0.99$; (b) $k_1/k_2 = 0.2$. The backscattered waves are shown by *solid curves* and the transmitted waves by *dash-dot curves*

$$k_1 \int d\widehat{\Omega}\, \sigma_{11}^{(q/12+\infty)}(\widehat{\Omega}|\widehat{\Omega}') = k_2 \sigma_{21}^{(12)}(Z=0|\widehat{\Omega}') \ . \tag{5.2.68}$$

This implies that the diffusion approximation is not good enough for the term $\sigma_{11}^{(q/12+\infty)}$, in spite of its success for the other terms, including the right-hand sides of the expressions (5.2.52a,b).

Formally, a symmetrical (reciprocal) expression can be obtained by making an asymptotic evaluation of the integral representation (5.1.30c) for $|\lambda_z|$, $|\lambda_z'| \sim \infty$ under the diffusion condition that the change of $S_{22}^{(q/12+\infty)}(\widehat{\Omega}, z|\widehat{\Omega}', z')$ be negligibly small within the distance γ_2^{-1}. To evaluate the integral by using (5.2.33) for $S_A^{(q/12+23)}$, we first disregard the discontinuity at $z = z'$ so that we obtain an expression to the first order of ∂_z ($\partial'_z = 0$), as

$$S_{+2,+2}^{(q/12+\infty)}(\widehat{\Omega}|\widehat{\Omega}') = (\lambda_z\lambda'_z)^{-1}[1 - (\mathrm{i}\lambda_z)^{-1}\partial_z] \times S_{22}^{(q/12+\infty)}(\widehat{\Omega}, z(=0)|\widehat{\Omega}', z' = -0) \quad (5.2.69a)$$

$$= |\Omega_z\Omega'_z| \mathcal{I}_{22}^{(q/12+\infty)}(\widehat{\Omega}, z = 0|\widehat{\Omega}', z' = -0)\ , \quad (5.2.69b)$$

where the last expression from (5.1.30a) is required to be consistent. Here, $S_{22}^{(q/12+\infty)}$ on the right-hand side of (5.2.69a) is given from (2.5.7), by

$$S_{22}^{(q/12+\infty)} \sim (4\pi)^{-1}\gamma_2 f_A(\widehat{\Omega}, \mathrm{i}\partial_z) S_A^{(q/12+\infty)}(z|z'), \quad z > z'\ , \quad (5.2.70)$$

where, from (2.5.8),

$$f_A(\widehat{\Omega}, \mathrm{i}\partial_z) = (\gamma_2 + \Omega_z\partial_z)\phi_A(\widehat{\Omega}, \mathrm{i}\partial_z) = 1 - 3\overline{D}_2\Omega_z\partial_z\ , \quad (5.2.71a)$$

$$\overline{D}_2 = D_2 - (3\gamma_2)^{-1} = a_1 D_2\ ; \quad (5.2.71b)$$

and λ_z and λ'_z are given by (5.1.24a,b). Hence (5.2.69a) agrees, in fact, with (5.2.69b) when using the diffusion expression (5.2.42); this holds true including the case of $\boldsymbol{\lambda} \neq 0$, as may be shown by substituting the $\boldsymbol{\lambda}$-dependent expressions (5.1.24a,b) in the first factor $(\lambda_z\lambda'_z)^{-1}$ of (5.2.69a).

On the other hand, if we took into account the discontinuity of the integrand at $z = z'$, expression (5.2.69a) would be replaced by

$$S_{+2,+2}^{(q/12+\infty)}(\widehat{\Omega}|\widehat{\Omega}') = (\lambda_z\lambda'_z)^{-1}[1 + \mathrm{i}(\lambda_z - \lambda'_z)^{-1}\partial_z] \times S_{22}^{(q/12+\infty)}(\widehat{\Omega}, z(=0)|\widehat{\Omega}', z' = -0)\ , \quad (5.2.72)$$

which is invariant against the interchange of λ_z and $-\lambda'_z$, therefore satisfying the reciprocity. But, it is not consistent with the basic optical relation (5.2.68) necessary to ensure power conservation of the entire system; methods basically equivalent to this method were used in [5.2–4].

Summarizing, the diffusion equation is an equation for the coefficient $S_A^{(q/12+23)}(z|z')$ in the expression (5.2.8) for $\mathcal{I}_{22}^{(q/12+23)}$, which is an asymptotic expression in the same sense as expression (5.2.69a) is asymptotic, so that, whenever using the solution, it gives the asymptotic answer directly, without any need of further asymptotic evaluation according to (5.2.72).

5.2.5 Boundary Condition when Media are Random on Both Sides of the Boundary

The boundary equation can be written in a more general form so that the equation is also applicable to the case in which media are random on both sides of a boundary, with the cross section of the general form $\sigma_{ab}(\hat{\Omega}|\hat{\Omega}')$. We designate the eigenfunctions of the diffusion term in space k_b by $\phi_A^{(b)}(\hat{\Omega},\hat{\lambda})$ and $\overline{\phi}_A^{(b)}(\hat{\Omega},\hat{\lambda})$, and introduce operators $\langle\sigma_{ab}(\partial_n)\rangle$ and $p_b^{\pm}(\partial_n)$, defined by

$$\begin{aligned}\langle\sigma_{ab}(\partial_n)\rangle =&(4\pi)^{-1}\int_{(+a)} d\hat{\Omega}\int_{(-b)} d\hat{\Omega}'\,\sigma_{ab}(\hat{\Omega}|\hat{\Omega}')\\ &\times\phi_A^{(b)}(\hat{\Omega}',\mathrm{i}\hat{\partial}_n^{(b)})\ ,\end{aligned} \tag{5.2.73}$$

$$p_b^{\pm}(\partial_n) =(4\pi)^{-1}\int_{(\pm b)} d\hat{\Omega}\,\Omega_n^{(b)}\phi_A^{(b)}(\hat{\Omega}',\mathrm{i}\hat{\partial}_n^{(b)})\ . \tag{5.2.74}$$

Here, $\hat{\partial}_n^{(b)} = \hat{\boldsymbol{n}}^{(b)}\partial_n^{(b)}$, where $\partial_n^{(b)} = \hat{\boldsymbol{n}}^{(b)}\cdot\partial/\partial\hat{\boldsymbol{\rho}}$, is the differential operator directed normally to the boundary toward the space k_b, and the integration ranges $(\pm b)$ designate the half solid angles of $\Omega_n^{(b)} = \hat{\boldsymbol{n}}^{(b)}\cdot\hat{\Omega} \gtrless 0$, respectively. Hence, when using (5.2.6) with $\Omega_z\partial_z \to \Omega_n^{(b)}\partial_n^{(b)}$, $\gamma_2 \to \gamma_b$, and $D_2 \to D_b$, $p_b^{\pm}(\partial_n)$ is reduced to

$$p_b^{\pm}(\partial_n) = (2\gamma_b)^{-1}(\pm\tfrac{1}{2} - D_b\partial_n^{(b)})\ , \tag{5.2.75}$$

in the same form as (5.2.20).

Now the generalized version of boundary equation (5.2.19) is written as

$$p_a^{+}(\partial_n)S_A^{(a)} = \sum_b\langle\sigma_{ab}(\partial_n)\rangle S_A^{(b)}\ . \tag{5.2.76}$$

Here, $S_A^{(a)}$ denotes the boundary-value (and its normal derivative) on the side of space k_a, and the left-hand side means the power scattered toward the space k_a by the boundary. From (5.2.73,74) we can write $\langle\sigma_{ab}(\partial_n)\rangle$ and $p_a^{\pm}(\partial_n)$ in the form

$$\langle\sigma_{ab}(\partial_n)\rangle = \langle\sigma_{ab}\rangle_0 + \langle\sigma_{ab}\rangle_1\partial_n^{(b)}\ , \tag{5.2.77}$$

$$p_a^{\pm}(\partial_n) = p_{a,0}^{\pm} - p_{a,1}^{\pm}\partial_n^{(a)}\ , \tag{5.2.78}$$

similar to (5.2.21c,20), respectively.

In 2×2 matrix form, (5.2.76–78) can be simply written as

$$p^{+}(\partial_n)S_A = \langle\sigma(\partial_n)\rangle S_A\ , \tag{5.2.79}$$

$$\langle\sigma(\partial_n)\rangle = \langle\sigma\rangle_0 + \langle\sigma\rangle_1\partial_n\ , \tag{5.2.80}$$

$$p^{\pm}(\partial_n) = p_0^{\pm} - p_1^{\pm}\partial_n\ , \tag{5.2.81}$$

where $p^{\pm}(\partial_n)$ is regarded as a diagonal matrix with the elements $p_a^{\pm}(\partial_n)$. Hence (5.2.79), upon substitution of (5.2.80,81), becomes written in the final form,

$$\partial_n S_A = Z S_A \ , \tag{5.2.82a}$$

or, more explicitly,

$$\partial_n^{(a)} S_A^{(a)} = \sum_b Z_{ab} S_A^{(b)} \ . \tag{5.2.82b}$$

Here, Z is a 2×2 matrix, defined by

$$Z = \left(p_1^+ + \langle\sigma\rangle_1\right)^{-1}\left(p_0^+ - \langle\sigma\rangle_0\right) \ . \tag{5.2.83}$$

Equations (5.2.82,83) obviously correspond to (5.2.23,24), respectively, and they are, in fact, equivalent to each other in the case of a semi-infinite random layer.

To confirm the consistency of the boundary equation (5.2.76) with power conservation, we observe that the $\left\langle\sigma_{ab}(\partial_n)\right\rangle$'s are subject to a constraint resulting from optical relation (5.1.3),

$$\sum_a k_a \left\langle\sigma_{ab}(\partial_n)\right\rangle = -k_b p_b^-(\partial_n) \ . \tag{5.2.84}$$

Hence, using (5.2.76),

$$\begin{aligned}\sum_a k_a p_a^+(\partial_n) S_A^{(a)} &= \sum_{a,b} k_a \left\langle\sigma_{ab}(\partial_n)\right\rangle S_A^{(b)} \\ &= -\sum_b k_b p_b^-(\partial_n) S_A^{(b)} \ ,\end{aligned} \tag{5.2.85}$$

which can be written, in term of the notation

$$p_a(\partial_n) = p_a^+(\partial_n) + p_a^-(\partial_n) \ , \tag{5.2.86}$$

as

$$\sum_a k_a p_a(\partial_n) S_A^{(a)} = 0 \ . \tag{5.2.87}$$

The sum means the total (angle-averaged) power away from both sides of the boundary, as is clear from the definition (5.2.74).

The function $S_A^{(a)}$ in each space is governed by the diffusion equation from (2.5.16) with $\widehat{\boldsymbol{p}}_A$ of (2.5.15b), i.e., with $\widehat{\boldsymbol{p}}_A \rightarrow \widehat{\boldsymbol{p}}_A^{(a)}$, by

$$\frac{\partial}{\partial\widehat{\boldsymbol{\rho}}} \cdot \widehat{\boldsymbol{p}}_A^{(a)}(\mathrm{i}\partial/\partial\widehat{\boldsymbol{\rho}}) S_A^{(a)}(\widehat{\boldsymbol{\rho}}|\widehat{\boldsymbol{\rho}}') = \delta(\widehat{\boldsymbol{\rho}} - \widehat{\boldsymbol{\rho}}') \ , \tag{5.2.88}$$

with $A(\mathrm{i}\partial/\partial\widehat{\boldsymbol{\rho}}) \simeq 1$ on the right-hand side; $K_a^{(q)}(\widehat{\boldsymbol{\Omega}}|\widehat{\boldsymbol{\Omega}}')$ can even be rotationally variant. Here, the component of $\widehat{\boldsymbol{p}}_A^{(a)}$ in the direction $\widehat{\boldsymbol{n}}^{(a)}$ is given by $p_a(\partial_n)$ of (5.2.86), and can also be written in the form

$$\widehat{\boldsymbol{n}}^{(a)} \cdot \widehat{\boldsymbol{p}}_A^{(a)}(\mathrm{i}\partial/\partial\widehat{\boldsymbol{\rho}}) = p_a(\partial_n) = p_{a,0} - p_{a,1}\partial_n^{(a)} \ . \tag{5.2.89}$$

$p_{a,0}$ is generally dependent on the horizontal $\boldsymbol{\lambda} = \mathrm{i}\partial/\partial\boldsymbol{\rho}$, although it is identically zero when the scattering cross section is rotationally invariant.

6. Fixed Scatterer

Basic equation for a system of a random medium with boundaries and a fixed scatterer embedded in it, can be obtained from equations of Chap. 4 by simply replacing one of the boundaries with the scatterer [6.1].

6.1 Basic Equations

When a fixed scatterer, described by $q^{(\alpha)}(\hat{\boldsymbol{x}})$ with the center at $\hat{\boldsymbol{x}} = \hat{\boldsymbol{x}}_\alpha (= \hat{\boldsymbol{\rho}}_\alpha)$, is embedded in a random medium, $q(\hat{\boldsymbol{x}})$, the deterministic Green's function, $g^{(\alpha)}(\hat{\boldsymbol{x}}|\hat{\boldsymbol{x}}')$, is governed by the wave equation,

$$[L - q(\hat{\boldsymbol{x}}) - q^{(\alpha)}(\hat{\boldsymbol{x}})]\, g^{(\alpha)}(\hat{\boldsymbol{x}}|\hat{\boldsymbol{x}}') = \delta(\hat{\boldsymbol{x}} - \hat{\boldsymbol{x}}') \ , \tag{6.1.1}$$

or, in matrix form,

$$[L - q - q^{(\alpha)}]\, g^{(\alpha)} = 1 \ . \tag{6.1.2}$$

A basic assumption implied here is that $q(\hat{\boldsymbol{x}})$ and $q^{(\alpha)}(\hat{\boldsymbol{x}})$ can both be nonzero at the same place. Equation (6.1.2) has the same form as (4.1.9a) with $B^{(12)} \to q^{(\alpha)}$, so that the resulting statistical Green's functions can also be written by the same equations simply by replacing the superscript $(12) \to (\alpha)$. This is true even when the present system includes the boundary, S_{12}, with $B^{(12)} \to B^{(12)} + q^{(\alpha)}$, or with the superscript $(12) \to (12 + \alpha)$ in those equations.

To obtain the first order Green's function, $G^{(\alpha)} = \langle g^{(\alpha)} \rangle$, from (6.1.2), we observe that the effective medium is changed from $M = M^{(q)}$ to $M^{(\alpha)} \equiv M^{(q)} + \Delta M^{(\alpha)}$ by a still unknown amount, $\Delta M^{(\alpha)}$, due to the presence of $q^{(\alpha)}$,

$$\langle q g^{(\alpha)} \rangle = (M^{(q)} + \Delta M^{(\alpha)})\, G^{(\alpha)} \ . \tag{6.1.3}$$

Now the average of (6.1.2) can be written as

$$(L - M - q^{(\alpha + \Delta\alpha)})\, G^{(\alpha)} = 1 \ , \tag{6.1.4}$$

with an effective scatterer, $q^{(\alpha+\Delta\alpha)}$, given by

$$q^{(\alpha+\Delta\alpha)} = q^{(\alpha)} + \Delta M^{(\alpha)} \ . \tag{6.1.5}$$

Hence the solution is, with the notation $G^{(0)}$ for $G = G_M$,

$$G^{(\alpha)} = G^{(0)} + G^{(0)}T^{(\alpha)}G^{(0)} \ , \tag{6.1.6a}$$

$$(L - M)G^{(0)} = 1 \ , \tag{6.1.6b}$$

in terms of the scattering matrix, $T^{(\alpha)}$ of $q^{(\alpha+\Delta\alpha)}$, defined by

$$\begin{aligned} T^{(\alpha)} &= q^{(\alpha+\Delta\alpha)}\left[1 + G^{(0)}T^{(\alpha)}\right] \\ &= \left[1 - q^{(\alpha+\Delta\alpha)}G^{(0)}\right]^{-1} q^{(\alpha+\Delta\alpha)} \ . \end{aligned} \tag{6.1.7}$$

To find $\Delta M^{(\alpha)}$, we write the deterministic $g^{(\alpha)}$ in terms of $G^{(\alpha)}$, using (6.1.2–5),

$$g^{(\alpha)} = G^{(\alpha)}\left[1 + (q - M - \Delta M^{(\alpha)})g^{(\alpha)}\right] \ , \tag{6.1.8}$$

where

$$\left\langle (q - M - \Delta M^{(\alpha)})g^{(\alpha)} \right\rangle = 0 \ , \tag{6.1.9}$$

in the same fashion as (4.1.17a,b). Hence, to the ladder approximation of q where $\langle q \rangle = 0$, we obtain

$$\left\langle q g^{(\alpha)} \right\rangle \simeq \left\langle q G^{(\alpha)} q \right\rangle G^{(\alpha)} \ , \tag{6.1.10}$$

in the same fashion as having obtained (3.2.4), showing from (6.1.3) that

$$M = M^{(q)} = \left\langle q G^{(0)} q \right\rangle \ , \tag{6.1.11a}$$

$$\Delta M^{(\alpha)} = \left\langle q G^{(0)} T^{(\alpha)} G^{(0)} q \right\rangle \ , \tag{6.1.11b}$$

as a consequence of (6.1.6). On the other hand, when the medium is composed of independent particles, (6.11a,b) are replaced with

$$M = n \int d\hat{\boldsymbol{x}}_a \left\langle T_a^M \right\rangle' \ , \tag{6.1.12a}$$

$$\Delta M^{(\alpha)} = n \int d\hat{\boldsymbol{x}}_a \left\langle T_a^M G^{(0)} T^{(\alpha)} G^{(0)} T_a^M \right\rangle' \ , \tag{6.1.12b}$$

in terms of the notations of (2.2.14b), and are derived from

$$M^{(\alpha)} = M + \Delta M^{(\alpha)} = n \int d\hat{\boldsymbol{x}}_a \left\langle T_a^{M+\alpha} \right\rangle' \ , \tag{6.1.12c}$$

where $T_a^{M+\alpha}$ is given by T_a^M, with the replacement of $M \rightarrow M + q^{(\alpha+\Delta\alpha)}$, using formula (2.2.9a)

$$\begin{aligned} T_a^{b+C} &= a\left(1 + G_{(b+C)} T_a^{b+C}\right) \\ &= T_a^b\left(1 + G_b T_C^b G_b T_a^{b+C}\right) \\ &\simeq T_a^b + T_a^b G_b T_C^b G_b T_a^b \ . \end{aligned} \tag{6.1.12d}$$

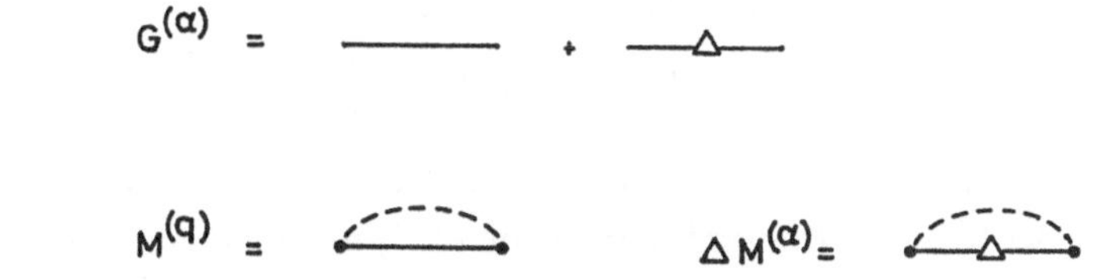

Fig. 6.1a. Schematic diagrams of $G^{(\alpha)}$, $M^{(q)}$, and $\Delta M^{(\alpha)}$, given by (6.1.6,11a,b), respectively. Here $G^{(0)}$, q, and $T^{(\alpha)}$ are represented by a *solid line*, *filled circle*, and *triangle*, respectively, and are connected in the order of their matrix multiplication; $\langle q \dots q \rangle$ is represented by a *dashed line* connecting the q's

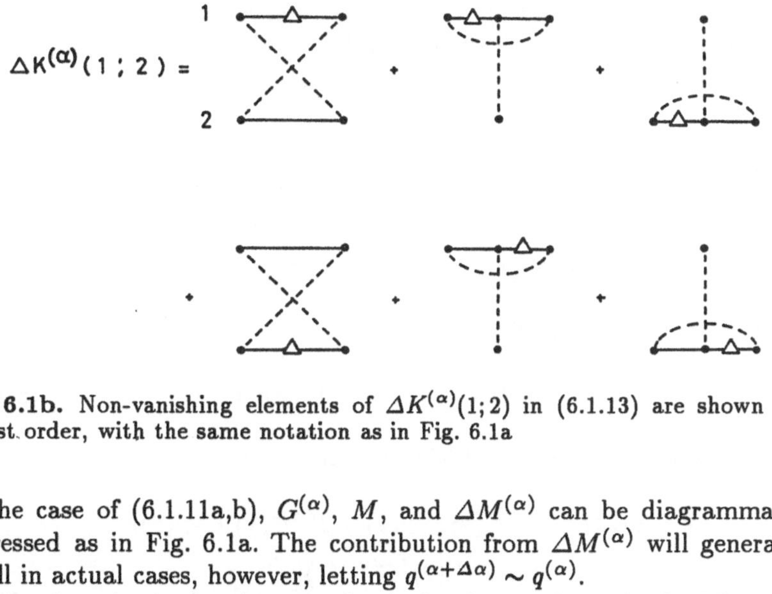

Fig. 6.1b. Non-vanishing elements of $\Delta K^{(\alpha)}(1;2)$ in (6.1.13) are shown to the lowest order, with the same notation as in Fig. 6.1a

In the case of (6.1.11a,b), $G^{(\alpha)}$, M, and $\Delta M^{(\alpha)}$ can be diagrammatically expressed as in Fig. 6.1a. The contribution from $\Delta M^{(\alpha)}$ will generally be small in actual cases, however, letting $q^{(\alpha+\Delta\alpha)} \sim q^{(\alpha)}$.

The situation is also the same for the incoherent factor in the BS equation, which is changed from $K(1;2)$ to

$$K^{(\alpha)}(1;2) = K(1;2) + \Delta K^{(\alpha)}(1;2) \ , \tag{6.1.13}$$

and is defined by the same equation as (4.1.19), just as $M^{(\alpha)}$ is defined by (6.1.3). Here it is remarked that, although the change $\Delta K^{(\alpha)}$ is generally small with the diagram of Fig. 6.1b, it is the only term responsible for the possible enhanced backscattering by the fixed scatterer.

When there are two fixed scatterers, the term $q^{(\alpha+\Delta\alpha)} \sim q^{(\alpha)}$ in the wave equation (6.1.4) is replaced by $q^{(\alpha_1)} + q^{(\alpha_2)}$, and the solution, $G^{(\alpha_1+\alpha_2)}$, can be written according to (2.2.5–9), as

$$G^{(\alpha_1+\alpha_2)} = G^{(\alpha_1)}\left(1 + T^{(\alpha_2/\alpha_1)}G^{(\alpha_1)}\right) , \tag{6.1.14}$$

$$T^{(\alpha_2/\alpha_1)} = \left(1 - T^{(\alpha_2)}G^{(0)}T^{(\alpha_1)}G^{(0)}\right)^{-1}T^{(\alpha_2)} , \tag{6.1.15}$$

in terms of the scattering matrices, $T^{(\alpha_1)}$ and $T^{(\alpha_2)}$, of the respective scatterers, or by (6.1.6), with the replacement of $T^{(\alpha)}$, by the resultant

$$T^{(\alpha_1+\alpha_2)} = T^{(\alpha_1)} + \left(1 + T^{(\alpha_1)}G^{(0)}\right)T^{(\alpha_2/\alpha_1)}\left(G^{(0)}T^{(\alpha_1)} + 1\right) . \tag{6.1.16}$$

Here, $G^{(0)}$ is a short-range function of the order of the wave coherence distance, and $T^{(\alpha_2/\alpha_1)}$ tends to $T^{(\alpha_2)}$ as the distance between the scatterers increases, yielding

$$T^{(\alpha_1+\alpha_2)} \sim T^{(\alpha_1)} + T^{(\alpha_2)} . \tag{6.1.17}$$

It may be remarked that one of the scatterers, say $q^{(\alpha_2)}$, may be regarded as a boundary, say $B^{(12)}$ of the boundary S_{12}, so that $T^{(\alpha_2)}$ also, as the 2×2 scattering matrix with the elements $\langle T_{ab}^{(12)} \rangle$, a, $b = 1, 2$, of (4.1.51a).

a) Case of an Unbounded Random Medium

In the case of one fixed scatterer, the propagator for the coherent wave, say $U^{(\alpha)}$, is given by

$$U^{(\alpha)}(1;2) = G^{(\alpha)*}(1)G^{(\alpha)}(2) , \tag{6.1.18}$$

and can be written, using (6.1.6), as

$$U^{(\alpha)}(1;2) = U(1;2) + U(1;2)V^{(\alpha)}(1;2)U(1;2) , \tag{6.1.19}$$

in the same form as $U^{(C)}$ of (4.1.50a,b), with

$$V^{(\alpha)}(1;2) = T^{(\alpha)*}(1)T^{(\alpha)}(2) + T^{(\alpha)*}(1)\left[G^{(0)}(2)\right]^{-1} + T^{(\alpha)}(2)\left[G^{(0)*}(1)\right]^{-1} . \tag{6.1.20}$$

Thus the BS equation for $I^{(q+\alpha)}(1;2)$ is obtained, by the procedure leading to (4.1.18) for $I^{(q+12)}$,

$$I^{(q+\alpha)}(1;2) = U^{(\alpha)}(1;2)\left[1 + K(1;2)I^{(q+\alpha)}\right] , \tag{6.1.21}$$

to the approximation of $K^{(\alpha)}(1;2) \simeq K(1;2)$.

It is presently convenient to use expression (4.2.5) with $\sigma^{(12)} \to V^{(\alpha)}$ (in view of the definition (4.1.48a) for $\sigma^{(12)}$), to write the solution of (6.1.21) as

$$I^{(q+\alpha)} = I^{(0q)} + I^{(0q)} V^{(\alpha/q)} I^{(0q)} \ . \tag{6.1.22}$$

Here, from (4.2.2) with $\mathcal{I}^{(0q)} = U S^{(0q)} U$,

$$V^{(\alpha/q)} = \left(1 - V^{(\alpha)} \mathcal{I}^{(0q)}\right)^{-1} V^{(\alpha)} \ . \tag{6.1.23}$$

Equation (6.1.23) can be interpreted as an effective $V^{(\alpha)}$ under the effect of medium fluctuation. When the scatterer is sufficiently small compared to the wave coherence distance, the optical cross section of $V^{(\alpha)}$, say $V^{(\alpha)}(\widehat{\Omega}|\widehat{\Omega}')$, will be shown later to be given by

$$V^{(\alpha)}(\widehat{\Omega}|\widehat{\Omega}') = \sigma^{(\alpha)}(\widehat{\Omega}|\widehat{\Omega}') - \gamma^{(\alpha)}(\widehat{\Omega})\delta^2(\widehat{\Omega} - \widehat{\Omega}') \ , \tag{6.1.24a}$$

in terms of the conventional cross section, $\sigma^{(\alpha)}(\widehat{\Omega}|\widehat{\Omega}')$, and the total cross section, $\gamma^{(\alpha)}(\widehat{\Omega})$ (including an effect of the medium fluctuation); hence,

$$\int d\widehat{\Omega}\, V^{(\alpha)}(\widehat{\Omega}|\widehat{\Omega}') \sim 0 \ . \tag{6.1.24b}$$

It may be remarked here that the cross section becomes critically negative in the shadow direction of the scatterer, and condition (6.1.24b) says that the total cross section of $V^{(\alpha)}$ is almost zero, as it should be in view of nearly the same amount of the scattered power as the incident, even in the present case of single scattering. While the effective cross section, $V^{(\alpha/q)}(\widehat{\Omega}|\widehat{\Omega}')$ of $V^{(\alpha/q)}$, is subject to (as long as the enhanced backscattering is negligible)

$$\int d\widehat{\Omega}\, V^{(\alpha/q)}(\widehat{\Omega}|\widehat{\Omega}') = 0 \ , \tag{6.1.25}$$

implying that the incident power after the multiple scattering with the random medium, is completely scattered out without being absorbed by the scatterer.

b) Case of Two Fixed Scatterers

When there are two fixed scatterers having their own scattering matrices, $V^{(\alpha_1)}$ and $V^{(\alpha_2)}$, then the resultant scattering matrix, $V^{(\alpha_1+\alpha_2)}$, can be written according to (6.1.20) with $T^{(\alpha)} \to T^{(\alpha_1+\alpha_2)}$ of (6.1.16), in the form

$$V^{(\alpha_1+\alpha_2)} = V^{(\alpha_1)} + {}'V^{(\alpha_2)} \ , \tag{6.1.26}$$

in the same fashion as (4.2.55b) for $V^{(C)}$. Here, ${}'V^{(\alpha_2)}$ includes the entire effect of the (coherent) multiple scattering between the two scatterers, and tends to $V^{(\alpha_2)}$ as the distance between the scatterers increases. Now the coherence function, $I^{(q+\alpha_1+\alpha_2)}$, is again given by (6.1.22) with the replacement of $V^{(\alpha/q)}$ with $V^{(\alpha_1+\alpha_2/q)}$, which can be obtained from formulas (4.2.98a,b) with $V^{(C_1)} \to V^{(\alpha_1)}$ and $V^{(C_2)} \to {}'V^{(\alpha_2)}$, as

$$V^{(\alpha_1+\alpha_2/q)} = V^{(\alpha_1/q)} + \left(1 + V^{(\alpha_1/q)}\mathcal{I}^{(0q)}\right) \times {}'V^{(\alpha_2/q/\alpha_1)}\left(\mathcal{I}^{(0q)}V^{(\alpha_1/q)} + 1\right) , \qquad (6.1.27)$$

which tends to $V^{(\alpha_1+\alpha_2)}$ as $K^{(q)} \to 0$. Here, as the separation of the scatterers increases, all the factors $\mathcal{I}^{(0q)}$ connecting $V^{(\alpha_1/q)}$ and ${}'V^{(\alpha_2/q)}$ in (6.1.27) become negligible, and the second term is reduced to $V^{(\alpha_2/q)}$. Hence,

$$V^{(\alpha_1+\alpha_2/q)} \sim V^{(\alpha_1/q)} + V^{(\alpha_2/q)} , \qquad (6.1.28a)$$

in the region where

$$\left|V^{(\alpha_1/q)}\mathcal{I}^{(0q)}V^{(\alpha_2/q)}\right| \ll \left|V^{(\alpha_2/q)}\right| . \qquad (6.1.28b)$$

c) Case of a Fixed Scatterer Embedded in a Semi-Infinite Random Layer

We now suppose that the fixed scatterer, $q^{(\alpha)}$, is embedded in a semi-infinite random layer with a rough boundary, S_{12}, as illustrated in Fig. 6.2. Then the previous equations for two boundaries in Sect. 4.2 are available as they are, by regarding one of the boundaries, say S_{23} with the scattering matrix $\sigma^{(23)}$, as the scatterer with $V^{(\alpha)}$. Hence the solution of the BS equation, $I^{(q+12+\alpha)}$, can be written in the form of (4.2.46a),

$$I^{(q+12+\alpha)} = I^{(0q)} + I^{(0q)}\sigma^{(12+\alpha/q)}I^{(0q)} . \qquad (6.1.29)$$

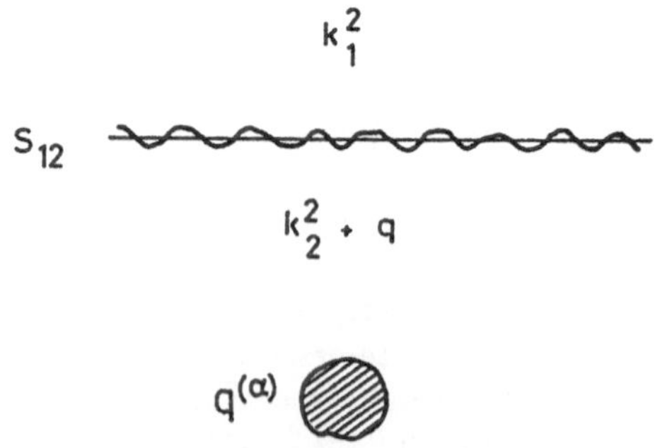

Fig. 6.2. Fixed scatterer embedded in a semi-infinite random medium for (6.1.29)

Here,

$$\sigma^{(12+\alpha)} = \sigma^{(12)} + V_{(12)}^{(\alpha)} , \qquad (6.1.30)$$

which is given by (4.2.60) for $\sigma_C^{(12)}$ with $\sigma_{(23)}^{(12)} \to V_{(12)}^{(\alpha)}$ and $'V^{(23)} \to {}'V^{(\alpha)}$. Hence $\sigma^{(12+\alpha/q)}$ is obtained from (4.2.44) with $\sigma^{(23)} \to V_{(12)}^{(\alpha)}$, in the form

$$\sigma^{(12+\alpha/q)} = \sigma^{(12/q)} + V^{(\alpha/q+12)} , \tag{6.1.31a}$$

where

$$V^{(\alpha/q+12)} = \left(1 + \sigma^{(12/q)}\mathcal{I}^{(0q)}\right) V^{(\alpha/q/12)} \left(\mathcal{I}^{(0q)}\sigma^{(12/q)} + 1\right) , \tag{6.1.31b}$$

with $V^{(\alpha/q/12)}$ given by (4.2.41), with $\sigma^{(23)} \to V_{(12)}^{(\alpha)}$.

Thus (6.1.29) can also be written as

$$I^{(q+12+\alpha)} = I^{(q+12)} + I^{(0q)} V^{(\alpha/q+12)} I^{(0q)} , \tag{6.1.32}$$

so that the second term gives the entire scattered wave by the scatterer. More specifically, when $q_1 = 0$, and the scatterer is embedded in the space of $q_2 \neq 0$, $z < 0$,

$$I_{11}^{(q+12+\alpha)} = I_{11}^{(q+12)} + U_1 V_{11}^{(\alpha/q+12)} U_1 , \tag{6.1.33a}$$

where, from (6.1.31b),

$$\begin{aligned} V_{11}^{(\alpha/q+12)} &= V_{11}^{(\alpha/q/12)} \\ &+ V_{12}^{(\alpha/q/12)} \mathcal{I}_2^{(0q)} \sigma_{21}^{(12/q)} + \sigma_{12}^{(12/q)} \mathcal{I}_2^{(0q)} V_{21}^{(\alpha/q/12)} \\ &+ \sigma_{12}^{(12/q)} \mathcal{I}_2^{(0q)} V_{22}^{(\alpha/q/12)} \mathcal{I}_2^{(0q)} \sigma_{21}^{(12/q)} , \end{aligned} \tag{6.1.33b}$$

with $\sigma_{21}^{(12/q)}$ given by (4.2.31a).

Here, when the boundary is perfectly smooth so that $\sigma^{(12)} = V^{(12)}$ and $V_{(12)}^{(\alpha)} = {}'V^{(\alpha)}$ in (6.1.30), using (4.2.41) yields

$$V_{22}^{(\alpha/q/12)} = \left[1 - {}'V_{22}^{(\alpha/q)} \mathcal{I}_2^{(0q)} V_{22}^{(12/q)} \mathcal{I}_2^{(0q)}\right]^{-1} {}'V_{22}^{(\alpha/q)} \tag{6.1.34a}$$

$$\sim {}'V_{22}^{(\alpha/q)} , \tag{6.1.34b}$$

where the last is valid either when $V_{22}^{(12)} = 0$ (the condition of no reflection), or when the incoherent multiple reflection through $\mathcal{I}_2^{(0q)}$, between the scatterer and the boundary is negligible; and further, ${}'V_{22}^{(\alpha/q)} \sim V_{22}^{(\alpha/q)}$ when the scatterer is separated far enough from the boundary, as compared with the wave coherence distance, say $\gamma_2 L \gg 1$. The situation is also the same for the other elements of $V^{(\alpha/q/12)}$. To write ${}'V^{(\alpha)}$ specifically, we observe that

$$V^{(12+\alpha)}(1;2) \equiv V^{(12)}(1;2) + {}'V^{(\alpha)}(1;2) \tag{6.1.35a}$$

$$\simeq T^{(12+\alpha)*}(1) T^{(12+\alpha)}(2) , \tag{6.1.35b}$$

where, by using formula (6.1.16), $T^{(12+\alpha)}$ can be divided into two parts as

$$T^{(12+\alpha)} = T^{(12)} + {}'T^{(\alpha)} , \tag{6.1.36}$$

with

$$'T^{(\alpha)} = (1 + T^{(12)}G^{(0)})T^{(\alpha/12)}(G^{(0)}T^{(12)} + 1) , \tag{6.1.37a}$$

$$T^{(\alpha/12)} = [1 - T^{(\alpha)}G^{(0)}T^{(12)}G^{(0)}]^{-1}T^{(\alpha)} . \tag{6.1.37b}$$

Hence,

$$V^{(12)}(1;2) = T^{(12)*}(1)T^{(12)}(2) , \tag{6.1.38a}$$

$$'V^{(\alpha)}(1;2) = {}'T^{(\alpha)*}(1)'T^{(\alpha)}(2) , \tag{6.1.38b}$$

on neglect of the interference terms. Here, since the only non-vanishing element of $T^{(\alpha)}$ is $T_{22}^{(\alpha)}$,

$$'T_{11}^{(\alpha)} = T_{12}^{(12)}G_2^{(0)}T_{22}^{(\alpha/12)}G_2^{(0)}T_{21}^{(12)} , \tag{6.1.39a}$$

$$'T_{12}^{(\alpha)} = T_{12}^{(12)}G_2^{(0)}T_{22}^{(\alpha/12)}(G_2^{(0)}T_{22}^{(12)} + 1) , \tag{6.1.39b}$$

$$'T_{22}^{(\alpha)} = (1 + T_{22}^{(12)}G_2^{(0)})T_{22}^{(\alpha/12)}(G_2^{(0)}T_{22}^{(12)} + 1) ; \tag{6.1.39c}$$

and, as $\gamma_2 L \to \infty$, $'T^{(\alpha)} \to T^{(\alpha)}$ with the only non-vanishing element, $T_{22}^{(\alpha)}$.

When $\gamma_2 L \gg 1$, the first three terms in (6.1.33b) are negligible in view of (6.1.39a–c) and (6.1.38b), showing that the non-vanishing element of $'V^{(\alpha)}$ is then given by

$$'V_{22}^{(\alpha)}(1;2) \sim V_{22}^{(\alpha)}(1;2) = T_{22}^{(\alpha)*}(1)T_{22}^{(\alpha)}(2) . \tag{6.1.40}$$

Hence, to the approximation of (6.1.34b), (6.1.33b) gives

$$V_{11}^{(\alpha/q+12)} \simeq V_{12}^{(12/q)}I_2^{(0q)}V_{22}^{(\alpha/q)}I_2^{(0q)}V_{21}^{(12/q)} , \tag{6.1.41}$$

On the other hand, when $\gamma_2 L \ll 1$, the first term of (6.1.33b) is dominant, which is mostly of coherent characteristic and, using an equation similar to (6.1.34b), is given by

$$V_{11}^{(\alpha/q+12)} \simeq V_{11}^{(\alpha/q/12)} \simeq {}'V_{11}^{(\alpha/q)} \simeq {}'V_{11}^{(\alpha)} , \tag{6.1.42}$$

where, from (6.1.38b,39a),

$$'V_{11}^{(\alpha)} = V_{12}^{(12)}U_2|T_{22}^{(\alpha/12)}|^2U_2V_{21}^{(12)} , \tag{6.1.43}$$

in terms of $V^{(12)}$ of (6.1.38a).

6.2 Power Equations and Optical Relations

We first divide $U^{(\alpha)}$ and $I^{(q+\alpha)}$ of (6.1.19,22) into two parts:

$$U^{(\alpha)} = U + \Delta U^{(\alpha)} \ , \tag{6.2.1}$$

$$I^{(q+\alpha)} = I^{(0q)} + \Delta I^{(\alpha)} \ , \tag{6.2.2}$$

and investigate power equations for the changes $\Delta U^{(\alpha)}$ and $\Delta I^{(\alpha)}$ caused by the scatterer. These are obtained from power equations (4.1.36) for $U^{(C)}$, and (4.2.107) for $I^{(0q)}$, by the replacement of $U^{(C)} \rightarrow U^{(\alpha)}$ and $I^{(0q)} \rightarrow I^{(q+\alpha)}$, respectively. Hence,

$$\left(\frac{\partial}{\partial \hat{\boldsymbol{x}}} \cdot \hat{\boldsymbol{\alpha}} + \Gamma\right) \Delta U^{(\alpha)}(\hat{\boldsymbol{x}}) + \Delta \Gamma^{(\alpha)} U^{(\alpha)}(\hat{\boldsymbol{x}}) = \Delta^2 G^{(\alpha)}(\hat{\boldsymbol{x}}) \ , \tag{6.2.3}$$

$$\frac{\partial}{\partial \hat{\boldsymbol{x}}} \cdot \hat{\boldsymbol{\alpha}} \Delta I^{(\alpha)}(\hat{\boldsymbol{x}}) = \Delta^2 G^{(\alpha)}(\hat{\boldsymbol{x}}) \ , \tag{6.2.4}$$

where

$$\Delta \Gamma^{(\alpha)}(1;2) = (2\mathrm{i})^{-1} \left[\Delta M^{(\alpha)*}(1) - \Delta M^{(\alpha)}(2)\right] \ , \tag{6.2.5}$$

$$\Delta^2 G^{(\alpha)}(1;2) = (2\mathrm{i})^{-1} \left[G^{(0)*} T^{(\alpha)*} G^{(0)*}(1) - G^{(0)} T^{(\alpha)} G^{(0)}(2)\right] \ . \tag{6.2.6}$$

Here, using

$$\Delta I^{(\alpha)} = I^{(0q)} V^{(\alpha/q)} I^{(0q)} \ , \tag{6.2.7}$$

$$\Delta U^{(\alpha)} = U V^{(\alpha)} U \ , \tag{6.2.8}$$

(6.2.4) can be expressed by

$$\Delta G^{(0)}(\hat{\boldsymbol{x}}) V^{(\alpha/q)} I^{(0q)} = \Delta^2 G^{(\alpha)}(\hat{\boldsymbol{x}}) \ , \tag{6.2.9}$$

with the aid of (4.2.107); and, in the same fashion, (6.2.3) is expressed by

$$\left[\Delta G^{(0)}(\hat{\boldsymbol{x}}) V^{(\alpha)} + \Delta \Gamma^{(\alpha')}(\hat{\boldsymbol{x}})\right] U = \Delta^2 G^{(\alpha)}(\hat{\boldsymbol{x}}) \ , \tag{6.2.10}$$

where

$$\Delta \Gamma^{(\alpha')} = \Delta \Gamma^{(\alpha)}(1 + U V^{(\alpha)}) \ . \tag{6.2.11}$$

Here, (6.2.9,10) have the same right-hand sides, which means an interaction term of the scatterer with the source, and tends to zero as the distance increases; while, on the left-hand sides,

$$\Delta G^{(0)}(\hat{\boldsymbol{x}}|1;2) = \delta(\hat{\boldsymbol{x}}|1;2)(2\mathrm{i})^{-1} \left[G^{(0)*}(1) - G^{(0)}(2)\right] \ , \tag{6.2.12}$$

from (2.3.16b), with $G \to G^{(0)}$ when $M = M^{(q)}$ [see also (6.1.6)].

Thus (6.2.9,10) provide local optical relations for $V^{(\alpha/q)}$ and $V^{(\alpha)}$, respectively, yielding

$$\Delta G^{(0)}(\hat{\boldsymbol{x}})V^{(\alpha/q)} = 0 \ , \tag{6.2.13}$$

$$\Delta G^{(0)}(\hat{\boldsymbol{x}})V^{(\alpha)} = -\Delta\Gamma^{(\alpha')}(\hat{\boldsymbol{x}}) \ , \tag{6.2.14}$$

similar to (2.3.17) [or (4.1.39b)] for $K^{(q)}$ and $\Gamma^{(q)}$; and, from (6.2.4),

$$\frac{\partial}{\partial \hat{\boldsymbol{x}}} \cdot \hat{\boldsymbol{\alpha}} \Delta I^{(\alpha)}(\hat{\boldsymbol{x}}) = 0 \ , \tag{6.2.15}$$

being divergence-free everywhere in the far region where the source term is negligible. Notice here that the right-hand side of (6.2.13) for $V^{(\alpha/q)}$ is zero, whereas that of (6.2.14) for $V^{(\alpha)}$ is $-\Delta\Gamma^{(\alpha')}$, implying that the additional part of $V^{(\alpha/q)}$ due to the medium fluctuation, i.e., $V^{(\alpha/q)} - V^{(\alpha)}$, contributes to the total cross section by the amount $+\Delta\Gamma^{(\alpha')}$, so that $V^{(\alpha/q)}$ makes no absorption of the incident power at all as the whole, Here, $\Delta\Gamma^{(\alpha')}$ is caused by the medium fluctuation, and is a contribution purely from the effective change of the scatterer, in view of (6.2.11,5) and (6.1.5).

6.3 Optical Cross Section and Shadowing Effect

Equation (6.2.7) for $\Delta I^{(\alpha)}$ can be written in optical form, according to the procedure of (2.3.33–35):

$$\begin{aligned} \Delta I^{(\alpha)}(\hat{\boldsymbol{\Omega}},\hat{\boldsymbol{\rho}}|\hat{\boldsymbol{\Omega}}',\hat{\boldsymbol{\rho}}') = \int d\hat{\boldsymbol{\rho}}''\, d\hat{\boldsymbol{\Omega}}'' \int d\hat{\boldsymbol{\rho}}'''\, d\hat{\boldsymbol{\Omega}}'''\; I^{(0q)}(\hat{\boldsymbol{\Omega}},\hat{\boldsymbol{\rho}}|\hat{\boldsymbol{\Omega}}'',\hat{\boldsymbol{\rho}}'') \\ \times V^{(\alpha/q)}(\hat{\boldsymbol{\Omega}}'',\hat{\boldsymbol{\rho}}''|\hat{\boldsymbol{\Omega}}''',\hat{\boldsymbol{\rho}}''') I^{(0q)}(\hat{\boldsymbol{\Omega}}''',\hat{\boldsymbol{\rho}}'''|\hat{\boldsymbol{\Omega}}',\hat{\boldsymbol{\rho}}') \ . \end{aligned} \tag{6.3.1}$$

With the same notations, (6.2.13) is written as

$$\int d\hat{\boldsymbol{\Omega}}\, V^{(\alpha/q)}(\hat{\boldsymbol{\Omega}},\hat{\boldsymbol{\rho}}|\hat{\boldsymbol{\Omega}}',\hat{\boldsymbol{\rho}}') = 0 \ ; \tag{6.3.2}$$

a similar optical relation is also obtained from (6.2.14) for $V^{(\alpha)}$ with a minor nonzero term on the right-hand side. Here, $V^{(\alpha/q)} \sim V^{(\alpha)}$ when $\Delta M^{(\alpha)} \sim 0$ so that $q^{(\alpha+\Delta\alpha)} \sim q^{\alpha}$ and $\Delta\Gamma^{(\alpha')} \sim 0$; and when the scatterer is sufficiently small compared with the wave coherence distance, the $V^{(\alpha)}$ has the form

$$V^{(\alpha)}(\hat{\boldsymbol{\Omega}},\hat{\boldsymbol{\rho}}|\hat{\boldsymbol{\Omega}}',\hat{\boldsymbol{\rho}}') \simeq V^{(\alpha)}(\hat{\boldsymbol{\Omega}}|\hat{\boldsymbol{\Omega}}')\delta(\hat{\boldsymbol{\rho}} - \hat{\boldsymbol{\rho}}_{\alpha})\delta(\hat{\boldsymbol{\rho}}' - \hat{\boldsymbol{\rho}}_{\alpha}) \ , \tag{6.3.3}$$

where $\widehat{\rho}_\alpha$ is the center coordinates of the scatterer. Hence, upon substitution of (6.3.3), relation (6.3.2) should lead, approximately, to the optical relation (6.1.24b) for $V^{(\alpha)}(\widehat{\Omega}|\widehat{\Omega}')$.

The power equation (6.2.15) for $\Delta I^{(\alpha)}$ is also optically written as

$$\frac{\partial}{\partial\widehat{\rho}}\cdot\int d\widehat{\Omega}\,\widehat{\Omega}\Delta I^{(\alpha)}(\widehat{\Omega},\widehat{\rho}|\widehat{\Omega}',\widehat{\rho}')=0\ , \tag{6.3.4}$$

which is consistent with (6.3.1), as can be shown by using optical relation (6.3.2) and

$$\frac{\partial}{\partial\widehat{\rho}}\cdot\int d\widehat{\Omega}\,\widehat{\Omega}I^{(0q)}(\widehat{\Omega},\widehat{\rho}|\widehat{\Omega}',\widehat{\rho}')=\delta(\widehat{\rho}-\widehat{\rho}')\ , \tag{6.3.5}$$

which is an optical version of (4.2.107).

To prove expression (6.1.24a) for $V^{(\alpha)}(\widehat{\Omega}|\widehat{\Omega}')$, we show here that the Fourier transform of $V^{(\alpha)}$, from (6.1.20), is

$$\begin{aligned}\widetilde{V}^{(\alpha)}(\widehat{\boldsymbol{u}}|\widehat{\boldsymbol{u}}')&=\left|\widetilde{T}^{(\alpha)}(\widehat{\boldsymbol{u}}|\widehat{\boldsymbol{u}}')\right|^2+(2\pi)^3\delta(\widehat{\boldsymbol{u}}-\widehat{\boldsymbol{u}}')\\&\times\left[\widetilde{T}^{(\alpha)*}(\widehat{\boldsymbol{u}}|\widehat{\boldsymbol{u}})\widetilde{G}^{(0)^{-1}}(\widehat{\boldsymbol{u}})+\widetilde{T}^{(\alpha)}(\boldsymbol{u}|\widehat{\boldsymbol{u}})\widetilde{G}^{(0)*^{-1}}(\widehat{\boldsymbol{u}})\right]\ ,\end{aligned} \tag{6.3.6}$$

where $\widehat{\boldsymbol{\lambda}}=\widehat{\boldsymbol{\lambda}}'=0$, and leads to the optical expression (6.1.24a) with

$$\sigma^{(\alpha)}(\widehat{\Omega}|\widehat{\Omega}')=(4\pi)^{-2}\left|\widetilde{T}^{(\alpha)}(\widehat{\boldsymbol{u}}=k\widehat{\Omega}|\widehat{\boldsymbol{u}}'=k\widehat{\Omega}')\right|^2\ , \tag{6.3.7a}$$

$$\gamma^{(\alpha)}(\widehat{\Omega})=(2ik)^{-1}\left(\widetilde{T}^{(\alpha)*}-\widetilde{T}^{(\alpha)}\right)\left(\widehat{\boldsymbol{u}}=k\widehat{\Omega}|\widehat{\boldsymbol{u}}'=k\widehat{\Omega}\right)\ . \tag{6.3.7b}$$

Here, the term (6.3.7a) is obtained directly from the first term of (6.3.6), according to the procedure of (2.3.33–35). For the first term in [ ] of the second term of (6.3.6), we first observe that it is nonzero only when $\widehat{\boldsymbol{u}}=k\widehat{\Omega}$ and $\widehat{\boldsymbol{u}}'=k\widehat{\Omega}'$ are in the same direction; and then that its total scattered power due to this term is given by [refer to (2.3.17)]

$$\int d\widehat{\boldsymbol{x}}\,\Delta G^{(0)}(\widehat{\boldsymbol{x}}|1;2)T^{(\alpha)*}(1)\left[G^{(0)}(2)\right]^{-1}\ . \tag{6.3.8}$$

This is expressed in terms of the corresponding Fourier transform (6.3.6) as

$$\begin{aligned}&\Delta\widetilde{G}^{(0)}(\widehat{\boldsymbol{u}}')\widetilde{T}^{(\alpha)*}(\widehat{\boldsymbol{u}}'|\widehat{\boldsymbol{u}}')\left[\widetilde{G}^{(0)}(\widehat{\boldsymbol{u}}')\right]^{-1}\\&=(2i)^{-1}\left[\widetilde{G}^{(0)*}(\widehat{\boldsymbol{u}}')/\widetilde{G}^{0}(\widehat{\boldsymbol{u}}')-1\right]\widetilde{T}^{(\alpha)*}(\widehat{\boldsymbol{u}}'|\widehat{\boldsymbol{u}}')\\&=-(2i)^{-1}\widetilde{T}^{(\alpha)*}(\widehat{\boldsymbol{u}}'|\widehat{\boldsymbol{u}}')\ ,\end{aligned} \tag{6.3.9}$$

for any incident wave of $\widehat{\boldsymbol{u}}'$ subject to

$$1/\widetilde{G}^{(0)}(\widehat{\boldsymbol{u}}') = (\widehat{\boldsymbol{u}}')^2 - k^2 = 0 \ .$$

The same is also true for the second term in [ ] of (6.3.6), whose contribution is the complex conjugate of (6.3.9). Hence the sum is given, in view of (6.3.7b), by $-k\gamma^{(\alpha)}(\widehat{\boldsymbol{\Omega}}')$, and agrees with that which should be from the second term of (6.1.24a). A more direct proof can also be given. Note that the cross section (6.1.24a) becomes critically negative in the shadow direction, meaning absorption of the wave, instead of scattering, and this comes from the interference terms of $V^{(\alpha)}$ (6.1.20).

6.4 Observation of a Fixed Scatterer Embedded in a Semi-Infinite Random Layer

The entire contribution from the fixed scatterer, $q^{(\alpha)}$, to $I_{11}^{(q+12+\alpha)}$ in this case (Fig. 6.2), say $\Delta I_{11}^{(\alpha/q+12)}$, is given by the second term of (6.1.33a), i.e.,

$$\Delta I_{11}^{(\alpha/q+12)} = U_1 V_{11}^{(\alpha/q+12)} U_1 \ . \tag{6.4.1}$$

Here, when the source and the observer are both separated far enough from the boundary, its intensity is given by the asymptotic expression of a form

$$\begin{aligned}\Delta I_{11}^{(\alpha/q+12)}(\widehat{\boldsymbol{\rho}}|\widehat{\boldsymbol{\rho}}') &\equiv \int d\widehat{\boldsymbol{\Omega}}\, d\widehat{\boldsymbol{\Omega}}'\, \Delta I_{11}^{(\alpha/q+12)}(\widehat{\boldsymbol{\Omega}}, \widehat{\boldsymbol{\rho}}|\widehat{\boldsymbol{\Omega}}', \widehat{\boldsymbol{\rho}}') \\ &\sim \int d\overline{\boldsymbol{\rho}}\, |\widehat{\boldsymbol{\rho}} - \overline{\boldsymbol{\rho}}|^{-2} V_{11}^{(\alpha/q+12)}(\widehat{\boldsymbol{\Omega}}|\overline{\boldsymbol{\rho}}|\widehat{\boldsymbol{\Omega}}')|\overline{\boldsymbol{\rho}} - \widehat{\boldsymbol{\rho}}'|^{-2} \ ,\end{aligned} \tag{6.4.2}$$

similar to (5.1.28), with the relative coordinates $\overline{\boldsymbol{\rho}}$ and $\Delta\boldsymbol{\rho}$ of (5.1.20), and the unit vectors $\widehat{\boldsymbol{\Omega}}$ and $\widehat{\boldsymbol{\Omega}}'$ in the directions of $\widehat{\boldsymbol{\rho}} - \overline{\boldsymbol{\rho}}$ and $\overline{\boldsymbol{\rho}} - \widehat{\boldsymbol{\rho}}'$, respectively. Here, the right-hand side is a surface integral over the boundary, wherein $V_{11}^{(\alpha/q+12)}(\widehat{\boldsymbol{\Omega}}|\overline{\boldsymbol{\rho}}|\widehat{\boldsymbol{\Omega}}')$ means the scatterer's cross section per unit area at $\overline{\boldsymbol{\rho}}$ on S_1, and is given by (6.1.33b) in the original matrix form; when the boundary is perfectly smooth, it is reduced to (6.1.41) and (6.1.42,43), depending on whether $\gamma_2 L \gg 1$ and $\gamma_2 L \ll 1$, respectively. Here, since $d\overline{\boldsymbol{\rho}} = |\Omega_z|^{-1} |\widehat{\boldsymbol{\rho}} - \overline{\boldsymbol{\rho}}|^2 d\widehat{\boldsymbol{\Omega}}$, (6.4.2) can be written as

$$\Delta I_{11}^{(\alpha/q+12)}(\widehat{\boldsymbol{\rho}}|\widehat{\boldsymbol{\rho}}') \sim \int d\widehat{\boldsymbol{\Omega}}\, |\Omega_z|^{-1} V_{11}^{(\alpha/q+12)}(\widehat{\boldsymbol{\Omega}}|\overline{\boldsymbol{\rho}}|\widehat{\boldsymbol{\Omega}}')|\overline{\boldsymbol{\rho}} - \widehat{\boldsymbol{\rho}}'|^{-2} \ , \tag{6.4.3}$$

in which the integrand means the wave flux per unit solid angle in the direction $\widehat{\boldsymbol{\Omega}}$.

When $\gamma_2 L \gg 1$, and by applying formulas (4.2.4a,b) to $V^{(12/q)}$, (6.1.41) can be written as

$$V_{11}^{(\alpha/q+12)} = V_{12}^{(12)} \mathcal{I}_{22}^{(q/12)} V_{22}^{(\alpha/q)} \mathcal{I}_{22}^{(q/12)} V_{21}^{(12)} \ . \tag{6.4.4}$$

Hence we obtain the optical expression of $V_{11}^{(\alpha/q+12)}$ in (6.4.2,3) in the form

$$V_{11}^{(\alpha/q+12)}(\widehat{\Omega}|\overline{\rho}|\widehat{\Omega}') = V_{12}^{(12)}(\widehat{\Omega}) \mathcal{I}_{22}^{(\alpha+q/12)}(\widehat{\Omega}|\overline{\rho}|\widehat{\Omega}') V_{21}^{(12)}(\widehat{\Omega}') \ . \tag{6.4.5}$$

Here, from (5.2.53),

$$V_{12}^{(12)}(\widehat{\Omega}) = \left|\Omega_z \langle R_{12}^{(12)}(\widehat{\Omega})\rangle^2\right| \ , \tag{6.4.6}$$

and, when the scatterer is located at $\widehat{\rho}_\alpha = (\rho_\alpha, -L)$ with $V^{(\alpha)}$ of (6.3.3), we obtain, on referring to (5.1.29a),

$$\begin{aligned} \mathcal{I}_{22}^{(\alpha+q/12)}(\widehat{\Omega}|\overline{\rho}|\widehat{\Omega}') = & \int d\widehat{\Omega}''\, d\widehat{\Omega}''' \int_{-\infty}^{\infty} d(\Delta\rho) \\ & \times \mathcal{I}_{22}^{(q/12)}(\widehat{\Omega}, z=0|\rho-\rho_\alpha|\widehat{\Omega}'', z''=-L) V^{(\alpha)}(\widehat{\Omega}''|\widehat{\Omega}''') \\ & \times \mathcal{I}_{22}^{(q/12)}(\widehat{\Omega}''', z'''=-L|\rho_\alpha-\rho'|\widehat{\Omega}', z'=0) \ , \end{aligned} \tag{6.4.7}$$

where, presently,

$$\rho = \overline{\rho} + \Delta\rho/2, \qquad \rho' = \overline{\rho} - \Delta\rho/2 \ . \tag{6.4.8}$$

We can finally write (6.4.7) in the form

$$\begin{aligned} \mathcal{I}_{22}^{(\alpha+q/12)}(\widehat{\Omega}|\overline{\rho}|\widehat{\Omega}') = & \int d\widehat{\Omega}''\, d\widehat{\Omega}'''\, \mathcal{I}_L^{(q/12)}(\widehat{\Omega}, \widehat{\Omega}'|\widehat{\Omega}'', \widehat{\Omega}''') \\ & \times V^{(\alpha)}(\widehat{\Omega}''|\widehat{\Omega}''') \ . \end{aligned} \tag{6.4.9}$$

Here,

$$\begin{aligned} \mathcal{I}_L^{(q/12)}(\widehat{\Omega}, \widehat{\Omega}'|\widehat{\Omega}'', \widehat{\Omega}''') = & \, 4(2\pi)^{-2} \int d\boldsymbol{\lambda} \exp[-i2\boldsymbol{\lambda}\cdot(\overline{\rho}-\rho_\alpha)] \\ & \times \mathcal{I}_{22}^{(q/12)}(\widehat{\Omega}, 0|\boldsymbol{\lambda}|\widehat{\Omega}'', -L) \mathcal{I}_{22}^{(q/12)}(\widehat{\Omega}''', -L|-\boldsymbol{\lambda}|\widehat{\Omega}', 0) \ , \end{aligned} \tag{6.4.10}$$

in which the Fourier transform, $\mathcal{I}_{22}^{(q/12)}(\widehat{\Omega}, z|\boldsymbol{\lambda}|\widehat{\Omega}', z')$, with the Fourier variable, $\boldsymbol{\lambda}$, as a parameter may be obtained according to (4.1.62), in terms of $\mathcal{I}_2^{(0q)}$ as the boundary-value solution of the transport equation (5.1.5), subjected to (5.1.7a) at $z = 0$.

Here, the $\mathcal{I}_{22}^{(q/12)}$'s involved in (6.4.10) are connected to $S_{22}^{(q/12)}(\widehat{\Omega}, z|\boldsymbol{\lambda}|\widehat{\Omega}', z')$, e.g., through the relation

$$\mathcal{I}_{22}^{(q/12)}(\widehat{\Omega}, 0|\boldsymbol{\lambda}|\widehat{\Omega}'', -L) = |\Omega_z \Omega_z''|^{-1} S_{0,-L}^{(q/12)}(\widehat{\Omega}|\boldsymbol{\lambda}|\widehat{\Omega}'') \ , \tag{6.4.11}$$

where

$$S_{0,-L}^{(q/12)}(\widehat{\boldsymbol{\Omega}}|\boldsymbol{\lambda}|\widehat{\boldsymbol{\Omega}}'') = \int_{-\infty}^{0} dz \int_{-L}^{0} dz'' \exp\{i[\lambda_z z - \lambda_z''(z''+L)]\} \times S_{22}^{(q/12)}(\widehat{\boldsymbol{\Omega}}, z|\boldsymbol{\lambda}|\widehat{\boldsymbol{\Omega}}'', z'') . \tag{6.4.12}$$

In this expression Ω_z, $\Omega_z'' > 0$, in the same fashion as (5.1.32), and means the optical cross section (per unit area) of the present random volume (distributed over the range $z < 0$) is for the wave incident at $z = -L$ in the direction $\widehat{\boldsymbol{\Omega}}''$ and emitted at $z = 0$ in the direction $\widehat{\boldsymbol{\Omega}}$.

7. Forward Scattering Approximation

In turbulent air, the refractive index has a fluctuation of the order of magnitude 10^{-6} and the correlation distance of several meters. Therefore, the medium can be assumed to be subject to Gaussian statistics in most cases, fulfilling the condition described in Sect. 1.3. Also, the forward scattering approximation is possible in the case of light wave propagation through air where the wave is scattered mostly in the forward direction. This enables us to write the moment equations of wave functions, as treated by various authors [7.1–6], in a form similar to the Schrödinger equation in many body problems, and to utilize some of the operator methods introduced therein to obtain the solutions.

7.1 Moment Equations of a Light Wave in a Turbulent Medium

When a medium changes slightly and also slowly enough in space so that the change is negligible within a distance of the order of the wave length, then the wave function of a beam wave with the main direction of propagation along the z-axis, say, changes mostly by the phase factor $\exp(-ikz)$. Therefore, the wave equations (2.1.7a,b) are considerably simplified by the replacement of

$$\psi(\hat{x}) \to \psi(\hat{x})\exp(-ikz), \qquad j(\hat{x}) \to j(\hat{x})\exp(-ikz) \ , \tag{7.1.1}$$

which yields the wave equations of the form

$$[L_x^* - q(\hat{x})]\,\psi^*(\hat{x}) = j^*(\hat{x}) \ , \tag{7.1.2a}$$
$$[L_y - q(\hat{y})]\,\psi(\hat{y}) = j(\hat{y}) \ . \tag{7.1.2b}$$

In this chapter, the coordinate vector of the complex conjugate wave function is denoted by $\hat{x} = (\vec{x}, z)$ with the two-dimensional coordinate vector, $\vec{x} = (x_1, x_2)$, orthogonal to the z-axis, while the coordinate vector of the original wave function is denoted by $\hat{y} = (\vec{y}, z)$ with $\vec{y} = (y_1, y_2)$; and

$$L_x^* = -2ik\frac{\partial}{\partial z} - \left(\frac{\partial}{\partial \vec{x}}\right)^2, \qquad L_y = +2ik\frac{\partial}{\partial z} - \left(\frac{\partial}{\partial \vec{y}}\right)^2, \tag{7.1.3}$$

upon neglect of the $(\partial/\partial z)^2$ terms. Here, comparing with the original wave equations (2.1.7), the differences are only in the replacement of L by the $\widehat{x}$ operator, L_x^*, for the complex conjugate wave function; and by the $\widehat{y}$ operator, L_y, for the original wave function. Therefore, the basic equations in Sects. 2.1,2 for a medium of independent particles, also remain unchanged for the present wave equations. For example, the first order Green's functions are still given by (2.1.9a,b), hence, from (2.2.1),

$$[L_y - \widehat{q}(\widehat{y})]\,G(\widehat{y}|\widehat{y}') = \delta(\widehat{y} - \widehat{y}') \;, \tag{7.1.4}$$

where, since the present random medium is subject to Gaussian statistics, the operator, $\widehat{q}(\widehat{y})$, is given according to (1.3.13), by

$$\widehat{q}(\widehat{y}) = C(\widehat{y}) + \int d\widehat{y}'\, D(\widehat{y} - \widehat{y}')\frac{\delta}{\delta C(\widehat{y}')} \;, \tag{7.1.5a}$$

$$D(\widehat{y} - \widehat{y}') = \langle q(\widehat{y})q(\widehat{y}')\rangle \;; \tag{7.1.5b}$$

the second order Green's functions are similarly given by (2.1.10a–c). In this section, the moment equations are derived and written in a simple form for use in the following sections [7.5].

From (2.1.9a), we can write

$$G(\widehat{y}|\widehat{y}') = \widehat{G}_q(\widehat{y}|\widehat{y}')Z_0 \;. \tag{7.1.6}$$

Here,

$$\delta\widehat{G}_q(\widehat{y}|\widehat{y}') = \int d\widehat{y}''\, \widehat{G}_q(\widehat{y}|\widehat{y}'')\delta C(\widehat{y}'')\widehat{G}(\widehat{y}''|\widehat{y}') \;, \tag{7.1.7a}$$

so that

$$\frac{\delta}{\delta C(\widehat{y}'')}\widehat{G}_q(\widehat{y}|\widehat{y}') = \widehat{G}_q(\widehat{y}|\widehat{y}'')\widehat{G}_q(\widehat{y}''|\widehat{y}') \;. \tag{7.1.7b}$$

Hence, using (7.1.5a,b),

$$\widehat{q}(\widehat{y})G(\widehat{y}|\widehat{y}')\big|_{C=0} = \int d\widehat{y}''\, D(\widehat{y} - \widehat{y}'')\widehat{G}_q(\widehat{y}|\widehat{y}'')G(\widehat{y}''|\widehat{y}') \tag{7.1.8}$$

$$\equiv \int d\widehat{y}''\, M(\widehat{y}|\widehat{y}'')G(\widehat{y}''|\widehat{y}') \;, \tag{7.1.9}$$

where, in the limit, $C(\widehat{y}) = 0$, in which $G(\widehat{y}|\widehat{y}'') = G(\widehat{y} - \widehat{y}'')$,

$$M(\widehat{y}|\widehat{y}'') \simeq D(\widehat{y} - \widehat{y}'')G(\widehat{y} - \widehat{y}'') \;, \tag{7.1.10}$$

In the last derivation, the approximation has been made based on the observation that the factor $\widehat{G}_q(\widehat{y}|\widehat{y}'')$ in (7.1.8) has a negligible correlation with the wave incident at $\widehat{y}''$, either because q is small enough so that its effect is negligible over the entire nonzero range of the factor $D(\hat{\boldsymbol{y}}-\hat{\boldsymbol{y}}'')$, or because the wave is scattered mostly in the forward direction, seldom being backscattered to the same path as once passed by the incident wave. Thus (7.1.4) becomes written, in matrix form, as

$$(L-M)G=1 \ ; \tag{7.1.11a}$$

in the same way

$$(L^*-M^*)G^*=1 \ . \tag{7.1.11b}$$

Also, for the second order moments of the wave functions, defined by

$$M_{11}(\hat{\boldsymbol{x}};\hat{\boldsymbol{y}})=\langle\psi^*(\hat{\boldsymbol{x}})\psi(\hat{\boldsymbol{y}})\rangle \ , \tag{7.1.12a}$$

$$M_{20}(\hat{\boldsymbol{x}}_1,\hat{\boldsymbol{x}}_2)=\langle\psi^*(\hat{\boldsymbol{x}}_1)\psi^*(\hat{\boldsymbol{x}}_2)\rangle \ , \tag{7.1.12b}$$

$$M_{02}(\hat{\boldsymbol{y}}_1,\hat{\boldsymbol{y}}_2)=\langle\psi(\hat{\boldsymbol{y}}_1)\psi(\hat{\boldsymbol{y}}_2)\rangle \ , \tag{7.1.12c}$$

we can obtain the equations is the same fashion. For example, from wave equations (7.1.2a,b),

$$\langle[L_x^*-q(\hat{\boldsymbol{x}})]\psi^*(\hat{\boldsymbol{x}})\psi(\hat{\boldsymbol{y}})\rangle=j^*(\hat{\boldsymbol{x}})\langle\psi(\hat{\boldsymbol{y}})\rangle \ . \tag{7.1.13}$$

Here, for the left-hand side, using formula (1.3.10) we obtain

$$\langle q(\hat{\boldsymbol{x}})\psi^*(\hat{\boldsymbol{x}})\psi(\hat{\boldsymbol{y}})\rangle=\bar{q}(\hat{\boldsymbol{x}})\langle\psi^*(\hat{\boldsymbol{x}})\psi(\hat{\boldsymbol{y}})\rangle \tag{7.1.14a}$$

$$\simeq\int d\hat{\boldsymbol{x}}'\,D(\hat{\boldsymbol{x}}-\hat{\boldsymbol{x}}')G^*(\hat{\boldsymbol{x}}|\hat{\boldsymbol{x}}')\langle\psi^*(\hat{\boldsymbol{x}}')\psi(\hat{\boldsymbol{y}})\rangle$$

$$+\int d\hat{\boldsymbol{y}}'\,D(\hat{\boldsymbol{x}}-\hat{\boldsymbol{y}}')G(\hat{\boldsymbol{y}}|\hat{\boldsymbol{y}}')\langle\psi^*(\hat{\boldsymbol{x}})\psi(\hat{\boldsymbol{y}}')\rangle \ , \tag{7.1.14b}$$

with the aid of

$$\frac{\delta}{\delta C(\hat{\boldsymbol{x}}')}\psi^*(\hat{\boldsymbol{x}})=G_q^*(\hat{\boldsymbol{x}}|\hat{\boldsymbol{x}}')\psi^*(\hat{\boldsymbol{x}}'),\qquad \text{etc.} \ , \tag{7.1.15}$$

from (7.1.7b). In terms of M of (7.1.10), and a new function, $\Lambda(\hat{\boldsymbol{x}},\hat{\boldsymbol{y}})$, defined by

$$\Lambda(\hat{\boldsymbol{x}}-\hat{\boldsymbol{y}},\hat{\boldsymbol{y}}-\hat{\boldsymbol{y}}')=D(\hat{\boldsymbol{x}}-\hat{\boldsymbol{y}}')G(\hat{\boldsymbol{y}}-\hat{\boldsymbol{y}}') \tag{7.1.16a}$$

with the relation

$$M(\hat{\boldsymbol{y}}-\hat{\boldsymbol{y}}')=\Lambda(0,\hat{\boldsymbol{y}}-\hat{\boldsymbol{y}}') \tag{7.1.16b}$$

(7.1.3) can be written as:

$$L_x^*M_{11}(\hat{\boldsymbol{x}};\hat{\boldsymbol{y}})-\int d\hat{\boldsymbol{x}}'\,\Lambda^*(0,\hat{\boldsymbol{x}}-\hat{\boldsymbol{x}}')M_{11}(\hat{\boldsymbol{x}}';\hat{\boldsymbol{y}})$$

$$-\int d\hat{\boldsymbol{y}}'\Lambda(\hat{\boldsymbol{x}}-\hat{\boldsymbol{y}},\hat{\boldsymbol{y}}-\hat{\boldsymbol{y}}')M_{11}(\hat{\boldsymbol{x}};\hat{\boldsymbol{y}}')=j^*(\hat{\boldsymbol{x}})\langle\psi(\hat{\boldsymbol{y}})\rangle \ . \tag{7.1.17}$$

Here, in view of the wave function redefined by (7.1.1), the change of M_{11} is negligible within a distance of the order of the medium correlation distance, so that the factors M_{11} in the integrands of (7.1.17) can be regarded as being constant over the whole range of integration. Hence, with a function $\Lambda(\widehat{\boldsymbol{x}}-\widehat{\boldsymbol{y}})$ defined by

$$\Lambda(\widehat{\boldsymbol{x}}-\widehat{\boldsymbol{y}})=\int_{-\infty}^{\infty} d\widehat{\boldsymbol{y}}'\,\Lambda(\widehat{\boldsymbol{x}}-\widehat{\boldsymbol{y}},\widehat{\boldsymbol{y}}-\widehat{\boldsymbol{y}}')\ , \tag{7.1.18}$$

(7.1.17) can be approximated by

$$\left[L_x^*-\Lambda^*(0)-\Lambda(\widehat{\boldsymbol{x}}-\widehat{\boldsymbol{y}})\right]M_{11}(\widehat{\boldsymbol{x}};\widehat{\boldsymbol{y}})=j^*(\widehat{\boldsymbol{x}})\langle\psi(\widehat{\boldsymbol{y}})\rangle\ . \tag{7.1.19}$$

Equation (7.1.19) is written with respect to the coordinates $\widehat{\boldsymbol{x}}$, with the other coordinates, $\widehat{\boldsymbol{y}}$, fixed. For the latter, we similarly obtain the following equation:

$$\left[L_y-\Lambda(0)-\Lambda^*(\widehat{\boldsymbol{y}}-\widehat{\boldsymbol{x}})\right]M_{11}(\widehat{\boldsymbol{x}};\widehat{\boldsymbol{y}})=j(\widehat{\boldsymbol{y}})\langle\psi^*(\widehat{\boldsymbol{x}})\rangle\ . \tag{7.1.20}$$

Also, for the higher order moment functions, $M_{\mu\nu}$, defined by

$$\begin{aligned}M_{\mu\nu}&(\widehat{\boldsymbol{x}}_1,\dots,\widehat{\boldsymbol{x}}_\mu;\widehat{\boldsymbol{y}}_1,\dots,\widehat{\boldsymbol{y}}_\nu)\\&=\langle\psi^*(\widehat{\boldsymbol{x}}_1)\dots\psi^*(\widehat{\boldsymbol{x}}_\mu)\psi(\widehat{\boldsymbol{y}}_1)\dots\psi(\widehat{\boldsymbol{y}}_\nu)\rangle\ ,\end{aligned} \tag{7.1.21}$$

we can obtain equations in the same fashion and they are summarized as follows:

$$\begin{aligned}&\left[L_{x_1}^*-\Lambda^*(0)-\sum_{m\neq1}^{\mu}\Lambda(\widehat{\boldsymbol{x}}_1-\widehat{\boldsymbol{x}}_m)-\sum_{n=1}^{\nu}\Lambda(\widehat{\boldsymbol{x}}_1-\widehat{\boldsymbol{y}}_n)\right]\\&\qquad\times M_{\mu\nu}(\widehat{\boldsymbol{x}}_1,\dots,\widehat{\boldsymbol{x}}_m,\dots,\widehat{\boldsymbol{x}}_\mu;\widehat{\boldsymbol{y}}_1,\dots,\widehat{\boldsymbol{y}}_n,\dots,\widehat{\boldsymbol{y}}_\nu)\\&\quad=j^*(\widehat{\boldsymbol{x}}_1)M_{\mu-1,\nu}(\widehat{\boldsymbol{x}}_2,\dots,\widehat{\boldsymbol{x}}_\mu;\widehat{\boldsymbol{y}}_1,\dots,\widehat{\boldsymbol{y}}_\nu)\ ,\end{aligned} \tag{7.1.22a}$$

$$\begin{aligned}&\left[L_{y_1}-\Lambda(0)-\sum_{n\neq1}^{\nu}\Lambda(\widehat{\boldsymbol{y}}_1-\widehat{\boldsymbol{y}}_n)-\sum_{m=1}^{\mu}\Lambda^*(\widehat{\boldsymbol{y}}_1-\widehat{\boldsymbol{x}}_m)\right]\\&\qquad\times M_{\mu\nu}(\widehat{\boldsymbol{x}}_1,\dots,\widehat{\boldsymbol{x}}_m,\dots,\widehat{\boldsymbol{x}}_\mu;\widehat{\boldsymbol{y}}_1,\dots,\widehat{\boldsymbol{y}}_n,\dots,\widehat{\boldsymbol{y}}_\nu)\\&\quad=j(\widehat{\boldsymbol{y}}_1)M_{\mu,\nu-1}(\widehat{\boldsymbol{x}}_1,\dots,\widehat{\boldsymbol{x}}_\mu;\widehat{\boldsymbol{y}}_2,\dots,\widehat{\boldsymbol{y}}_\nu)\ .\end{aligned} \tag{7.1.22b}$$

Here, the factor $G(\widehat{\boldsymbol{y}}-\widehat{\boldsymbol{y}}'')$ in (7.1.10) can be approximated by the Green's function in free space, say $G_0(\widehat{\boldsymbol{y}}-\widehat{\boldsymbol{y}}'')$, which is the solution of

$$\left[2\mathrm{i}k\frac{\partial}{\partial z}-\left(\frac{\partial}{\partial\vec{y}}\right)^2\right]G(\widehat{\boldsymbol{y}}-\widehat{\boldsymbol{y}}'')=\delta(\widehat{\boldsymbol{y}}-\widehat{\boldsymbol{y}}'')\ , \tag{7.1.23a}$$

in view of (7.1.3), and is given by the Fourier representation,

$$G_0(\hat{\boldsymbol{y}}) = (2\pi)^{-3} \int d\hat{\boldsymbol{\lambda}}\, \widetilde{G}_0(\hat{\boldsymbol{\lambda}}) \exp(-\mathrm{i}\hat{\boldsymbol{\lambda}} \cdot \hat{\boldsymbol{y}}) \tag{7.1.23b}$$

with

$$\widetilde{G}_0(\hat{\boldsymbol{\lambda}}) = \left(2k\lambda_z + \vec{\lambda}^2 + \mathrm{i}\varepsilon\right)^{-1}, \quad \varepsilon = +0 \ , \tag{7.1.23c}$$

where $\hat{\boldsymbol{\lambda}} = (\vec{\lambda}, \lambda_z)$; hence,

$$G_0(\hat{\boldsymbol{y}}) = \begin{cases} (4\pi z)^{-1} \exp\left[-\mathrm{i}k\vec{y}^{\,2}(2z)^{-1}\right], & z > 0 \\ 0, & z < 0 \ . \end{cases} \tag{7.1.23d}$$

Use of (7.1.18,16a) leads to an integral representation of the function, $\Lambda(\hat{\boldsymbol{x}})$, as

$$\Lambda(\hat{\boldsymbol{x}}) = (2\pi)^{-3} \int d\hat{\boldsymbol{\lambda}} \exp(-\mathrm{i}\hat{\boldsymbol{\lambda}} \cdot \hat{\boldsymbol{x}}) \widetilde{D}(\hat{\boldsymbol{\lambda}}) \widetilde{G}_0(-\hat{\boldsymbol{\lambda}}) \ . \tag{7.1.24}$$

Here, since the Fourier transform, $\widetilde{D}(\hat{\boldsymbol{\lambda}})$, is appreciable only within a range of $|\hat{\boldsymbol{\lambda}}| \ll k$, the term $\vec{\lambda}^2$ in (7.1.23c) is negligible for $\widetilde{G}_0(-\hat{\boldsymbol{\lambda}})$ in (7.1.24), yielding the Fourier inversion,

$$\begin{aligned} &(2\pi)^{-3} \int d\hat{\boldsymbol{\lambda}} \exp(-\mathrm{i}\hat{\boldsymbol{\lambda}} \cdot \hat{\boldsymbol{x}}) \widetilde{G}_0(-\hat{\boldsymbol{\lambda}}) \\ &\simeq \delta(\vec{x})(2\pi)^{-1} \int d\lambda_z \frac{\exp(-\mathrm{i}\lambda_z z)}{-2k\lambda_z + \mathrm{i}\varepsilon} \\ &= -\mathrm{i}(2k)^{-1}\delta(\vec{x})\delta_-(z) \ , \end{aligned} \tag{7.1.25}$$

where $\delta_-(z) = 1$ or 0, depending on $z \gtrless 0$, respectively. Hence (7.1.24) is found to be given by

$$\begin{aligned} \Lambda(\hat{\boldsymbol{x}}) &= -\mathrm{i}(2k)^{-1} \int_z^\infty dz'\, D(\vec{x}, z') \\ &= -\Lambda^*(\hat{\boldsymbol{x}}) \ , \end{aligned} \tag{7.1.26}$$

which is a purely imaginary function.

Thus the moment equations can be written in compact form by the introduction of a constant, γ, and a function, $V(\hat{\boldsymbol{x}})$, defined by

$$\gamma = \mathrm{i}\left(2k^2\right)^{-1}\Lambda(0) = \left(4k^3\right)^{-1} \int_0^\infty dz'\, D(0, z') > 0 \ , \tag{7.1.27a}$$

$$V(\widehat{\boldsymbol{x}}) = \mathrm{i}k^{-2}[\Lambda(0) - \Lambda(\widehat{\boldsymbol{x}})]$$
$$= (2k^3)^{-1}\left[\int_0^{\infty} dz' \left\{D(0,z') - D(\vec{\boldsymbol{x}},z')\right\} + \int_0^{z} dz' D(\vec{\boldsymbol{x}},z')\right] , \qquad (7.1.27b)$$

both of which have no physical dimension. Hence, using (7.1.16b,18) for M, (7.1.11a,b) are written as

$$\left[k^{-1}\frac{\partial}{\partial z} + \frac{\mathrm{i}}{2}\left(k^{-1}\frac{\partial}{\partial \vec{\boldsymbol{y}}}\right)^2 + \gamma\right]G(\widehat{\boldsymbol{y}} - \widehat{\boldsymbol{y}}') = -\mathrm{i}(2k^2)^{-1}\delta(\widehat{\boldsymbol{y}} - \widehat{\boldsymbol{y}}') , \qquad (7.1.28a)$$

$$\left[k^{-1}\frac{\partial}{\partial z} - \frac{\mathrm{i}}{2}\left(k^{-1}\frac{\partial}{\partial \vec{\boldsymbol{x}}}\right) + \gamma\right]G^*(\widehat{\boldsymbol{x}} - \widehat{\boldsymbol{x}}') = +i(2k^2)^{-1}\delta(\widehat{\boldsymbol{x}} - \widehat{\boldsymbol{x}}') . \qquad (7.1.28b)$$

In the same fashion, (7.1.19) becomes

$$\left[k^{-1}\frac{\partial}{\partial z} - \frac{\mathrm{i}}{2}\left(k^{-1}\frac{\partial}{\partial \vec{\boldsymbol{x}}}\right)^2 + \frac{1}{2}V(\widehat{\boldsymbol{x}} - \widehat{\boldsymbol{y}})\right]M_{11}(\widehat{\boldsymbol{x}}, \widehat{\boldsymbol{y}})$$
$$= \mathrm{i}(2k^2)^{-1}j^*(\widehat{\boldsymbol{x}})\langle\psi(\widehat{\boldsymbol{y}})\rangle , \qquad (7.1.29a)$$

while the equation of $M_{11}(\widehat{\boldsymbol{x}};\widehat{\boldsymbol{y}})$, with respect to another coordinate $\widehat{\boldsymbol{y}} = (\vec{\boldsymbol{y}}, z)$ for fixed $\widehat{\boldsymbol{x}}$, is obtained from (7.1.20) as

$$\left[k^{-1}\frac{\partial}{\partial z} + \frac{\mathrm{i}}{2}\left(k^{-1}\frac{\partial}{\partial \vec{\boldsymbol{y}}}\right)^2 + \frac{1}{2}V(\widehat{\boldsymbol{y}} - \widehat{\boldsymbol{x}})\right]M_{11}(\widehat{\boldsymbol{x}}; \widehat{\boldsymbol{y}})$$
$$= -\mathrm{i}(2k^2)^{-1}j(\widehat{\boldsymbol{y}})\langle\psi^*(\widehat{\boldsymbol{x}})\rangle . \qquad (7.1.29b)$$

Consequently, when the z components of the coordinates $\widehat{\boldsymbol{x}} = (\vec{\boldsymbol{x}}, z)$ and $\widehat{\boldsymbol{y}} = (\vec{\boldsymbol{y}}, z)$ are set as equal, the equation with respect to the common z is given by the sum of the two equations (7.1.29a,b); Hence, using the notations

$$M_{11}(\vec{\boldsymbol{x}}; \vec{\boldsymbol{y}}; z) = M_{11}(\widehat{\boldsymbol{x}}; \widehat{\boldsymbol{y}}) , \qquad (7.1.30a)$$
$$V(\vec{\boldsymbol{x}} - \vec{\boldsymbol{y}}) = V(\widehat{\boldsymbol{x}} - \widehat{\boldsymbol{y}}) , \qquad (7.1.30b)$$

whenever the coordinates $\widehat{\boldsymbol{x}}$ and $\widehat{\boldsymbol{y}}$ have the same z components, we obtain

$$\left[k^{-1}\frac{\partial}{\partial z} + \frac{\mathrm{i}}{2k^2}\left\{\left(\frac{\partial}{\partial \vec{\boldsymbol{y}}}\right)^2 - \left(\frac{\partial}{\partial \vec{\boldsymbol{x}}}\right)^2\right\} + V(\vec{\boldsymbol{x}} - \vec{\boldsymbol{y}})\right]M_{11}(\vec{\boldsymbol{x}}; \vec{\boldsymbol{y}}; z)$$
$$= \mathrm{i}(2k^2)^{-1}\left[j^*(\widehat{\boldsymbol{x}})\langle\psi(\widehat{\boldsymbol{y}})\rangle - j(\widehat{\boldsymbol{y}})\langle\psi^*(\widehat{\boldsymbol{x}})\rangle\right] . \qquad (7.1.31)$$

The same is also true for the moment equations (7.1.22a,b) of arbitrary order, and, when all the other coordinates are fixed, the equation for the coordinates $\widehat{\boldsymbol{x}}_1 = (\vec{\boldsymbol{x}}_1, z_1)$ becomes

$$\left[k^{-1}\frac{\partial}{\partial z_1}-\frac{i}{2}\left(k^{-1}\frac{\partial}{\partial \vec{x}_1}\right)^2+(\mu-\nu)\gamma-\sum_{m\neq 1}^{\mu}\frac{1}{2}V(\hat{x}_1-\hat{x}_m)\right.$$
$$\left.+\sum_{n=1}^{\nu}\frac{1}{2}V(\hat{x}_1-\hat{y}_n)\right]M_{\mu\nu}(\hat{x}_1,\ldots,\hat{x}_\mu;\hat{y}_1,\ldots,\hat{y}_\nu)$$
$$=i(2k^2)^{-1}j^*(\hat{x}_1)M_{\mu-1,\nu}(\hat{x}_2,\ldots,\hat{x}_\mu;\hat{y}_1,\ldots,\hat{y}_\nu)\ ; \qquad (7.1.32a)$$

while the corresponding equation for the coordinates $\hat{y}_1=(\vec{y}_1,z_1)$ becomes

$$\left[k^{-1}\frac{\partial}{\partial z_1}+\frac{i}{2}\left(k^{-1}\frac{\partial}{\partial \vec{y}_1}\right)^2+(\nu-\mu)\gamma-\sum_{n\neq 1}^{\nu}\frac{1}{2}V(\hat{y}_1-\hat{y}_n)\right.$$
$$\left.+\sum_{m=1}^{\mu}\frac{1}{2}V(\hat{y}_1-\hat{x}_m)\right]M_{\mu\nu}(\hat{x}_1,\ldots,\hat{x}_\mu;\hat{y}_1,\ldots,\hat{y}_\nu)$$
$$=-i(2k^2)^{-1}j(\hat{y}_1)M_{\mu,\nu-1}(\hat{x}_1,\ldots,\hat{x}_\mu;\hat{y}_2,\ldots,\hat{y}_\nu)\ . \qquad (7.1.32b)$$

Therefore, when the z coordinates are the same for all the coordinates so that $\hat{x}_m=(\vec{x}_m,z)$, $m=1, 2, \ldots, \mu$, and $\hat{y}_n=(\vec{y}_n,z)$, $n=1, 2, \ldots, \nu$, then, the equation for the moment, say $M_{\mu\nu}(\vec{x}_1,\ldots,\vec{x}_\mu;\vec{y}_1,\ldots,\vec{y}_\nu;z)$, with respect to the common z, is given by the sum of equations (7.1.32a,b). Hence,

$$\left[k^{-1}\frac{\partial}{\partial z}+\frac{i}{2k^2}\left\{\sum_{n=1}^{\nu}\left(\frac{\partial}{\partial \vec{y}_n}\right)^2-\sum_{m=1}^{\mu}\left(\frac{\partial}{\partial \vec{x}_m}\right)^2\right\}+(\nu-\mu)^2\gamma\right.$$
$$+\sum_{m=1}^{\mu}\sum_{n=1}^{\nu}V(\vec{x}_m-\vec{y}_n)-\sum_{m>l=1}^{\mu}V(\vec{x}_m-\vec{x}_l)$$
$$\left.-\sum_{n>S=1}^{\nu}V(\vec{y}_n-\vec{y}_S)\right]M_{\mu\nu}(\vec{x}_1,\ldots,\vec{x}_\mu;\vec{y}_1,\ldots,\vec{y}_\nu;z)$$
$$=i(2k^2)^{-1}\left[\sum_{x_1}j^*(\hat{x}_1)M_{\mu-1,\nu}(\vec{x}_2,\ldots,\vec{x}_\mu;\vec{y}_1,\ldots,\vec{y}_\nu;z)\right.$$
$$\left.-\sum_{y_1}j(\hat{y}_1)M_{\mu,\nu-1}(\vec{x}_1,\ldots,\vec{x}_\mu;\vec{y}_2,\ldots,\vec{y}_\nu;z)\right]\ . \qquad (7.1.33)$$

Here, in the particular case, $\mu=\nu$, it is convenient to write the equations in terms of the (two-dimensional) relative coordinates, $\vec{r}_n$ and $\vec{\rho}_n$, $n=1,2,\ldots,\nu$, defined by

$$\vec{r}_n=\vec{y}_n-\vec{x}_n, \qquad \vec{\rho}_n=2^{-1}(\vec{y}_n+\vec{x}_n)\ . \qquad (7.1.34)$$

In the region free from the source where $j(\hat{y})=0$, (7.1.31), for example, becomes

$$\left[k^{-1}\frac{\partial}{\partial z}+\frac{i}{k^2}\frac{\partial}{\partial \vec{r}}\cdot\frac{\partial}{\partial \vec{\rho}}+V(\vec{r})\right]M_{11}(\vec{r};\vec{\rho};z)=0\ , \tag{7.1.35}$$

while the equation for M_{22} can be written as

$$\left[k^{-1}\frac{\partial}{\partial z}+\frac{i}{k^2}\left(\frac{\partial}{\partial \vec{r}_1}\cdot\frac{\partial}{\partial \vec{\rho}_1}+\frac{\partial}{\partial \vec{r}_2}\cdot\frac{\partial}{\partial \vec{\rho}_2}\right)+V(\vec{r}_1)+V(\vec{r}_2)\right.$$
$$\left.+V_I(\vec{r}_1,\vec{r}_2,\vec{\rho}_{12})\right]M_{11}(\vec{r}_1,\vec{r}_2;\vec{\rho}_1,\vec{\rho}_2;z)=0\ , \tag{7.1.36}$$

Here, $\vec{\rho}_{12}=\vec{\rho}_1-\vec{\rho}_2$ and

$$\begin{aligned}V_I(\vec{r}_1,\vec{r}_2,\vec{\rho}_{12}) &= V(\vec{x}_1-\vec{y}_2)+V(\vec{y}_1-\vec{x}_2)\\ &\quad -V(\vec{x}_1-\vec{x}_2)-V(\vec{y}_1-\vec{y}_2)\\ &= V\left[\vec{\rho}_{12}-2^{-1}(\vec{r}_1+\vec{r}_2)\right]+V\left[\vec{\rho}_{12}+2^{-1}(\vec{r}_1+\vec{r}_2)\right]\\ &\quad -V\left[\vec{\rho}_{12}-2^{-1}(\vec{r}_1-\vec{r}_2)\right]-V\left[\vec{\rho}_{12}+2^{-1}(\vec{r}_1-\vec{r}_2)\right]\ .\end{aligned} \tag{7.1.37}$$

Thus, when $\mu=\nu$, in the region free from the wave source, Eq. (7.1.33) can be written as [7.5]

$$\left[k^{-1}\frac{\partial}{\partial z}+\sum_{n=1}^{\nu}\left\{\frac{i}{k^2}\frac{\partial}{\partial \vec{r}_n}\cdot\frac{\partial}{\partial \vec{\rho}_n}+V(\vec{r}_n)\right\}\right.$$
$$\left.+\sum_{n>m=1}^{\nu}V_I(\vec{r}_m,\vec{r}_n,\vec{\rho}_{mn})\right]$$
$$\times M_{\nu\nu}(\vec{r}_1,\ldots,\vec{r}_\nu;\vec{\rho}_1,\ldots,\vec{\rho}_\nu;z)=0\ , \tag{7.1.38}$$

in which the medium appears only through the function, $V(\vec{r})$, given, according to (7.1.27b,30b), by

$$V(\vec{r})=(2k^3)^{-1}\int_0^\infty dz\left[D(0,z)-D(\vec{r},z)\right]\ , \tag{7.1.39}$$

in terms of the correlation function, $D(\hat{x})=D(\vec{x},z)$.

7.1.1 Turbulent Medium of Kolmogorov Spectrum

In Sect. 7.1, we have seen that when the forward scattering approximation is possible, the moment equations for $M_{\mu\nu}$ become considerably simple, particularly in the case $\mu=\nu$ in which the equations are given by (7.1.38), and the random medium is involved only through the basic function, $V(\vec{r})$, which is defined by (7.1.39) in terms of the medium correlation function, $D(\hat{x})$.

The medium correlation function can be given in terms of the conventional spectrum density function, $\phi_\varepsilon(\boldsymbol{\lambda})$, by

$$D(\hat{x} - \hat{x}') = k^4 \langle \varepsilon(\hat{x})\varepsilon(\hat{x}')\rangle = k^4 \int d\hat{\boldsymbol{\lambda}}\, \phi_\varepsilon(\hat{\boldsymbol{\lambda}}) \exp\left[-i\hat{\boldsymbol{\lambda}}(\hat{x} - \hat{x}')\right] \ , \quad (7.1.40a)$$

and, based on the Kolmogorov theory of turbulence, $\phi_\varepsilon(\boldsymbol{\lambda})$ has often been represented by a modified (Von Karman) spectrum function [7.6,7],

$$\phi_\varepsilon(\hat{\boldsymbol{\lambda}}) = 4\phi_n(\hat{\boldsymbol{\lambda}}) = 4 \times 0.033\, C_n^2 \frac{\exp\left[-(\hat{\boldsymbol{\lambda}}/k_m)^2\right]}{(\hat{\boldsymbol{\lambda}}^2 + l_M^{-2})^{11/6}} \ . \quad (7.1.40b)$$

Here, ϕ_n is the spectrum density function of the refractive index, and C_n^2 is a structure constant which represents the intensity of medium fluctuation and has the order of magnitude of 10^{-16} to $10^{-12} m^{-2/3}$; $l_m = 5.92/k_m$ and l_M are the respective minimum and maximum lengths associated with the spatial size of the medium fluctuation, and have the magnitude of the order of $l_m \sim 10^{-3}\, m$ and $l_M \sim 10\, m$.

By using (7.1.40a,b), the function $V(\vec{r})$, defined by (7.1.39), is shown to be given by

$$V(\vec{r}) = 2\pi k \int_{-\infty}^{\infty} d\vec{\lambda}\, \phi_n(\vec{\lambda}, 0)\left[1 - \exp(-i\vec{\lambda} \cdot \vec{r})\right] \quad (7.1.41)$$

$$\sim \begin{cases} \beta' |k\vec{r}|^2, & |\vec{r}| \ll l_m \quad (7.1.42a) \\ \beta |k\vec{r}|^{5/3}, & l_M \gg |\vec{r}| \gg l_m \ , \quad (7.1.42b) \end{cases}$$

where

$$\beta' = 1.65\, k^{-1} l_m^{-1/3} C_n^2, \qquad \beta = 1.46\, k^{-2/3} C_n^2 \ . \quad (7.1.43)$$

Hence, in the range where expression (7.1.42a) holds, the function V_I of (7.1.37) becomes independent of the coordinates $\vec{\rho}_{12}$, as given by

$$V_I(\vec{r}_1, \vec{r}_2, \vec{\rho}_{12}) = 2\beta' k^2 \vec{r}_1 \cdot \vec{r}_2 \ ; \quad (7.1.44)$$

and thereby all medium terms involved in the moment equation (7.1.38) are unified to be given by

$$\sum_{j=1}^{\nu} V(\vec{r}_j) + \sum_{i>j=1}^{\nu} V_I(\vec{r}_i, \vec{r}_j, \vec{\rho}_{ij})$$

$$= \beta' \left(\sum_{j=1}^{\nu} k\vec{r}_j\right)^2 = \beta' \left[\sum_{j=1}^{\nu} k(\vec{y}_j - \vec{x}_j)\right]^2 \ , \quad (7.1.45)$$

which is simple enough to obtain exact solutions of the moment equations for all orders (Sect. 7.2.4).

In a pure Kolmogorov medium where $l_m = 0$ and $l_M = \infty$, it is mathematically convenient to write $V(\vec{r})$ as [7.6, 8, 9]

$$V(\vec{r}) = C|k\vec{r}|^{\alpha} \ , \tag{7.1.46}$$

$$\begin{aligned} C = &-\pi 2^{-\alpha-1}\Gamma(1+\alpha)\Gamma^{-2}(1+\alpha/2) \\ &\times \cot(\pi\alpha/2)k^{1-\alpha}C_n^2 \ . \end{aligned} \tag{7.1.47}$$

Here, $2 > \alpha > 1$, and $\alpha = 5/3$ for the pure Kolmogorov medium; even when $\alpha = 2$, the expression (7.1.46) is available with $C = \beta'$ of (7.1.43).

7.2 Solutions of the Moment Equations

In the case of a plane wave, (7.1.28a,35) yield the solutions

$$\langle\psi(\hat{y})\rangle = \exp(-\gamma kz)\langle\psi(z=0)\rangle, \qquad z > 0 \ ,$$

$$M_{11}(\vec{r};\vec{\rho};z) = \exp[-V(\vec{r})kz]M_{11}(z=0) \ .$$

Here, $V(0) = 0$ from (7.1.41), and $V(\vec{r})$ has expressions (7.1.42a,b) which are derived based on the Kolmogorov theory of turbulence. The above equations show that $\langle\psi(\hat{y})\rangle$ decreases exponentially with the coherence distance, $(k\gamma)^{-1}$, and $\langle\psi^*\psi\rangle\,|_{\vec{r}=0}$ is independent of the distance, being entirely not affected by the medium fluctuation.

On the other hand, for the moment M_{22} and the higher order, solving the moment equations becomes quite involved, even for plane and spherical wave incidences, unless expression (7.1.42a) can be assumed over the entire $\vec{r}$ range, and the solutions have been obtained mostly by numerical methods except for a few special cases. In Fig. 7.1 are shown numerical results given in [7.10] for cases of plane and spherical waves in three-dimensional space, based on a pure Kolmogorov spectrum given by (7.1.40b) when $l_M = \infty$ and $l_m = 0$, and obtained by using approximate expressions (7.2.60,61). Here, each curve of the variance, $\sigma_I^2 = \langle I^2\rangle/\langle I\rangle^2 - 1$, versus a numerical distance, $\tilde{\zeta}_k$, shows a maximum and saturation phenomenon, tending to 1 as $\tilde{\zeta}_k \to \infty$. An exact integral representation of M_{22} can be obtained, on the other hand, when the wave is propagated through a thin turbulent layer or phase screen (Sect. 7.2.2).

7.2.1 Basic Equations

To treat the moment equation (7.1.38), it is convenient to introduce the dimensionless coordinates, $\boldsymbol{r}$, $\boldsymbol{\rho}$ and ζ, defined by

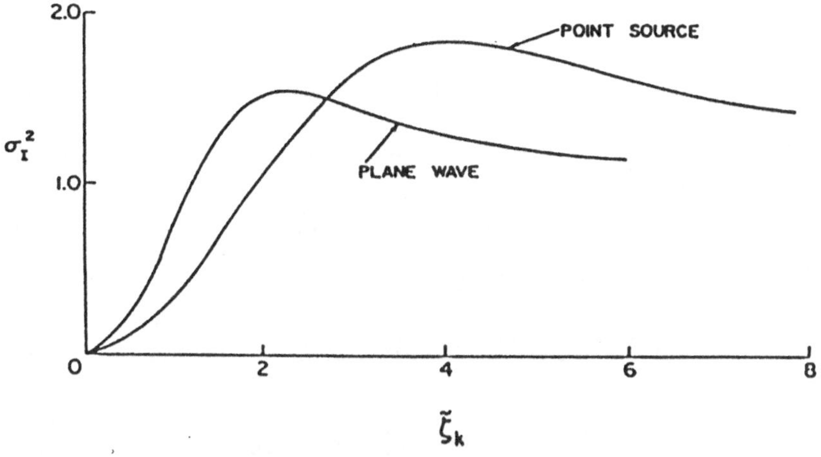

Fig. 7.1. Scintillation indices as a function of dimensionless distance for plane-wave and point-source initial conditions propagating through a pure Kolmogorov medium of (7.1.42b) (taken from [7.10])

$$\boldsymbol{r} = (k/L)^{1/2}\vec{r}, \qquad \boldsymbol{\rho} = (k/L)^{1/2}\vec{\rho}, \qquad \zeta = z/L , \tag{7.2.1}$$

where L is the distance of the wave propagation; and also a medium function, $\overline{V}(\boldsymbol{r})$, defined by

$$\overline{V}(\boldsymbol{r}) = kLV\left[(L/k)^{1/2}\boldsymbol{r}\right] \tag{7.2.2}$$

$$\simeq \tfrac{1}{2}\phi|\boldsymbol{r}|^{\alpha} . \tag{7.2.3}$$

Here, the last is an expression available in the important ranges (7.1.42a,b) of a Kolmogorov medium, and, using (7.1.47) for $2 > \alpha > 1$,

$$\phi = -\pi 2^{-\alpha}\Gamma(1+\alpha)\Gamma^{-2}(1+\alpha/2)\cot(\pi\alpha/2)L^{1+\alpha/2}k^{2-\alpha/2}C_n^2 , \tag{7.2.4a}$$

while, when $\alpha = 2$,

$$\phi = 2\beta'(kL)^2 = 3.3\, kL^2 l_m^{-1/3} C_n^2 . \tag{7.2.4b}$$

Hence the moment equation (7.1.35) can be written, for example, as

$$\frac{\partial}{\partial\zeta}M_{11}(\zeta) = (T_1 - V_1)M_{11}(\zeta) , \tag{7.2.5}$$

where

$$T_1 = -\mathrm{i}\frac{\partial}{\partial \boldsymbol{r}_1}\cdot\frac{\partial}{\partial \boldsymbol{\rho}_1}, \qquad V_1 = \overline{V}(\boldsymbol{\rho}_1) , \tag{7.2.6}$$

and the coordinates other then ζ have been suppressed. We write the solution of (7.2.5) in the form

$$M_{11}(\zeta) = \exp(T_1\zeta)U_1(\zeta)M_{11}(0) \ . \tag{7.2.7}$$

The factor $U_1(\zeta)$ is an operator to express an effect of the medium fluctuation and is a solution of

$$\frac{\partial}{\partial\zeta}U_1(\zeta) = -\widehat{V}_1(\zeta)U_1(\zeta) \ , \tag{7.2.8}$$

in terms of the notation [cf. (1.1.9)],

$$\widehat{V}_1(\zeta) = \exp(-T_1\zeta)\overline{V}(\boldsymbol{r}_1)\exp(T_1\zeta) = \overline{V}[\widehat{\boldsymbol{r}}_1(\zeta)] \ , \tag{7.2.9}$$

with

$$\widehat{\boldsymbol{r}}_1(\zeta) = \exp(-T_1\zeta)\boldsymbol{r}_1\exp(T_1\zeta) \ . \tag{7.2.10}$$

This is reduced, with the convention $[A, B] \equiv AB - BA$, to

$$\begin{aligned}\widehat{\boldsymbol{r}}_1(\zeta) &= \boldsymbol{r}_1 - \zeta[T_1, \boldsymbol{r}_1] + \tfrac{1}{2}\zeta^2[T_1, [T_1, \boldsymbol{r}_1]] - \cdots \\ &= \boldsymbol{r}_1 + \mathrm{i}\zeta\frac{\partial}{\partial\boldsymbol{\rho}_1} \ ,\end{aligned} \tag{7.2.11}$$

where the third and higher order terms of the series are identically zero (1.1.13).

The solution of (7.2.8), subject to $U_1(0) = 1$, is obviously

$$U_1(\zeta) = \exp\left[-\int_0^\zeta d\zeta' \, \widehat{V}_1(\zeta')\right] \ . \tag{7.2.12}$$

Hence, in terms of the solution in free space, $M_{11}^0(\zeta = 1)$ given by

$$M_{11}^0(1) = \exp(T_1)M_{11}(0) \ , \tag{7.2.13}$$

Eq. (7.2.7) at $\zeta = 1$ can be written as

$$\begin{aligned}M_{11}(1) &= \exp(T_1)U_1(1)\exp(-T_1)M_{11}^0(1) \\ &= \exp\left[-\int_0^1 d\zeta \, \widehat{V}_1(\zeta - 1)\right]M_{11}^0(1) \ ,\end{aligned} \tag{7.2.14}$$

as a consequence of definition (7.2.9). Since the expression (7.2.14) is only a functional of the commutable operators, $\widehat{\boldsymbol{r}}_1(\zeta)$, $1 \geq \zeta \geq 0$, evaluation of $M_{11}(1)$ is straightforward by using the Fourier representation of $M_{11}^0(1)$ with respect to $\boldsymbol{\rho}_1$ (7.2.31).

In the same way, for $M_{22}(\zeta)$, the moment equation (7.1.38) can be written in the form

$$\frac{\partial}{\partial \zeta} M_{22}(\zeta) = (T_1 + T_2 - V_1 - V_2 - V_{12}) M_{22}(\zeta) \ . \tag{7.2.15}$$

Here, T_j and V_j are defined by (7.2.6), and

$$V_{12} = V_{21} = \overline{V}_I(\boldsymbol{r}_1, \boldsymbol{r}_2, \boldsymbol{\rho}_{12}), \qquad \boldsymbol{\rho}_{12} = \boldsymbol{\rho}_1 - \boldsymbol{\rho}_2 \ , \tag{7.2.16}$$

where $\overline{V}_I$ is defined in the same way as V_I is by (7.1.37), in terms of $\boldsymbol{r}_1$, $\boldsymbol{r}_2$, $\boldsymbol{\rho}_{12}$, and $\overline{V}(\boldsymbol{r})$. Hence, writing the solution of (7.2.15) in the form

$$M_{22}(\zeta) = \exp\left[(T_1 + T_2)\zeta\right] U_{12}(\zeta) M_{22}(0) \ , \tag{7.2.17}$$

similar to (7.2.7), the operator, $U_{12}(\zeta)$, is a solution of

$$\frac{\partial}{\partial \zeta} U_{12}(\zeta) = -\widehat{V}^{(2)}(\zeta) U_{12}(\zeta) \ , \tag{7.2.18}$$

with

$$\widehat{V}^{(2)}(\zeta) = \widehat{V}_1(\zeta) + \widehat{V}_2(\zeta) + \widehat{V}_{12}(\zeta) \ . \tag{7.2.19}$$

Here, $\widehat{V}_j(\zeta)$ is defined by (7.2.9), and

$$\begin{aligned} \widehat{V}_{12}(\zeta) &= \exp\left[-(T_1 + T_2)\zeta\right] \overline{V}_I(\boldsymbol{r}_1, \boldsymbol{r}_2, \boldsymbol{\rho}_{12}) \exp\left[(T_1 + T_2)\zeta\right] \\ &= \overline{V}_I\left[\widehat{\boldsymbol{r}}_1(\zeta), \widehat{\boldsymbol{r}}_2(\zeta), \widehat{\boldsymbol{\rho}}_{12}(\zeta)\right] \ , \end{aligned} \tag{7.2.20}$$

where $\widehat{\boldsymbol{\rho}}_{12} = \widehat{\boldsymbol{\rho}}_1 - \widehat{\boldsymbol{\rho}}_2$, and $\widehat{\boldsymbol{\rho}}_j(\zeta)$ is defined by

$$\widehat{\boldsymbol{\rho}}_j(\zeta) = \exp(-T_j\zeta)\boldsymbol{\rho}_j \exp(T_j\zeta) \tag{7.2.21}$$

$$= \boldsymbol{\rho}_j + \mathrm{i}\zeta \frac{\partial}{\partial \boldsymbol{r}_j}, \qquad j = 1, 2 \ , \tag{7.2.22}$$

in exactly the same manner as $\widehat{\boldsymbol{r}}_j(\zeta)$ is defined by (7.2.10,11).

The ζ-dependent operators, $\widehat{\boldsymbol{r}}_j(\zeta)$ and $\widehat{\boldsymbol{\rho}}_j(\zeta')$, are both Hermitian and not commutable when $\zeta \neq \zeta'$, subject to the commutation relation

$$\left[\widehat{\boldsymbol{r}}_i(\zeta), \widehat{\boldsymbol{\rho}}_j(\zeta')\right] = \mathrm{i}(\zeta - \zeta')\delta_{ij} \ ; \tag{7.2.23}$$

whereas,

$$\left[\widehat{\boldsymbol{r}}_i(\zeta), \widehat{\boldsymbol{r}}_j(\zeta')\right] = \left[\widehat{\boldsymbol{\rho}}_i(\zeta), \widehat{\boldsymbol{\rho}}_j(\zeta')\right] = 0 \ . \tag{7.2.24}$$

Hence the solution of (7.2.18), subject to $U_{12} = 1$, should be written in the form

$$U_{12}(\zeta = 1) = P \exp\left[-\int_0^1 d\zeta \, \widehat{V}^{(2)}(\zeta)\right] \ , \tag{7.2.25}$$

with an ordering symbol, P, defined so that for any operators, $A(\zeta)$ and $B(\zeta')$,

$$P[A(\zeta)B(\zeta')] = \begin{cases} A(\zeta)B(\zeta'), & \zeta > \zeta' \\ B(\zeta')A(\zeta), & \zeta < \zeta' \end{cases} . \tag{7.2.26}$$

From (7.2.17, 25), $M_{22}(1)$ can be written, in terms of the free space solution, $M_{22}^0(1) = \exp(T_1 + T_2)M_{22}(0)$, as

$$M_{22}(1) = P\exp\left[-\int_0^1 d\zeta\, \widehat{V}^{(2)}(\zeta - 1)\right] M_{22}^0(1) , \tag{7.2.27}$$

in the same form as (7.2.14) for $M_{11}(1)$.

It is now straightforward to generalize expression (7.2.27) to $M_{\nu\nu}(1)$ of arbitrary order, ν, as the solution of the moment equation,

$$\frac{\partial}{\partial\zeta} M_{\nu\nu}(\zeta) = \left[\sum_{j=1}^{\nu}(T_j - V_j) - \sum_{i>j=1}^{\nu} V_{ij}\right] M_{\nu\nu}(\zeta) , \tag{7.2.28}$$

from (7.1.38); hence,

$$M_{\nu\nu}(1) = P\exp\left[-\int_0^1 d\zeta\, \widehat{V}^{(\nu)}(\zeta - 1)\right] M_{\nu\nu}^0(1) , \tag{7.2.29}$$

where

$$V^{(\nu)} = \sum_{j=1}^{\nu} V_j + \sum_{i>j=1}^{\nu} V_{ij} . \tag{7.2.30}$$

a) Average Wave Intensity

Here we evaluate the average intensity by using (7.2.14) for $M_{11}(1)$, which can be written, in terms of the Fourier transform $\widetilde{M}_{11}^0(\boldsymbol{s}_1;\boldsymbol{\lambda}_1;\zeta = 1)$ of $M_{11}^0(\boldsymbol{r}_1;\boldsymbol{\rho}_1;\zeta = 1)$ with respect to $\boldsymbol{r}_1$ and $\boldsymbol{\rho}_1$, as

$$M_{11}(\boldsymbol{r}_1;\boldsymbol{\rho}_1;\zeta = 1) = (2\pi)^{-4}\int d\boldsymbol{s}_1 d\boldsymbol{\lambda}_1 \exp[-i(\boldsymbol{s}_1\cdot\boldsymbol{r}_1 + \boldsymbol{\lambda}_1\cdot\boldsymbol{\rho}_1)]$$
$$\times \exp\left\{-\int_0^1 d\zeta\, \overline{V}[\boldsymbol{r}_1 + (\zeta - 1)\boldsymbol{\lambda}_1]\right\} \widetilde{M}_{11}^0(\boldsymbol{s}_1;\boldsymbol{\lambda}_1;\zeta = 1) . \tag{7.2.31}$$

When the incident wave is a collimated beam wave with the minimum radius b at $z = 0$, i.e., with the dimensionless radius $\overline{b} = (k/L)^{1/2}b$ at $z = 0$,

$$\widetilde{M}_{11}^0(\boldsymbol{s}_1;\boldsymbol{\lambda}_1;\zeta) = I_0(2\pi\overline{b}^2)^2 \exp[i\boldsymbol{s}_1\cdot\boldsymbol{\lambda}_1\zeta - \overline{b}^2(\boldsymbol{s}_1^2 + \tfrac{1}{4}\boldsymbol{\lambda}_1^2)] ; \tag{7.2.32}$$

which is reduced to:

(i) Case of a Spherical Wave ($b = 0$).

$$\widetilde{M}_{11}^0(\boldsymbol{s}_1;\boldsymbol{\lambda}_1;\zeta = 0) = (4\pi L)^{-2}(2\pi)^2 ; \tag{7.2.33}$$

(ii) Case of a Plane Wave ($b = \infty$).

$$\widetilde{M}_{11}^{0}(\boldsymbol{s}_1; \boldsymbol{\lambda}_1; \zeta = 0) = I_0 (2\pi)^4 \delta(\boldsymbol{s}_1)\delta(\boldsymbol{\lambda}_1) \ . \tag{7.2.34}$$

When $\boldsymbol{r}_1 = 0$, (7.2.31,32) give an exact integral representation of the intensity, i.e.,

$$M_{11}(\boldsymbol{r}_1 = 0; \boldsymbol{\rho}_1; z = 1) = I_0 \bar{b}^2 (4\pi)^{-1} \int d\boldsymbol{\lambda}_1 \exp(-\mathrm{i}\boldsymbol{\lambda}_1 \cdot \boldsymbol{\rho}_1)$$
$$\times \exp\left\{ -\int_0^1 d\zeta \, \overline{V}[(\zeta - 1)\boldsymbol{\lambda}_1] - \tfrac{1}{4}(\bar{b}^2 + 1/\bar{b}^2)\boldsymbol{\lambda}_1^2 \right\} \ . \tag{7.2.35}$$

Here, when using expression (7.2.3),

$$\int_0^1 d\zeta \, \overline{V}[(\zeta - 1)\boldsymbol{\lambda}] = \tfrac{1}{2}(1 + \alpha)^{-1} \phi |\boldsymbol{\lambda}|^{\alpha} \ . \tag{7.2.36}$$

Hence, upon introduction of new parameters, f, X, and Y, defined by

$$f^2 = 1 + (\bar{b})^{-4} \ , \tag{7.2.37a}$$
$$X = 2(\bar{b}f)^{-1} |\boldsymbol{\rho}| \ , \tag{7.2.37b}$$
$$Y = 2^{-1}(1 + \alpha)^{-1} \phi (2/\bar{b}f)^{\alpha} \ , \tag{7.2.37c}$$

and also a change of the variable of integration by $t = (\bar{b}f/2)^2 \boldsymbol{\lambda}^2$, (7.2.35) is reduced to the integral,

$$M_{11}(\zeta = 1) = I_0 f^{-2} \int_0^1 dt \, J_0(X t^{1/2}) \exp(-Y t^{\alpha/2} - t) \ , \tag{7.2.38}$$

where $J_0(x)$ is the Bessel function of zeroth order. When $\alpha = 5/3$, it gives the mean wave intensity for a medium of the pure Kolmogorov spectrum, and the integral is tabulated as a function of X and Y in [7.11,6]. While, when $\alpha = 2$, (7.2.38) is reduced to

$$M_{11} = I_0 f^{-2} (1 + \sigma_E)^{-1} \exp[-(1 + \sigma_E)^{-1} (\bar{b}f)^{-2} |\boldsymbol{\rho}|^2] \ , \tag{7.2.39}$$

where

$$\sigma_E = Y \,\big|_{\alpha=2} = \tfrac{2}{3} \phi (\bar{b}f)^{-2} \ , \tag{7.2.40}$$

showing that the average intensity keeps its original Gaussian form when $\alpha = 2$, whereas this is not the case of $\alpha = 5/3$ [7.11–13].

7.2.2 Phase Screen

When a turbulent layer with an infinitesimal width, say $\Delta\zeta$, is located at ζ, the only operators involved in the expression (7.2.27) are $\widehat{\boldsymbol{r}}_j(\zeta-1)$, $j=1$, 2, and $\widehat{\boldsymbol{\rho}}_{12}(\zeta-1)$, which are commutable with each other, in view of (7.2.23), and, therefore, can be treated in just the same way as ordinary (commutable) variables are, enabling $M_{22}(1)$ to be Fourier represented with respect to the operator, $\widehat{\boldsymbol{\rho}}_{12}(\zeta-1)$; that is, with the notation

$$\widehat{V}^{(2)}(\zeta)=V^{(2)}\left[\widehat{\boldsymbol{r}}_1(\zeta),\widehat{\boldsymbol{r}}_2(\zeta),\widehat{\boldsymbol{\rho}}_{12}(\zeta)\right] , \tag{7.2.41}$$

similar to (7.2.20), we obtain the representation

$$M_{22}(1)=\exp\left\{-\Delta\zeta V^{(2)}\left[\widehat{\boldsymbol{r}}_1(\zeta-1),\widehat{\boldsymbol{r}}_2(\zeta-1),\widehat{\boldsymbol{\rho}}_{12}(\zeta-1)\right]\right\}M_{22}^{0}(1) \tag{7.2.42}$$

$$=(2\pi)^{-2}\int d\boldsymbol{\lambda}'d\boldsymbol{\rho}'\exp\left\{-\Delta\zeta V^{(2)}\left[\widehat{\boldsymbol{r}}_1(\zeta-1),\widehat{\boldsymbol{r}}_2(\zeta-1),\boldsymbol{\rho}'\right]\right\}$$
$$\times\exp\left\{\mathrm{i}\boldsymbol{\lambda}'\cdot\left[\boldsymbol{\rho}'-\widehat{\boldsymbol{\rho}}_{12}(\zeta-1)\right]\right\}M_{22}^{0}(1) \tag{7.2.43}$$

(as can be directly shown by making the $\boldsymbol{\lambda}'$ integration first to yield the factor $\delta\left[\boldsymbol{\rho}'-\widehat{\boldsymbol{\rho}}_{12}(\zeta-1)\right]$). Here the order of the exponential factors does not matter, and the integral can be evaluated in exactly the same manner as when deriving (7.2.35), in terms of the Fourier transform of $M_{22}^{0}(1)$.

To obtain a specific expression of (7.2.43), it is convenient to introduce the relative coordinates, $\boldsymbol{\rho}_{12}$ and $\overline{\boldsymbol{\rho}}$, defined by

$$\boldsymbol{\rho}_{12}=\boldsymbol{\rho}_1-\boldsymbol{\rho}_2,\qquad \overline{\boldsymbol{\rho}}=2^{-1}(\boldsymbol{\rho}_1+\boldsymbol{\rho}_2) , \tag{7.2.44}$$

and similar Fourier variables,

$$\boldsymbol{v}=\boldsymbol{s}_1-\boldsymbol{s}_2,\qquad \overline{\boldsymbol{v}}=2^{-1}(\boldsymbol{s}_1+\boldsymbol{s}_2) ; \tag{7.2.45}$$

and also

$$\boldsymbol{\nu}=2^{-1}(\boldsymbol{\lambda}_1-\boldsymbol{\lambda}_2),\qquad \overline{\boldsymbol{\nu}}=\boldsymbol{\lambda}_1+\boldsymbol{\lambda}_2 , \tag{7.2.46a}$$

$$\boldsymbol{\lambda}_1=\boldsymbol{\nu}+2^{-1}\overline{\boldsymbol{\nu}},\qquad \boldsymbol{\lambda}_2=-\boldsymbol{\nu}+2^{-1}\overline{\boldsymbol{\nu}} . \tag{7.2.46b}$$

Hence,

$$\boldsymbol{\lambda}_1\cdot\boldsymbol{\rho}_1+\boldsymbol{\lambda}_2\cdot\boldsymbol{\rho}_2=\boldsymbol{\nu}\cdot\boldsymbol{\rho}_{12}+\overline{\boldsymbol{\nu}}\cdot\overline{\boldsymbol{\rho}} , \tag{7.2.47}$$

$$\boldsymbol{\lambda}_1\cdot\boldsymbol{s}_1+\boldsymbol{\lambda}_2\cdot\boldsymbol{s}_2=\boldsymbol{\nu}\cdot\boldsymbol{v}+\overline{\boldsymbol{\nu}}\cdot\overline{\boldsymbol{v}} . \tag{7.2.48}$$

For example, in terms of $\boldsymbol{s}=(\boldsymbol{s}_1,\boldsymbol{s}_2)=(\boldsymbol{v},\overline{\boldsymbol{v}})$ and $\boldsymbol{\lambda}=(\boldsymbol{\lambda}_1,\boldsymbol{\lambda}_2)=(\boldsymbol{\nu},\overline{\boldsymbol{\nu}})$, (7.2.32) gives an expression for $\widetilde{M}_{22}^{0}(\zeta=1)$ of the form

$$\widetilde{M}_{22}^{0}(\boldsymbol{s};\boldsymbol{\lambda};\zeta=1)=\exp\left[\mathrm{i}\left(\boldsymbol{\nu}\cdot\boldsymbol{v}+\overline{\boldsymbol{\nu}}\cdot\overline{\boldsymbol{v}}\right)\right]\widetilde{M}_{22}^{0}(\boldsymbol{s};\boldsymbol{\lambda};\zeta=0)\ . \tag{7.2.49}$$

By introducing two more new variables, $\boldsymbol{\lambda}''$ and $\boldsymbol{\rho}''$, defined by

$$\boldsymbol{\lambda}''=\boldsymbol{\lambda}'+\boldsymbol{\nu},\qquad \boldsymbol{\rho}''=\boldsymbol{\rho}'-\zeta\boldsymbol{v}\ , \tag{7.2.50}$$

we find that the operators $\widehat{\boldsymbol{r}}_j(\zeta-1)$ and $\widehat{\boldsymbol{\rho}}_{12}(\zeta-1)$, involved in (7.2.43), can be replaced by their eigenvalues, $\overline{\boldsymbol{r}}'_j(\zeta)$ and $\overline{\boldsymbol{\rho}}'_{12}(\zeta)$, respectively, in view of (7.2.11,22), given by

$$\overline{\boldsymbol{r}}'_1(\zeta)=\boldsymbol{r}_1+(\zeta-1)\left(\boldsymbol{\lambda}''+2^{-1}\overline{\boldsymbol{\nu}}\right)\ , \tag{7.2.51a}$$

$$\overline{\boldsymbol{r}}'_2(\zeta)=\boldsymbol{r}_2+(\zeta-1)\left(-\boldsymbol{\lambda}''+2^{-1}\overline{\boldsymbol{\nu}}\right)\ , \tag{7.2.51b}$$

$$\overline{\boldsymbol{\rho}}'_{12}(\zeta)=\boldsymbol{\rho}_{12}+(\zeta-1)\boldsymbol{v}\ . \tag{7.2.51c}$$

Also, after substituting (7.2.49) into (7.2.43), when $\boldsymbol{r}_j=0$, the resultant phase term can be written as

$$\begin{aligned}\boldsymbol{\lambda}'\cdot\left[\boldsymbol{\rho}'-\overline{\boldsymbol{\rho}}'_{12}(\zeta)\right]&+\sum_{j=1}^{2}\boldsymbol{\lambda}_j\cdot(\boldsymbol{s}_j-\boldsymbol{\rho}_j)\\ &=\boldsymbol{\lambda}''\cdot\left(\boldsymbol{\rho}''+\boldsymbol{v}-\boldsymbol{\rho}_{12}\right)-\boldsymbol{\nu}\cdot\boldsymbol{\rho}''+\overline{\boldsymbol{\nu}}\cdot\left(\overline{\boldsymbol{v}}-\overline{\boldsymbol{\rho}}\right)\ .\end{aligned} \tag{7.2.52}$$

Thus, when $\boldsymbol{r}_j=0$, the result can be written [upon changing the variables of integration, $\boldsymbol{\lambda}'$ and $\boldsymbol{\rho}'$ to $\boldsymbol{\lambda}''$ and $\boldsymbol{\rho}''$ of (7.2.50), respectively] by a Fourier representation, with respect to $\boldsymbol{\rho}_{12}$, of the form

$$M_{22}\left(\boldsymbol{\rho}_{12},\overline{\boldsymbol{\rho}};\zeta=1\right)=(2\pi)^{-2}\int d\boldsymbol{\lambda}''\,\exp(-\mathrm{i}\boldsymbol{\lambda}''\cdot\boldsymbol{\rho}_{12})\widetilde{M}_{22}(\boldsymbol{\lambda}'',\overline{\boldsymbol{\rho}})\ . \tag{7.2.53}$$

Here, $\widetilde{M}_{22}(\boldsymbol{\lambda}'',\overline{\boldsymbol{\rho}})$ means a spectrum function at the center of coordinates, $\overline{\boldsymbol{\rho}}$, and is given by [7.14]

$$\begin{aligned}\widetilde{M}_{22}(\boldsymbol{\lambda}'',\overline{\boldsymbol{\rho}})&=(2\pi)^{-4}\int d\boldsymbol{\rho}''d\boldsymbol{v}d\overline{\boldsymbol{\nu}}\,\exp\left\{\mathrm{i}\left[\boldsymbol{\lambda}''\cdot(\boldsymbol{\rho}''+\boldsymbol{v})-\overline{\boldsymbol{\nu}}\cdot\overline{\boldsymbol{\rho}}\right]\right\}\\ &\times\exp\left\{-\Delta\zeta V^{(2)}\left[(\zeta-1)(\boldsymbol{\lambda}''+\boldsymbol{\nu}/2),(\zeta-1)(-\boldsymbol{\lambda}''+\overline{\boldsymbol{\nu}}/2),\boldsymbol{\rho}''+\zeta\boldsymbol{v}\right]\right\}\\ &\times\widetilde{M}_{22}^{0}(\boldsymbol{\rho}'';\boldsymbol{v},\overline{\boldsymbol{\nu}})\ .\end{aligned} \tag{7.2.54}$$

where, in the case of a collimated beam wave (7.2.32),

$$\widetilde{M}_{22}^{0}(\boldsymbol{\rho}'';\boldsymbol{v},\overline{\boldsymbol{\nu}})=(2\pi)^{-4}\int d\boldsymbol{\nu}d\overline{\boldsymbol{v}}\,\exp\left[-\mathrm{i}\left(\boldsymbol{\nu}\cdot\boldsymbol{\rho}''-\overline{\boldsymbol{\nu}}\cdot\overline{\boldsymbol{v}}\right)\right]\times\widetilde{M}_{22}^{0}(\boldsymbol{s},\boldsymbol{\lambda};\zeta=0) \tag{7.2.55a}$$

$$=I_0^2(\pi\overline{b}^2)^2\exp\left[-\frac{1}{2}\left(\frac{\boldsymbol{\rho}''}{\overline{b}}\right)^2-\frac{1}{2}\left(\overline{b}\boldsymbol{v}\right)^2-\frac{1}{8}\left(\overline{b}^2+\frac{1}{\overline{b}^2}\right)\overline{\boldsymbol{\nu}}^2\right]\ . \tag{7.2.55b}$$

7.2.3 Formal Generalization to a Turbulent Continuum

When, on the other hand, a turbulent medium is continuously distributed, we may assume, as an approximation, that the resultant effect of the medium fluctuation is given just by an integrated effect of phase screens distributed successively over the range, $1 \geq \zeta \geq 0$, so that $\widetilde{M}_{22}$, for example, is still given by (7.2.54), simply with the replacement of

$$\Delta\zeta \rightarrow \int_0^1 d\zeta \; . \tag{7.2.56}$$

That is, in the special cases of spherical and plane waves, in which from (7.2.33, 34),

$$\widetilde{M}_{22}^0(\boldsymbol{\rho}''; \boldsymbol{v}, \overline{\boldsymbol{\nu}}) = (2\pi)^4 \delta(\overline{\boldsymbol{\nu}}) \times \begin{cases} (4\pi L)^{-4}\delta(\boldsymbol{\rho}''), & b = 0 \\ I_0^2 \delta(\boldsymbol{v}), & b = \infty \; , \end{cases} \tag{7.2.57}$$

(7.2.54) leads to:

(i) Case of a Spherical Wave ($b = 0$).

$$\widetilde{M}_{22}(\boldsymbol{\lambda}'', \overline{\boldsymbol{\rho}}) \sim (4\pi L)^{-4} \int d\boldsymbol{v} \exp(\mathrm{i}\boldsymbol{\lambda}'' \cdot \boldsymbol{v}) \times \exp\left\{ -\int_0^1 d\zeta \, V^{(2)}[(\zeta - 1)\boldsymbol{\lambda}'', -(\zeta - 1)\boldsymbol{\lambda}'', \zeta\boldsymbol{v}] \right\} \; ; \tag{7.2.58}$$

(ii) Case of a Plane Wave ($b = \infty$).

$$\widetilde{M}_{22}(\boldsymbol{\lambda}'', \overline{\boldsymbol{\rho}}) \sim I_0^2 \int d\boldsymbol{\rho}'' \exp(\mathrm{i}\boldsymbol{\lambda}'' \cdot \boldsymbol{\rho}'') \times \exp\left\{ -\int_0^1 d\zeta \, V^{(2)}[(\zeta - 1)\boldsymbol{\lambda}'', -(\zeta - 1)\boldsymbol{\lambda}'', \boldsymbol{\rho}''] \right\} \; , \tag{7.2.59}$$

which are both equivalent to those previously obtained [7.15.16], and are also reproduced based on the two-scale method [7.17–20].

Hence, when using $\overline{V}(\boldsymbol{r})$ of (7.2.3), (7.2.53,59) for a plane wave, for example, lead to the expression

$$\begin{aligned} M_{22}(\boldsymbol{\rho}_{12}, \overline{\boldsymbol{\rho}}) \sim I_0^2 (2\pi)^{-2} \int d\boldsymbol{\lambda}'' d\boldsymbol{\rho}'' \exp\left[\mathrm{i}\boldsymbol{\lambda}'' \cdot (\boldsymbol{\rho}'' - \boldsymbol{\rho}_{12})\right] \\ \times \exp\Big\{ -\phi[(1+\alpha)^{-1}|\boldsymbol{\lambda}''|^{\alpha} + |\boldsymbol{\rho}''|^{\alpha}] \\ + \frac{1}{2}\phi \int_0^1 d\zeta \left[|\boldsymbol{\rho}'' + (\zeta - 1)\boldsymbol{\lambda}''|^{\alpha} + |\boldsymbol{\rho}'' - (\zeta - 1)\boldsymbol{\lambda}''|^{\alpha}\right] \Big\} \; ; \end{aligned} \tag{7.2.60}$$

while, in the case of a spherical wave, use of (7.2.58) similarly yields

$$M_{22}(\boldsymbol{\rho}_{12},\overline{\boldsymbol{\rho}}) \sim (4\pi L)^{-4}(2\pi)^{-2}\int d\boldsymbol{\lambda}''d\boldsymbol{v}\exp\left[\mathrm{i}\boldsymbol{\lambda}''\cdot(\boldsymbol{v}-\boldsymbol{\rho}_{12})\right]$$
$$\times\exp\left\{-\phi(1+\alpha)^{-1}\left(|\boldsymbol{\lambda}''|^{\alpha}+|\boldsymbol{v}|^{\alpha}\right)\right.$$
$$\left.+\frac{1}{2}\phi\int_0^1 d\zeta\left[|\zeta\boldsymbol{v}+(\zeta-1)\boldsymbol{\lambda}''|^{\alpha}+|\zeta\boldsymbol{v}-(\zeta-1)\boldsymbol{\lambda}''|^{\alpha}\right]\right\} . \qquad (7.2.61)$$

Note that in both (7.2.60,61), the exponential terms { } become zero when $\alpha = 2$.

Here, to the first order of ϕ, the integral in (7.2.60) can be analytically evaluated with the result, which, when $\alpha = 5/3$ and $\boldsymbol{\rho}_{12} = 0$, agrees with the conventional result by the Rytov approximation [7.6]. To briefly give the proof, we observe that (7.2.60) can be written to the first order of $\overline{V}_I$ as

$$\langle I^2\rangle/\langle I\rangle^2 = (2\pi)^{-2}\int d\boldsymbol{\lambda}''d\boldsymbol{\rho}''\exp(\mathrm{i}\boldsymbol{\lambda}''\cdot\boldsymbol{\rho}'')$$
$$\times\exp\left[-\phi(1+\alpha)^{-1}|\boldsymbol{\lambda}''|^{\alpha}\right]$$
$$\times\left\{1-\int_0^1 d\zeta\overline{V}_I\left[(\zeta-1)\boldsymbol{\lambda}'',-(\zeta-1)\boldsymbol{\lambda}'',\boldsymbol{\rho}''\right]\right\} . \qquad (7.2.62)$$

Here, the function, $\overline{V}_I(\boldsymbol{r}_1,\boldsymbol{r}_2,\boldsymbol{\rho}_{12})$, is defined by (7.2.16), and, as $|\boldsymbol{\rho}_{12}|\to\infty$, becomes the order of $|\boldsymbol{\rho}_{12}|^{\alpha-2}$, tending to zero in view of $2>\alpha>1$. Thus it can be Fourier represented with respect to $\boldsymbol{\rho}_{12}$ (although not for the function $\overline{V}(\boldsymbol{r})$), and, using (7.1.37) with expression (7.2.3) for $\overline{V}(\boldsymbol{r})$, we obtain the representation [7.8, 9]

$$\overline{V}_I(\boldsymbol{r}_1,\boldsymbol{r}_2,\boldsymbol{\rho}_{12}) = 2\phi N(\alpha)\int_{-\infty}^{\infty} d\boldsymbol{t}\,|\boldsymbol{t}|^{-\alpha-2}$$
$$\times\exp(-\mathrm{i}\boldsymbol{t}\cdot\boldsymbol{\rho}_{12})\sin(\boldsymbol{t}\cdot\boldsymbol{r}_1/2)\sin(\boldsymbol{t}\cdot\boldsymbol{r}_2/2) , \qquad (7.2.63a)$$

where

$$N(\alpha) = \pi^{-2}2^{\alpha}\Gamma^2(1+\alpha/2)\sin(\pi\alpha/2) . \qquad (7.2.63b)$$

Hence,

$$\overline{V}_I\left[(\zeta-1)\boldsymbol{\lambda}'',-(\zeta-1)\boldsymbol{\lambda}'',\boldsymbol{\rho}''\right] = -2\phi N(\alpha)\int_{-\infty}^{\infty} d\boldsymbol{t}\,|\boldsymbol{t}|^{-\alpha-2}$$
$$\times\exp(-\mathrm{i}\boldsymbol{t}\cdot\boldsymbol{\rho}'')\sin^2\left[(1-\zeta)\boldsymbol{t}\cdot\boldsymbol{\lambda}''/2\right] , \qquad (7.2.64)$$

so that, making the $\boldsymbol{\rho}''$ integration in (7.2.62), first, the result is reduced to

$$\sigma_I^2 \equiv \langle I^2\rangle/\langle I\rangle^2 - 1 = 2\phi N(\alpha)\int_0^1 d\zeta$$
$$\times\int_{-\infty}^{\infty} d\boldsymbol{t}\,|\boldsymbol{t}|^{-\alpha-2}\sin^2\left[(1-\zeta)\boldsymbol{t}^2/2\right]\exp\left[-\phi(1+\alpha)^{-1}|\boldsymbol{t}|^{\alpha}\right] , \qquad (7.2.65)$$

by virtue of the resulting factor, $\delta(\boldsymbol{\lambda}''-\boldsymbol{t})$. Here, since $\phi \ll 1$, the exponential term is negligible, and, by the replacement of $(1-\zeta)^{1/2}\boldsymbol{t} \to \boldsymbol{t}$, (7.2.65) is reduced to

$$\sigma_I^2 = 2\phi N(\alpha) \int_0^1 d\zeta\,(1-\zeta)^{\alpha/2} \int_{-\infty}^{\infty} d\boldsymbol{t}\,|\boldsymbol{t}|^{-\alpha-2}\sin^2(\boldsymbol{t}^2/2)\ , \qquad (7.2.66)$$

wherein the last integral becomes

$$2\pi \int_0^{\infty} dt\, t^{-\alpha-1}\sin^2(t^2/2) = (\pi/2)^2 \operatorname{cosec}(\pi\alpha/4)\Gamma^{-1}(1+\alpha/2)\ . \qquad (7.2.67)$$

Thus, with (7.2.63b), we obtain [7.8]

$$\sigma_I^2 = 2^{\alpha}\Gamma(1+\alpha/2)(1+\alpha/2)^{-1}\cos(\pi\alpha/4)\phi, \qquad \phi \ll 1\ , \qquad (7.2.68)$$

which, when $\alpha = 5/3$, gives the expression by the Rytov approximation.

Also, in the case of a spherical wave, the expression (7.2.61) leads to the same result as (7.2.66), except for the ζ integral, which is replaced by

$$\int_0^1 d\zeta\,\zeta^{\alpha/2}(1-\zeta)^{\alpha/2} = \Gamma^2(1+\alpha/2)/\Gamma(2+\alpha)\ ;$$

hence σ_I^2 of (7.2.68), say $\sigma_I^2\,|_{PL}$, is replaced by [7.1]

$$\sigma_I^2\,|_{SP} = (1+\alpha/2)\Gamma^2(1+\alpha/2)\Gamma^{-1}(2+\alpha)\sigma_I^2\,|_{PL}\ . \qquad (7.2.69)$$

Expressions (7.2.60,61) for M_{22} are both based on a pure Kolmogorov medium corresponding to the spectrum function (7.1.40b), with $l_M = \infty$ and $l_m = 0$, and are equivalent to those used to numerically obtain the curves of Fig. 7.1. While, to introduce the inner scale, l_m, of finite value, a simple model was proposed [7.21], which is equivalent to replacing (7.2.3) by

$$\overline{V}(\boldsymbol{r}) = \tfrac{1}{2}\phi\left[|\overline{l}_m^2 + \boldsymbol{r}^2|^{\alpha/2} - \overline{l}_m^{\alpha}\right], \qquad \alpha = 5/3\ , \qquad (7.2.70)$$

where $\overline{l}_m = (k/L)^{1/2} l_m$, and which is reduced to (7.2.3) with ϕ of (7.2.4a,b), respectively, depending on whether $\overline{l}_m \ll |\boldsymbol{r}|$ or $\overline{l}_m \gg |\boldsymbol{r}|$. Shown in Fig 7.2 are numerical results (in three-dimensional space) using (7.2.70,58,59) for several values of $\overline{l}_m$ as a parameter [7.10]. Here the variable, $\tilde{\zeta}_k$, taken in the abscissa, is $\phi^{1/(1+\alpha/2)}$ (except for a numerical factor of the order of unity), being proportional to the propogation distance, L; the curves demonstrate very large values of the scintillation index up to about four (see also experimental results of Figs. 7.6,7).

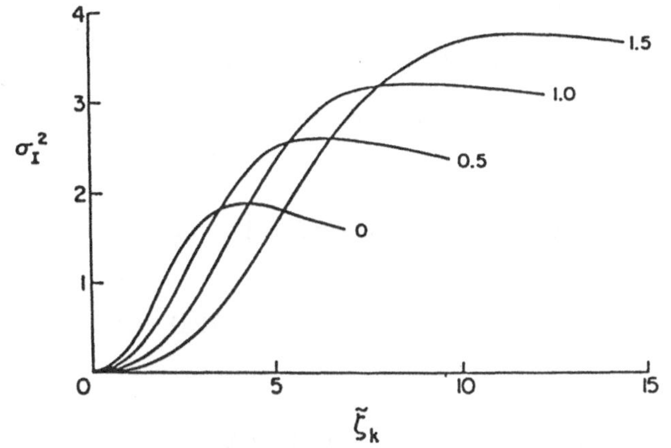

Fig. 7.2. Scintillation index versus distance for a point source in a Kolmogorov medium of (7.2.70). The numbers on the curves correspond to values of $\bar{l}_m = (k/L)^{1/2} l_m$. Thus each curve represents a constant value of the inner scale l_m (taken from [7.10])

Summarizing, the spectrum function $\widetilde{M}_{22}(\widehat{\boldsymbol{\lambda}}, \overline{\rho})$ of (7.2.54) has been obtained for a phase screen of width $\Delta\zeta$, and for any incident beam wave [including the collimated beam wave of (7.2.55a,b)], it has been intuitively generalized to a turbulent continuum by the replacement (7.2.56). The theoretical basis of this assumption is investigated in detail in Sect. 7.5, based on the Lagrange variational principle.

7.2.4 Exact Solutions of All Orders in a Special Case

The ν-th order moment with respect to the pair of wave functions, $\psi^*\psi$, has been given by (7.2.29,30) as the solution of equation (7.2.28). Here, the operators, $\widehat{V}_i(-\zeta)$ and $\widehat{V}_{ij}(-\zeta')$, are not mutually commutable for $\zeta \neq \zeta'$, in view of commutation relations (7.2.23), and must be ordered depending on their ζ values, according to the ordering P of (7.2.26).

However, when $V(\vec{r})$ can be given by (7.1.42a) over the entire $\vec{r}$ range, the total sum of the medium terms in the moment equation is given by (7.1.45), which is entirely free from the $\vec{\rho}$ coordinates. Consequently, the operators involved in (7.2.29) are only the $\widehat{\boldsymbol{r}}_i(-\zeta)$'s which are mutually commutable independent of their ζ values (7.2.24), so that the ordering P is not necessary. Thus, by using (7.1.45) which is presently written by

$$V^{(\nu)} = \frac{1}{2}\phi\left(\sum_{j=1}^{\nu} \boldsymbol{r}_j\right)^2 , \tag{7.2.71}$$

it is straightforward to find that, when $\boldsymbol{r}_1 = \boldsymbol{r}_2 = \ldots = \boldsymbol{r}_\nu = 0$, (7.2.29) is reduced to

$$M_{\nu\nu}(1) = \exp\left[6^{-1}\phi\left(\frac{\partial}{\partial\rho_1} + \frac{\partial}{\partial\rho_2} + \cdots \frac{\partial}{\partial\rho_\nu}\right)^2\right] M^0_{\nu\nu}(1) \ , \qquad (7.2.72a)$$

as a consequence of (7.2.11), with the Fourier representation of $M^0_{\nu\nu}(1)$ of the form

$$M^0_{\nu\nu}(1) = \prod_{j=1}^{\nu}(2\pi)^{-2}\int d\boldsymbol{\lambda}_j\,\widetilde{M}^0_{11}(\boldsymbol{\lambda}_j;\zeta=1)\exp(-\mathrm{i}\boldsymbol{\lambda}_j\cdot\boldsymbol{\rho}_j) \ . \qquad (7.2.72b)$$

Hence, when $\boldsymbol{\rho}_1 = \boldsymbol{\rho}_2 = \ldots = \boldsymbol{\rho}_\nu = \boldsymbol{\rho}$, (7.2.72a) can be rewritten by an integral as

$$M_{\nu\nu}(1) = \int d\boldsymbol{\lambda}\prod_{j=1}^{\nu}(2\pi)^{-2}\int d\boldsymbol{\lambda}_j\,\widetilde{M}^0_{11}(\boldsymbol{\lambda}_j;\zeta=1)$$
$$\times\,\delta\left(\boldsymbol{\lambda}-\sum_{a=1}^{\nu}\boldsymbol{\lambda}_a\right)\exp\left[-\mathrm{i}\boldsymbol{\lambda}\cdot\boldsymbol{\rho}-6^{-1}\phi\boldsymbol{\lambda}^2\right] \ . \qquad (7.2.73)$$

Upon substitution of the representation,

$$\delta\left(\boldsymbol{\lambda}-\sum_{a=1}^{\nu}\boldsymbol{\lambda}_a\right) = (2\pi)^{-2}\int d\boldsymbol{\rho}'\exp\left[\mathrm{i}\boldsymbol{\rho}'\cdot\left(\boldsymbol{\lambda}-\sum_{a=1}^{\nu}\boldsymbol{\lambda}_a\right)\right] \ . \qquad (7.2.74)$$

and after performing all the $\boldsymbol{\lambda}_j$ integrations with the aid of (7.2.72b), (7.2.73) becomes

$$M_{\nu\nu}(1) = \int d\boldsymbol{\rho}'\left[M^0_{11}(\boldsymbol{\rho}';\zeta=1)\right]^\nu$$
$$\times\,(2\pi)^{-2}\int d\boldsymbol{\lambda}\exp\left[\mathrm{i}(\boldsymbol{\rho}'-\boldsymbol{\rho})\cdot\boldsymbol{\lambda}-6^{-1}\phi\boldsymbol{\lambda}^2\right] \ . \qquad (7.2.75)$$

showing that

$$\langle I^\nu(\boldsymbol{\rho};\zeta=1)\rangle = \int d\boldsymbol{\rho}'\left[I^0(\boldsymbol{\rho}';\zeta=1)\right]^\nu P_S(\boldsymbol{\rho}-\boldsymbol{\rho}') \ . \qquad (7.2.76)$$

Here, $I^0(\boldsymbol{\rho};\zeta=1) = M^0_{11}(1)\,|_{\boldsymbol{r}=0}$ is the wave intensity in free space, and

$$P_S(\boldsymbol{\rho}) = (3/2\pi\phi)\exp\left[-(3/2\phi)\boldsymbol{\rho}^2\right] \ , \qquad (7.2.77a)$$

which satisfies the normalization

$$\int d\boldsymbol{\rho}\,P_S(\boldsymbol{\rho}) = 1 \qquad (7.2.77b)$$

and as $\phi \to 0$, $P_S(\boldsymbol{\rho}) \to \delta(\boldsymbol{\rho})$.

Here, writing (7.2.77a,b) in the original coordinates $\vec{\rho}$, the function $P_S(\rho)$ is changed to $P_S(\vec{\rho})$, as given by

$$P_S(\vec{\rho}) = (3k/2\pi\phi L)\exp\left[-(3k/2\phi L)\vec{\rho}^{\,2}\right] , \tag{7.2.78}$$

which is independent of the order, ν, and of the wave number, k, in view of

$$2\phi L/3k = 2.2\, L^3 l_m^{-1/3} C_n^2 , \tag{7.2.79}$$

from (7.2.4b). Thus (7.2.76) leads to the conclusion that $P_S(\vec{\rho})$ means the probability density function of finding the original wave, $I^0(\rho'; \zeta = 1)$, at a location displaced transversally by the distance, $\rho - \rho'$, without getting any distortion of the beam shape [7.22]. As a matter of fact, the above result is what is expected by assuming the model (7.1.42a); it corresponds to an atmospheric turbulence model consisting of thin refractive wedges of infinite extent, which gives rise to a random bending of the optical beam during propogation [7.23]. The exact agreement of the physical model (7.1.42a) and the theoretical result (7.2.76) suggests that the moment equations are highly reliable, not only for the first few orders, but for all orders, and are applicable to various problems.

When the incident beam wave is Gaussian, given by

$$M_{11}^0(\zeta = 1) = I_0 f^{-2} \exp\left[-(\bar{b}f)^{-2}\rho^2\right] , \tag{7.2.80a}$$

$$\widetilde{M}_{11}^0(\zeta = 1) = I_0 \bar{b}^2 \pi \exp\left[-(\bar{b}f/2)^2 \boldsymbol{\lambda}^2\right] \tag{7.2.80b}$$

from (7.2.39) with $\sigma_E = 0$; then, (7.2.76) gives [7.5, 24, 25]

$$M_{\nu\nu}(\zeta = 1) = (I_0 f^{-2})^\nu (1 + \sigma_E \nu)^{-1} \times \exp\left[-\nu(1 + \sigma_E\nu)^{-1}(\bar{b}f)^{-2}\rho^2\right] , \tag{7.2.81}$$

where $\boldsymbol{r}_1 = \boldsymbol{r}_2 = \ldots = \boldsymbol{r}_\nu = 0$, $\boldsymbol{\rho}_1 = \boldsymbol{\rho}_2 = \ldots = \boldsymbol{\rho}_\nu = \boldsymbol{\rho}$, and σ_E is defined by (7.2.40). When $\nu = 1$, the result agrees with (7.2.39).

7.3 High Order Intensity Moments and the Cluster Approximation

In a turbulent medium the function $V(\vec{r})$ is usually given by (7.1.42b) over the important range, and this makes solving the moment equations quite involved except for the first order equation. On the other hand, since the moment equations have a form similar to the Schrödinger equation in many body problems when the coordinate ζ is regarded as the time, we can also effectively utilize some of the techniques introduced therein for the present problems.

7.3.1 Intensity Fluctuation

The fourth order moment equation (7.2.15) can be written in the form

$$\frac{\partial}{\partial \zeta} M_{22}(\zeta) = (H_1 + H_2 - V_{12}) M_{22}(\zeta) \ . \tag{7.3.1}$$

Here,

$$H_j = T_j - V_j, \qquad V_j = \overline{V}(\boldsymbol{r}_j), \qquad j = 1,2 \ , \tag{7.3.2a}$$
$$V_{12} = \overline{V}_I(\boldsymbol{r}_1, \boldsymbol{r}_2, \boldsymbol{\rho}_{12}), \qquad \boldsymbol{\rho}_{12} = \boldsymbol{\rho}_1 - \boldsymbol{\rho}_2 \ . \tag{7.3.2b}$$

To obtain the solution, we write it here in the form

$$M_{22}(\zeta) = \exp\left[(H_1 + H_2)\zeta\right] S_{12}(\zeta) M_{22}(0) \ , \tag{7.3.3}$$

so that the operator, $S_{12}(\zeta)$, is the solution of

$$\frac{\partial}{\partial \zeta} S_{12}(\zeta) = -\widetilde{V}_{12}(\zeta) S_{12}(\zeta), \qquad S_{12}(0) = 1 \ , \tag{7.3.4}$$

where

$$\begin{aligned}\widetilde{V}_{12}(\zeta) &= \exp\left[-(H_1 + H_2)\zeta\right] \overline{V}_I(\boldsymbol{r}_1, \boldsymbol{r}_2, \boldsymbol{\rho}_{12}) \exp\left[(H_1 + H_2)\zeta\right] \\ &= \overline{V}_I(\widetilde{\boldsymbol{r}}_1(\zeta), \widetilde{\boldsymbol{r}}_2(\zeta), \widetilde{\boldsymbol{\rho}}_{12}(\zeta)) \ ,\end{aligned} \tag{7.3.5}$$

with

$$\widetilde{\boldsymbol{r}}_j(\zeta) = \exp(-H_j\zeta) \boldsymbol{r}_j \exp(H_j\zeta) \ , \tag{7.3.6a}$$
$$\widetilde{\boldsymbol{\rho}}_j(\zeta) = \exp(-H_j\zeta) \boldsymbol{\rho}_j \exp(H_j\zeta) \ . \tag{7.3.6b}$$

Here, as we shall see below, $\widetilde{V}_{12}(\zeta)$ and $\widetilde{V}_{12}(\zeta')$ are not commutable for $\zeta \neq \zeta'$, and, therefore, the solution of (7.3.4) should be written, as in (7.2.25), as

$$S_{12}(\zeta) = P \exp\left[-\int_0^\zeta d\zeta' \, \widetilde{V}_{12}(\zeta')\right] \tag{7.3.7}$$

$$= 1 - \int_0^\zeta d\zeta_1 \, \widetilde{V}_{12}(\zeta_1) + \int_0^\zeta d\zeta_2 \, \widetilde{V}_{12}(\zeta_2) \int_0^{\zeta_2} d\zeta_1 \, \widetilde{V}_{12}(\zeta_1) - \cdots \ , \tag{7.3.8}$$

with the ordering P defined by (7.2.26).

We introduce the notation

$$I_j(\zeta) = \psi_j^*(\zeta)\psi_j(\zeta) \equiv \psi^*(\boldsymbol{x}_j, \zeta)\psi(\boldsymbol{y}_j, \zeta) \ , \tag{7.3.9}$$

where $\boldsymbol{x}_j$ are $\boldsymbol{y}_j$ are to be expressed by the relative coordinates $\boldsymbol{r}_j$ and $\boldsymbol{\rho}_j$ of (7.1.34); and also, for any functional Q_{12} of the operators $\widetilde{\boldsymbol{r}}_j(\zeta)$ and $\widetilde{\boldsymbol{\rho}}_j(\zeta)$, $1 \geq \zeta \geq 0$, $j = 1, 2$, an average symbol $\langle Q_{12}\rangle_L$ is defined by

$$\langle Q_{12}\rangle_L = \frac{\langle 0|\exp(H_1 + H_2)Q_{12}|I\,\rangle}{\langle 0|\exp(H_1 + H_2)|I\,\rangle} \ . \tag{7.3.10a}$$

Here,

$$|I\,\rangle = I_1(0)I_2(0) = M_{11}^0(0) \ , \tag{7.3.10b}$$

and $\langle 0|$ represents a function, $\langle 0|\boldsymbol{r}_1, \boldsymbol{r}_2; \boldsymbol{\rho}_1, \boldsymbol{\rho}_2\rangle$, defined by

$$\langle 0|\boldsymbol{r}_1, \boldsymbol{r}_2; \boldsymbol{\rho}_1, \boldsymbol{\rho}_2\rangle = \delta(\boldsymbol{r}_1)\delta(\boldsymbol{r}_2)\delta(\boldsymbol{\rho} - \boldsymbol{\rho}_1)\delta(\boldsymbol{\rho} - \boldsymbol{\rho}_2) \ , \tag{7.3.10c}$$

so that the inner product of a function with $\langle 0|$ results in setting $\boldsymbol{r}_1 = \boldsymbol{r}_2 = 0$ and $\boldsymbol{\rho}_1 = \boldsymbol{\rho}_2 = \boldsymbol{\rho}$, of the function. In view of (7.2.5), we obtain

$$\langle 0|\exp(H_1 + H_2)|I\,\rangle = \langle I(\zeta = 1)\rangle^2 \ , \tag{7.3.11}$$

being the square of the mean intensity at $\zeta = 1$; and, also, by multiplication of (7.3.3) at $\zeta = 1$ to the left with $\langle 0|$,

$$\langle I^2(1)\rangle = \langle S_{12}(1)\rangle_L \langle I(1)\rangle^2 \ , \tag{7.3.12}$$

as a consequence of definition (7.3.10a). Hence,

$$m_2 \equiv \langle I^2(1)\rangle / \langle I(1)\rangle^2 = \langle S_{12}(1)\rangle_L \ , \tag{7.3.13}$$

which shows that $\langle S_{12}(1)\rangle_L$ means the normalized intensity moment, m_2. From (7.3.8),

$$\begin{aligned}\langle S_{12}(1)\rangle_L = 1 &- \int_0^1 d\zeta_1 \, \langle \widetilde{V}_{12}(\zeta_1)\rangle_L \\ &+ \int_0^1 d\zeta_2 \int_0^{\zeta_2} d\zeta_1 \, \langle \widetilde{V}_{12}(\zeta_2)\widetilde{V}_{12}(\zeta_1)\rangle_L - \cdots \ ,\end{aligned} \tag{7.3.14}$$

in which the $\langle\cdots\rangle_L$ can be regarded as ordinary averaging.

To find $\langle S_{12}(1)\rangle_L$, with reference to (7.3.7), we can alternatively put

$$S_{12}(\zeta) = S_{12}'(\zeta)\exp\left[-\int_0^\zeta d\zeta' \, \langle \widetilde{V}_{12}(\zeta')\rangle_L\right] \ , \tag{7.3.15}$$

which, upon the substitution into (7.3.4), yields the equation for $S'_{12}(\zeta)$ as

$$\frac{\partial}{\partial \zeta} S'_{12}(\zeta) = -\Delta \widetilde{V}_{12}(\zeta) S'_{12}(\zeta) \ , \tag{7.3.16}$$

where

$$\Delta \widetilde{V}_{12}(\zeta) = \widetilde{V}_{12}(\zeta) - \langle \widetilde{V}_{12}(\zeta) \rangle_L, \qquad \langle \Delta \widetilde{V}_{12}(\zeta) \rangle_L = 0 \ . \tag{7.3.17}$$

The solution is given by

$$S'_{12}(\zeta) = P \exp\left[-\int_0^{\zeta} d\zeta' \, \Delta \widetilde{V}_{12}(\zeta') \right] \ ; \tag{7.3.18}$$

hence,

$$\begin{aligned} \langle S_{12}(1) \rangle_L &= \exp\left[-\int_0^1 d\zeta \, \langle \widetilde{V}_{12}(\zeta) \rangle_L \right] \\ &\times \left[1 + \int_0^1 d\zeta_2 \int_0^{\zeta_2} d\zeta_1 \, \langle \Delta \widetilde{V}_{12}(\zeta_2) \Delta \widetilde{V}_{12}(\zeta_1) \rangle_L \right. \\ &\left. - \int_0^1 d\zeta_3 \int_0^{\zeta_3} d\zeta_2 \int_0^{\zeta_2} d\zeta_1 \, \langle \Delta \widetilde{V}_{12}(\zeta_3) \Delta \widetilde{V}_{12}(\zeta_2) \Delta \widetilde{V}_{12}(\zeta_1) \rangle_L + \cdots \right] \ , \end{aligned} \tag{7.3.19}$$

which can also be written by a cumulant series, as

$$\begin{aligned} \langle S_{12}(1) \rangle_L \equiv \exp[K_1(12)] = \exp&\left[-\int_0^1 d\zeta \, \langle \widetilde{V}_{12}(\zeta) \rangle_L \right. \\ &\left. + \int_0^1 d\zeta_2 \int_0^{\zeta_2} d\zeta_1 \, \langle \Delta \widetilde{V}_{12}(\zeta_2) \Delta \widetilde{V}_{12}(\zeta_1) \rangle_L - \cdots \right] \ . \end{aligned} \tag{7.3.20}$$

On the other hand, from (7.3.5), $\widetilde{V}_{12}(\zeta)$ is a function of $\widetilde{\boldsymbol{r}}_1(\zeta)$, $\widetilde{\boldsymbol{r}}_2(\zeta)$, and $\widetilde{\rho}_{12}(\zeta)$, which are given [in view of

$$\exp(H_1 \zeta) = \exp(T_1 \zeta) U_1(\zeta) \ , \tag{7.3.21}$$

from (7.2.5–7)] by

$$\begin{aligned} \widetilde{\rho}_j(\zeta) &= U_j^{-1}(\zeta) \exp(-T_j \zeta) \rho_j \exp(T_j \zeta) U_j(\zeta) \\ &= U_j^{-1}(\zeta) \widehat{\rho}_j(\zeta) U_j(\zeta) \ , \end{aligned} \tag{7.3.22}$$

in terms of $\widehat{\rho}_j(\zeta)$ of (7.2.21), and, similarly,

$$\widetilde{\boldsymbol{r}}_j(\zeta) = U_j^{-1}(\zeta) \widehat{\boldsymbol{r}}_j(\zeta) U_j(\zeta) = \widehat{\boldsymbol{r}}_j(\zeta) \ , \tag{7.3.23}$$

with $U_j(\zeta)$ being commutable with $\widehat{\boldsymbol{r}}_j(\zeta')$ (7.2.12,9). Equation (7.3.22) is further reduced, with $U_j(\zeta)$ of (7.2.12), to [cf. (1.1.13)]

$$\begin{aligned}\widetilde{\rho}_j(\zeta) &= \widehat{\rho}_j(\zeta) + \int_0^\zeta d\zeta' \, [\overline{V}[\widehat{\boldsymbol{r}}_j(\zeta')], \widehat{\rho}_j(\zeta)] \\ &= \widehat{\rho}_j(\zeta) + \mathrm{i} \int_0^\zeta d\zeta' \, (\zeta' - \zeta) \frac{\partial}{\partial \boldsymbol{r}_j} \overline{V}[\widehat{\boldsymbol{r}}_j(\zeta')] \quad . \qquad (7.3.24)\end{aligned}$$

On the other hand, directly in terms of $\widehat{V}_1$, $\widehat{V}_2$, and $\widehat{V}_{12}$, $\widetilde{V}_{12}(\zeta)$ is expressed by

$$\widetilde{V}_{12}(\zeta) = U_2^{-1}(\zeta) U_1^{-1}(\zeta) \widehat{V}_{12}(\zeta) U_1(\zeta) U_2(\zeta) \qquad (7.3.25)$$

$$\begin{aligned}&= \exp\left\{\int_0^\zeta d\zeta' \, [\widehat{V}_1(\zeta') + \widehat{V}_2(\zeta')]\right\} \\ &\quad \times \widehat{V}_{12}(\zeta) \exp\left\{-\int_0^\zeta d\zeta' \, [\widehat{V}_1(\zeta') + \widehat{V}_2(\zeta')]\right\} , \qquad (7.3.26)\end{aligned}$$

which can be utilized to evaluate

$$K_1(12) \sim -\int_0^1 d\zeta \, \langle \widetilde{V}_{12}(\zeta) \rangle_L \qquad (7.3.27)$$

from (7.3.20), and to obtain the result (7.2.68) directly.

It is actually easier, compared to $\langle Q_{12} \rangle_L$, to evaluate another average, $\langle Q_{12} \rangle_L'$, defined by

$$\langle Q_{12} \rangle_L' = \frac{\langle 0 | Q_{12} \exp(H_1 + H_2) | I \rangle}{\langle 0 | \exp(H_1 + H_2) | I \rangle} , \qquad (7.3.28)$$

with the relation

$$\langle Q_{12} \rangle_L = \langle \exp(H_1 + H_2) Q_{12} \exp(-H_1 - H_2) \rangle_L' \; ; \qquad (7.3.29)$$

and hence (7.3.27), e.g., can be written, in view of (7.3.6a,b), as

$$K_{12} \sim -\int_0^1 d\zeta \, \langle \widetilde{V}_{12}(\zeta - 1) \rangle_L' \; . \qquad (7.3.30)$$

Here in the case of a plane wave,

$$\exp(H_1 + H_2) | I \rangle = \exp[-\overline{V}(\boldsymbol{r}_1) - \overline{V}(\boldsymbol{r}_2)] \, | I \rangle \; , \qquad (7.3.31)$$

so that evaluation of (7.3.30) becomes straightforward by using (7.3.26) and the integral representation (7.2.63) for $\overline{V}_I$.

7.3.2 Cluster Approximation

It is practically impossible to solve the moment equations for all orders except in the special case of Sect. 7.2.4 where $V(\vec{r})$ is assumed to have the form (7.1.42a) over the entire $\vec{r}$ range. On the other hand, it has been shown experimentally that the wave intensity in a turbulent medium has a distribution close to the log-normal, and this implies that the high order moments can be expressed to a good accuracy by the first few low order moments. In this section we show how, based on the moment equations, the high order intensity moments can be systematically constructed in terms of the low order moments so that the intensity distribution is reduced to the log-normal in a limit [7.9].

Moment Eq. (7.2.28) for $M_{\nu\nu}$, $\nu \geq 2$, can be written as

$$\frac{\partial}{\partial \zeta} M_{\nu\nu}(\zeta) = \left[\sum_{j=1}^{\nu} H_j - \sum_{i>j=1}^{\nu} V_{ij}\right] M_{\nu\nu}(\zeta) \ , \tag{7.3.32}$$

in terms of H_j and $V_{ij} = V_{ji}$, defined by (7.3.2a,b). Here we write the solution in a form similar to (7.3.3):

$$M_{\nu\nu}(\zeta) = \exp\left[\sum_{j=1}^{\nu} H_j \zeta\right] S^{(\nu)}(\zeta)|I\rangle \ , \tag{7.3.33}$$

where

$$|I\rangle = \prod_{j=1}^{\nu} \psi^*(\boldsymbol{x}_j, \zeta = 0)\psi(\boldsymbol{y}_j, \zeta = 0) \ , \tag{7.3.34}$$

being the initial value at $\zeta = 0$. Here, since V_{ij} depends only on the variables, $\boldsymbol{r}_i$, $\boldsymbol{r}_j$, and $\boldsymbol{\rho}_{ij}$, the substitution in (7.3.32) yields the equation for $S^{(\nu)}(\zeta)$ as

$$\frac{\partial}{\partial \zeta} S^{(\nu)}(\zeta) = -\sum_{i>j=1}^{\nu} \widetilde{V}_{ij}(\zeta) S^{(\nu)}(\zeta) \ , \tag{7.3.35}$$

in terms of $\widetilde{V}_{ij}(\zeta)$, defined by (7.3.5). Hence the formal solution is given by

$$S^{(\nu)}(\zeta) = P \exp\left[-\sum_{i>j=1}^{\nu} \int_0^{\zeta} d\zeta' \, \widetilde{V}_{ij}(\zeta')\right] \tag{7.3.36a}$$

$$= P \prod_{i>j=1}^{\nu} S_{ij}(\zeta), \qquad P^2 = P \ , \tag{7.3.36b}$$

in terms of $S_{ij}(\zeta)$, defined by (7.3.7).

Now we introduce an "averaging" notation, $\langle Q\rangle_L$, similar to (7.3.10), defined for any operator Q, by

$$\langle Q\rangle_L = \frac{\langle 0|\exp(\sum_{j=1}^{\nu} H_j)Q|I\rangle}{\langle 0|\exp(\sum_{j=1}^{\nu} H_j)|I\rangle} , \tag{7.3.37}$$

where the multiplication with $\langle 0|$ yields the setting, $\boldsymbol{r}_1 = \boldsymbol{r}_2 = \ldots = \boldsymbol{r}_\nu = 0$ and $\boldsymbol{\rho}_1 = \boldsymbol{\rho}_2 = \ldots = \boldsymbol{\rho}_\nu = \boldsymbol{\rho}$. Hence, from (7.3.33,36b), we find an expression,

$$\begin{aligned} m_\nu &\equiv \langle I^\nu(1)\rangle/\langle I(1)\rangle^\nu = \langle S^{(\nu)}(1)\rangle_L \\ &= \left\langle P \prod_{i>j=1}^{\nu} S_{ij}(1)\right\rangle_L , \end{aligned} \tag{7.3.38}$$

for the normalized intensity moments, m_ν, in terms of the operators, $S_{ij}(1)$.

We first introduce a quantity, $K_1(12)$, defined by

$$\langle S_{12}(1)\rangle_L = \exp[K_1(12)] , \tag{7.3.39}$$

with the symmetry,

$$K_1(12) = K_1(23) = \ldots = K_1(ij) = K_1 , \tag{7.3.40}$$

which is obvious, by the definition. Then we define $K_2(12,23)$ by

$$\langle PS_{12}(1)S_{23}(1)\rangle_L = \exp[K_1(12) + K_1(23) + K_2(12,23)] , \tag{7.3.41}$$

with symmetry similar to (7.3.40). Here, we note, that when the two "pair bonds" are disconnected, e.g.,

$$\langle PS_{12}S_{34}\rangle_L = \langle S_{12}\rangle_L\langle S_{34}\rangle_L = \exp[K_1(12) + K_1(34)] , \tag{7.3.42}$$

it follows that $K_2(12,34) = 0$.

In the same way, $K_3(12,23,34)$ is defined by [7.26]

$$\begin{aligned} \langle PS_{12}S_{23}S_{34}\rangle_L = \exp[&K_1(12) + K_1(23) + K_1(34) \\ &+ K_2(12,23) + K_2(23,34) + K_3(12,23,34)] , \end{aligned} \tag{7.3.43}$$

in view of $K_2(12,34) = 0$; therefore, $K_3(12,23,34)$ is that term which is connected by the three bonds, 12, 23, and 34.

Generally, if $K(A)$ and $K(B)$, for any two sets A and B of P-ordered products of S_{ij}, $i \neq j = 1, 2, 3, \ldots$, are defined by

$$\langle PA\rangle_L = \exp[K(A)], \qquad \langle PB\rangle_L = \exp[K(B)] , \tag{7.3.44}$$

then, $K(A,B)$ is defined by

$$\langle PAB\rangle_L = \exp\left[K(A) + K(B) + K(A,B)\right] \ . \tag{7.3.45}$$

Therefore, $K(A,B) = 0$ whenever the sets A and B are not correlated; i.e.,

$$\langle PAB\rangle_L = \langle PA\rangle_L\langle PB\rangle_L = \exp\left[K(A) + K(B)\right] \ , \tag{7.3.46}$$

In this way, in terms of the K_ν 's, we can write the normalized intensity moment, m_ν of (7.3.38), as:

$$m_2 = \langle S_{12}\rangle_L = \exp[K_1(12)] \ , \tag{7.3.47a}$$

$$\begin{aligned} m_3 &= \langle PS_{12}S_{23}S_{31}\rangle_L \\ &= \exp\left[3K_1(12) + 3K_2(12,23) + K_3(12,23,31)\right] \ , \end{aligned} \tag{7.3.47b}$$

$$\begin{aligned} m_\nu &= \left\langle P \prod_{i>j=1}^{\nu} S_{ij} \right\rangle_L \\ &= \exp\Big[\tbinom{\nu}{2} K_1(12) + \tbinom{\nu}{3}\{3K_2(12,23) + K_3(12,23,31)\} \\ &\quad + \tbinom{\nu}{4}\{12K_3(12,23,34) + 3K_4(12,23,34,41) \\ &\quad + 6K_5(12,23,34,41,13) \\ &\quad + K_6(12,23,34,41,13,24)\} + \cdots\Big] \ , \end{aligned} \tag{7.3.47c}$$

where we have used the symmetries similar to (7.3.40), and also the rule that K_ν vanishes whenever the involved pair bonds are disconnected; the high order K_ν 's are not always uniquely determined, however [7.26]. Here, the essential point is to express m_ν in terms of the low order moments by using (7.3.47a–c), i.e.,

$$\ln(m_2) = K_1(12) \ , \tag{7.3.48a}$$

$$\ln(m_3/m_2^3) = 3K_2(12,23) + K_3(12,23,31) \ , \tag{7.3.48b}$$

$$\begin{aligned} \ln(m_4 m_2^6/m_3^4) &= 12K_3(12,23,34) \\ &\quad +3K_4(12,23,34,41) + \cdots \ , \quad \textit{etc.} \ , \end{aligned} \tag{7.3.48c}$$

so that (7.3.47c) can be written as

$$\begin{aligned} m_\nu = \exp\Big[&\tbinom{\nu}{2}\ln(m_2) + \tbinom{\nu}{3}\ln(m_3/m_2^3) \\ &+ \tbinom{\nu}{4}\ln(m_4 m_2^6/m_3^4) + \cdots\Big] \ ; \end{aligned} \tag{7.3.49}$$

and to expect a good convergence of the series.

However, it is generally very difficult to discuss convergence of series (7.3.49), which should depend on the particular natures of H_j and V_{ij} in (7.3.32), being, therefore, mostly anticipated by a physical intuition. In the present case, we shall keep only the first two terms of the series, and first compare the results with experimental values to see its validity. Thus we obtain an approximate expression of m_ν of the form

$$m_\nu = \exp\{2^{-1}\nu(\nu-1)K_1[1-(\nu-2)\Delta']\} \ . \tag{7.3.50a}$$

Here,

$$\begin{aligned} K_1 &= \ln m_2 \ , \\ \Delta' &= -[K_2(12,23)+3^{-1}K_3(12,23,31)]/K_1(12) \\ &= 1-3^{-1}\ln m_3/\ln m_2 \ . \end{aligned} \tag{7.3.50b}$$

Table 7.1. Comparison of the experimental values of $m_\nu = \langle I^\nu\rangle/\langle I\rangle^\nu$ observed by [7.27] for $\beta_0^2 = 0.31\,C_\varepsilon^2 k^{7/6} L^{11/6} \geq 25$ with the corresponding theoretical values given by (7.4.29). $K_1 = \ln m_2 = 0.7793$ whereas the parameter Δ has the two different values according to whether it is determined by the experimental value of m_3 or of m_4. The resulted errors of m_4 or m_3 are shown in parenthesis in %. The two values in the last line and column are the ratios of the predicted theoretical values to the log-normal one [7.9]

ν \ m_ν	Experimental	Theoretical & Error (%) $\Delta = 0.0262$	$\Delta = 0.0346$	Log-normal ($\Delta = 0$)	Experimental/ Log-normal
2	2.18	2.18	2.18	2.18	1.0
3	9.76	9.76	9.58 (1.8 %)	10.36	0.94
4	79.32	85.0 (7.2 %)	79.32	107.32	0.74
5	xxxxx	1374	1165	2424	(0.57; 0.48)

In Table 7.1, the experimental values of m_ν observed in [7.27] are shown along with the corresponding theoretical values (to be given by (7.4.29a,b), which actually give the same values as (7.3.50a,b) for the shown experimental values of $\Delta' \sim \Delta \ll 1$ and $\nu = 2 \sim 4$). Here, since the parameter Δ depends critically on the value of m_3, two values of Δ are prepared; one being determined by the values of m_3 and the other by that of m_4. The first value gives rise to the error of 7.2 % for m_4, whereas the second one gives rise to the error of only 1.8 % for m_3. Note that $\Delta' \sim \Delta > 0$, and this is also theoretically confirmed to the first order of $\phi \ll 1$, when the distance of wave propogation is short enough.

It is also worthwhile to check the third term of series (7.3.49), which has been completely neglected in the derivation of expression (7.3.50a,b). According to the experimental values given in Table 7.1,

$$\ln(m_4 m_2^6/m_3^4) = 4.3735 + 6 \times 0.7793 - 4 \times 2.2783 = -0.064 \ ,$$

which is just 0.7 % of the cancelled value. Therefore, the above value will certainly be within the range of experimental error, and the large cancellation will mean that the third term is really very small to be neglected; i.e.,

$$m_4 \sim m_3^4/m_2^6 \ , \tag{7.3.51}$$

which is exactly the relation given by the series (7.3.49) when neglecting the third and higher order terms. Thus the agreement of the theoretical and experimental values is very good, showing a rapid convergence of the cumulant series (7.3.49), as expected.

In the expression (7.3.50a,b) of m_ν, $K_1(12)$ is given by the cumulant series of (7.3.20), and $K_2(12,23)$ and $K_3(12,23,34)$ are

$$K_2(12,23) \sim 2\int_0^1 d\zeta_2 \int_0^{\zeta_2} d\zeta_1 \, \langle \Delta\widetilde{V}_{12}(\zeta_2)\Delta\widetilde{V}_{23}(\zeta_1)\rangle_L \ , \tag{7.3.52a}$$

$$\begin{aligned} K_3(12,23,34) \sim -6\int_0^1 d\zeta_3 \int_0^{\zeta_3} d\zeta_2 \int_0^{\zeta_2} d\zeta_1 \\ \times \langle \Delta\widetilde{V}_{12}(\zeta_3)\Delta\widetilde{V}_{23}(\zeta_2)\Delta\widetilde{V}_{34}(\zeta_1)\rangle_L \ , \end{aligned} \tag{7.3.52b}$$

being the non-vanishing terms to the lowest order of ϕ.

Finally, it may be remarked that the expression (7.3.50a,b) for m_ν is valid, independent of the medium spectrum as well as the initial wave-condition, which are all involved only through the first two moments, m_2 and m_3; this is true even of the aperture size of the receiver, as can be directly shown by redefining $\langle 0|$ in (7.3.37). The expression is therefore universal in that sense.

7.4 Probability Distribution Function of Intensity

It has been a tough theoretical problem to account for the experimental fact that the intensity distribution of a light wave in a turbulent medium is close to the Gaussian with respect to the logarithm of intensity, even in the region where the variance of intensity scintillation is saturated, as was mentioned in [7.6,23, 27, 28], and others. Of course, the theory should be given based on the moment equations of wave functions, which have been successful for solving problems not only of the low order moments, but also of the high orders in a special case (Sect. 7.2.4). Although it is generally impossible to solve the equations for all orders to find the probability distribution of intensity, $P(I)$, a reasonable distribution can be obtained by constructing a self-consistent m_ν based on the cluster approximation of Sect. 7.3.2. In this chapter, integral representation of $P(I)$ for given moments $\langle I^n \rangle$, $n = 1, 2, 3, \ldots$, is first obtained, which is similar to the Watson-Sommerfeld transformation for a diffracted wave by a spherical surface; this is then applied to the $\langle I^n \rangle$'s given by (7.2.81) and (7.4.29). The integrals can be exactly evaluated by the residual method, and thus the situation again becomes the same as in the residual series expression of the Watson integral. Finally, the results are compared to the experimentals, along with several other theories.

7.4.1 Rice–Nakagami Distribution

The complete statistical information of a wave at a point in a random medium can be provided by the characteristic function of the wave functions, defined by

$$Z(\bar{j}^*, \bar{j}) = \langle \exp(\bar{j}^* \psi^* + \bar{j}\psi) \rangle \tag{7.4.1}$$

$$= \sum_{m,n=0}^{\infty} (m!n!)^{-1} \langle \psi^{*m} \psi^n \rangle \bar{j}^{*m} \bar{j}^n \tag{7.4.2a}$$

$$= \exp\left[\sum_{m,n=0}^{\infty} (m!n!)^{-1} \kappa_{mn} \bar{j}^{*m} \bar{j}^n \right] . \tag{7.4.2b}$$

Here, the cumulants, κ_{mn}, are given in terms of the moments by

$$\begin{aligned}
&\kappa_{00} = 0, \qquad \kappa_{01} = \langle \psi \rangle, \qquad \kappa_{10} = \langle \psi^* \rangle \ , \\
&\kappa_{11} = \langle \psi^* \psi \rangle - \langle \psi^* \rangle \langle \psi \rangle, \qquad \kappa_{02} = \langle \psi^2 \rangle - \langle \psi \rangle^2 \ , \\
&\kappa_{21} = \langle (\psi^* - \langle \psi^* \rangle)^2 (\psi - \langle \psi \rangle) \rangle \\
&\quad = \langle \psi^{*2} \psi \rangle - \langle \psi^{*2} \rangle \langle \psi \rangle - 2 \langle \psi^* \rangle \langle \psi^* \psi \rangle + 2 \langle \psi^* \rangle^2 \langle \psi \rangle \ , \quad \ldots
\end{aligned} \tag{7.4.3}$$

Hence, if the first three terms are enough in the cumulant series (7.4.2b), i.e.,

$$Z(\bar{j}^*, \bar{j}) \sim \exp[\bar{j}^* \langle \psi^* \rangle + \bar{j} \langle \psi \rangle + \bar{j}^* \bar{j} \kappa_{11}] \ , \tag{7.4.4}$$

then, the wave intensity, $I = \psi^* \psi$, is proved to be subject to the Rice–Nakagami distribution.

The probability density function, $P(I)$, can be given by the integral,

$$P(I) = \langle \delta(I - \psi^* \psi) \rangle = \frac{1}{2\pi} \int_{-\infty}^{\infty} dt \exp(\mathrm{i}tI) f(t) \ , \tag{7.4.5}$$

in terms of the characteristic function of I, $f(t)$, defined by

$$\begin{aligned}
f(t) &= \langle \exp(-\mathrm{i}t \psi^* \psi) \rangle \\
&= \exp\left[-\mathrm{i}t \frac{\partial}{\partial \bar{j}^*} \frac{\partial}{\partial \bar{j}} \right] Z(\bar{j}^*, \bar{j}) \bigg|_{\bar{j}^* = \bar{j} = 0} \ ,
\end{aligned} \tag{7.4.6}$$

where use has been made of (7.4.1) in the last derivation. Here, when $Z(\overline{\jmath}^*, \overline{\jmath})$ is approximated by (7.4.4), evaluation of (7.4.6) yields

$$f(t) = (1 + \mathrm{i}t\kappa_{11})^{-1} \exp\left[-\mathrm{i}t(1 + \mathrm{i}\kappa_{11})^{-1} |\langle\psi\rangle|^2\right] \ . \tag{7.4.7}$$

The proof is given as follows: with $Z(\overline{\jmath}^*, \overline{\jmath})$ of (7.4.4), we obtain from (7.4.6) [7.29]

$$\begin{aligned} \frac{\partial}{\partial\langle\psi\rangle} f(t) &= \exp\left[-\mathrm{i}t \frac{\partial}{\partial \overline{\jmath}^*} \frac{\partial}{\partial \overline{\jmath}}\right] \overline{\jmath} Z(\overline{\jmath}^*, \overline{\jmath}) \\ &= \left(\overline{\jmath} - \mathrm{i}t \frac{\partial}{\partial \overline{\jmath}^*}\right) f(t) \ , \end{aligned}$$

by using formula (1.1.13) with the replacement of $C \rightarrow \overline{\jmath}$ and $\theta(\partial/\partial C) \rightarrow -\mathrm{i}t(\partial/\partial \overline{\jmath}^*)(\partial/\partial \overline{\jmath})$. To evaluate the right-hand side, we remark that

$$\begin{aligned} \frac{\partial}{\partial \overline{\jmath}^*} Z(\overline{\jmath}^*, \overline{\jmath}) &= \left[\langle\psi^*\rangle + \kappa_{11} \overline{\jmath}\right] Z(\overline{\jmath}^*, \overline{\jmath}) \\ &= \left[\langle\psi^*\rangle + \kappa_{11} \frac{\partial}{\partial\langle\psi\rangle}\right] Z(\overline{\jmath}^*, \overline{\jmath}) \ ; \end{aligned}$$

hence, for $\overline{\jmath}^* = \overline{\jmath} = 0$,

$$\frac{\partial}{\partial\langle\psi\rangle} f(t) = -\mathrm{i}t \left[\langle\psi^*\rangle + \kappa_{11} \frac{\partial}{\partial\langle\psi\rangle}\right] f(t) \ ,$$

or

$$\frac{\partial}{\partial\langle\psi\rangle} f(t) = -\mathrm{i}t\langle\psi^*\rangle(1 + \mathrm{i}t\kappa_{11})^{-1} f(t) \ ,$$

whose solution is

$$f(t) = f_0(t) \exp\left[-\mathrm{i}t\langle\psi^*\rangle\langle\psi\rangle(1 + \mathrm{i}t\kappa_{11})^{-1}\right] \ .$$

Here, $f_0(t)$ is the initial function at $\langle\psi\rangle = \langle\psi^*\rangle = 0$, hence,

$$\begin{aligned} f_0(t) &= \exp\left[-\mathrm{i}t \frac{\partial}{\partial \overline{\jmath}^*} \frac{\partial}{\partial \overline{\jmath}}\right] \exp\left(\kappa_{11} \overline{\jmath}^* \overline{\jmath}\right) \Bigg|_{\overline{\jmath}^* = \overline{\jmath} = 0} \\ &= \sum_{n=0}^{\infty} (n!)^{-2} (-\mathrm{i}t\kappa_{11})^n \left(\frac{\partial}{\partial \overline{\jmath}^*} \frac{\partial}{\partial \overline{\jmath}}\right)^n (\overline{\jmath}^* \overline{\jmath})^n \Bigg|_{\overline{\jmath}^* = \overline{\jmath} = 0} \\ &= \sum_{n=0}^{\infty} (-\mathrm{i}t\kappa_{11})^n = (1 + \mathrm{i}t\kappa_{11})^{-1} \ ; \end{aligned}$$

thus, (7.4.7) is proved.

To evaluate the integral (7.4.5) for $P(I)$, we change the variable of integration, t, to

$$s = (1 + \mathrm{i}t\kappa_{11})I^{1/2}|\langle\psi\rangle|^{-1} \ ; \tag{7.4.8a}$$

hence, with (7.4.7),

$$P(I) = \kappa_{11}^{-1} \exp\left[-\kappa_{11}^{-1}(I + |\langle\psi\rangle|^2)\right] \times (2\pi\mathrm{i})^{-1} \int_{-\mathrm{i}\infty+\varepsilon}^{\mathrm{i}\infty+\varepsilon} ds\, s^{-1} \exp\left[I^{1/2}|\langle\psi\rangle|\kappa_{11}^{-1}(s + s^{-1})\right] \ , \tag{7.4.8b}$$

where $\varepsilon = +0$. Here, the integral can be given in terms of the residue value at the pole, $s = 0$; hence, with the aid of formula

$$I_n(x) = (2\pi\mathrm{i})^{-1} \oint ds\, s^{n-1} \exp\left[2^{-1}x(s + s^{-1})\right] \ , \tag{7.4.9}$$

$I_n(x) = I_{-n}(x)$, being the modified Bessel function of order n, and we finally obtain the Rice–Nakagami distribution [7.30]

$$P(I) = \kappa_{11}^{-1} \exp\left[-\kappa_{11}^{-1}(I + |\langle\psi\rangle|^2)\right] I_0(2\kappa_{11}^{-1}|\langle\psi\rangle|I^{1/2}) \ . \tag{7.4.10}$$

As $\langle\psi\rangle \to 0$, the distribution tends to the exponential distribution, and as was remarked in [7.31] in connection with high order intensity moments in a turbulent medium.

7.4.2 Integral Representation of the Probability Density Function for Given $\langle I^n \rangle$, $n = 1, 2, 3, \ldots$

The characteristic function of intensity, $f(t)$, defined by (7.4.6), is formally expanded in a power series of t, as

$$f(t) = \sum_{n=0}^{\infty} (n!)^{-1}(-\mathrm{i}t)^n \langle I^n \rangle \ , \tag{7.4.11}$$

in terms of given $\langle I^n \rangle$. However, the direct substitution in the integrand of (7.4.5) is not possible, because it yields a divergent integral for each term of the series.

On the other hand, with the aid of the relation,

$$\mathrm{Res}[\Gamma(-\nu)]\,\big|_{\nu=n} = (-)^{n+1}(n!)^{-1}, \qquad n = 0, 1, 2, \ldots \ , \tag{7.4.12}$$

the series (7.4.11) can be replaced by the integral,

$$f(t) = \frac{1}{2\pi\mathrm{i}} \int_C d\nu\, \Gamma(-\nu)(\mathrm{i}t)^{\nu} \langle I^{\nu} \rangle \ , \tag{7.4.13}$$

where the path of integration, C, starts at $\nu = +\infty$ and goes on the infinitesimally lower side of the real axis; it then encircles the origin in the clockwise direction and returns to $+\infty$ on the infinitesimally upper side of the real axis, as illustrated in Fig 7.3.

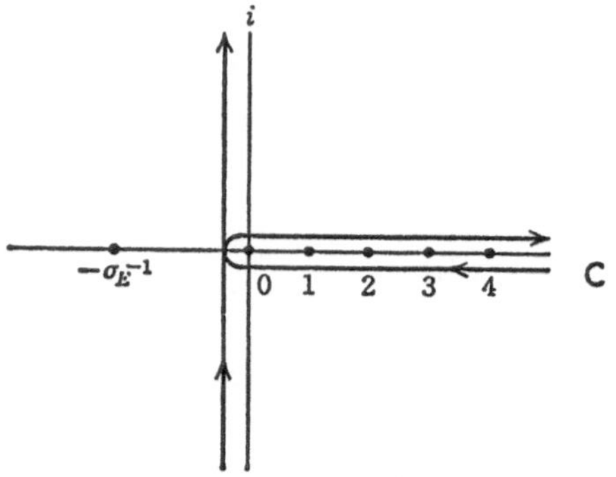

Fig. 7.3. The complex ν plane for (7.4.13) and (7.4.19)

We assume that $\langle I^\nu \rangle$ is analytic on the right half-plane of ν, and has the asymptotic form

$$\langle I^\nu \rangle \lesssim I_C^\nu, \qquad \mathrm{Re}\{\nu\} \sim +\infty \ , \tag{7.4.14}$$

which means that there is always a positive number, I_C, such that $\langle I^\nu \rangle / I_C^\nu < \infty$ as $\mathrm{Re}\{\nu\} \to +\infty$. Hence, as $|\nu| \to +\infty$ on the right half-plane of ν, the integrand of (7.4.13) tends to the order of magnitude,

$$\exp\left\{-(\nu + \tfrac{1}{2})\ln(-\nu) + \nu\left[1 + \ln(\mathrm{i}t) + \ln I_C\right]\right\} \ , \tag{7.4.15}$$

where $\ln(-\nu) = \mp \mathrm{i}\pi/2 + \ln|\nu|$ for $\nu \sim \pm \mathrm{i}\infty$. Therefore, when $|\arg(\mathrm{i}t)| \le \pi/2$, the integrand tends to zero as $|\nu| \to \infty$, and is analytic on the right half-plane of ν excluding the real axis. Thus (7.3.13) leads to the Barnes-type integral:

$$f(t) = \frac{1}{2\pi \mathrm{i}} \int_{-\mathrm{i}\infty - \varepsilon}^{\mathrm{i}\infty - \varepsilon} d\nu\, \Gamma(-\nu)(\mathrm{i}t)^\nu \langle I^\nu \rangle \ . \tag{7.4.16a}$$

$$|\arg(\mathrm{i}t)| \le \pi/2, \qquad |t| \ne 0 \ , \tag{7.4.16b}$$

where $\varepsilon = +0$ is an infinitesimal positive number.

If (7.4.16a) is substituted into (7.4.5), we notice that the integrand is singular at $t = 0$ through the factor $(\mathrm{i}t)^\nu$, $\mathrm{Re}\{\nu\} = -\varepsilon$, and, therefore, the path of integration of t in (7.4.5) has to be taken on the infinitesimally lower side of the real axis so that the conditions of (7.4.16b) are fulfilled. The relevant integral with respect to t becomes

$$(2\pi)^{-1} \int_{-\infty - \mathrm{i}\varepsilon}^{\infty - \mathrm{i}\varepsilon} dt\, (\mathrm{i}t)^\nu \exp(\mathrm{i}tI) \ . \tag{7.4.17}$$

where $I > 0$ and $\mathrm{Re}\{\nu\} = -\varepsilon$.

The path of integration of (7.4.17) can be deformed so that it starts at $+\mathrm{i}\infty$ on the imaginary axis, encircles the origin in the counter-clockwise direction, and returns to the starting point. Hence, rewritting the variable of integration, t, as $\mathrm{i}t/I$, the integral (7.4.17) becomes

$$I^{-\nu-1}(2\pi)^{-1}\mathrm{i}\int_{+\infty}^{(0+)} dt\,(-t)^{\nu}\mathrm{e}^{-t} = I^{-\nu-1}/\Gamma(-\nu)\ , \tag{7.4.18}$$

as a consequence of the Hankel contour integral for the Gamma function.

Thus, from (7.4.5,16–18), we obtain [7.5]

$$P(I) = \frac{1}{2\pi\mathrm{i}}\int_{-\mathrm{i}\infty-\varepsilon}^{\mathrm{i}\infty-\varepsilon} d\nu\, I^{-\nu-1}\langle I^{\nu}\rangle\ , \tag{7.4.19}$$

provided that $\langle I^{\nu}\rangle$ is analytic on the right half-plane of ν, and has the asymptotic form (7.4.14).

Equation (7.4.19) gives the distribution function $P(I)$ directly in terms of given $\langle I^{\nu}\rangle = \langle I^{n}\rangle\,\big|_{n=\nu}$, and is similar to the Watson-Sommerfeld transformation for a wave diffracted by a spherical surface, in that starting from the discrete values, $\langle I^{n}\rangle$, $n = 0, 1, 2, \ldots$, the function, $P(I)$, is represented by an integral with respect to the order variable, ν. Note that the intensity moments, $\langle I^{n}\rangle$, are defined only for the positive integral values of n, and are obtained from solutions of the moment equations. In the special case of model (7.1.42a), the exact solutions are given by (7.2.81) for all orders, which shows that $\langle I^{\nu}\rangle$ does indeed satisfy the asymptotic condition (7.4.14), resulting in $P(I)$ or $P(E)$ being given by (7.4.31) in terms of the residue value at a pole. Thus the situation again becomes the same as that of the residual series expression by Watson for the diffracted wave.

7.4.3 $\langle I^{\nu}\rangle$ as the Characteristic Function of $\ln(I)$

The integral representation (7.4.19) of $P(I)$ can be expressed in a more symmetrical form if the logarithm of intensity, $E = \ln(I)$, is introduced with the corresponding distribution function, $P(E)$, as given by

$$P(E) = P(I)\frac{dI}{dE} = \frac{1}{2\pi\mathrm{i}}\int_{-\mathrm{i}\infty-\varepsilon}^{\mathrm{i}\infty-\varepsilon} d\nu\, \exp(-\nu E)\langle I^{\nu}\rangle\ , \tag{7.4.20}$$

which has a form similar to the inverse Laplace transformation. Since, by assumption, $\langle I^{\nu}\rangle$ is analytic on the right half-plane of ν and has the asymptotic form (7.4.14), it immediately follows that

$$P(E) = 0, \qquad E > E_C \equiv \ln I_C\ ; \tag{7.4.21}$$

and also that, as in the Laplace transformation,

$$\langle I^\nu \rangle = \int_{-\infty}^{E_C} dE \exp(\nu E) P(E), \qquad \mathrm{Re}\{\nu\} \geq 0 \ , \tag{7.4.22}$$

which, conversely, can be regarded as an integral representation of $\langle I^\nu \rangle$, and, indeed, ensures that it is analytic on the right half-plane of ν, with the asymptotic form of (7.4.14).

Here, expansion of the right-hand side of (7.4.22) in a power series of ν yields

$$\langle I^\nu \rangle = \sum_{n=0}^{\infty} (n!)^{-1} \langle E^n \rangle \nu^n \ , \tag{7.4.23a}$$

$$\langle E^n \rangle = \int_{-\infty}^{E_C} dE\, E^n P(E) \ , \tag{7.4.23b}$$

which show that the expansion coefficients provide the moments of log-intensity, $\langle E^n \rangle$; i.e., $\langle I^\nu \rangle$ is precisely the characteristic function, $\langle \exp(\nu E) \rangle$, of $E = \ln(I)$. Hence the first three central moments can be directly obtained according to [cf. (7.4.3)]

$$\begin{aligned} \langle E \rangle &= \frac{\partial}{\partial \nu} \ln \langle I^\nu \rangle \Big|_{\nu=0} \ , \\ \langle (E - \langle E \rangle)^n \rangle &= \frac{\partial^n}{\partial \nu^n} \ln \langle I^\nu \rangle \Big|_{\nu=0} , \qquad n = 2, 3 \ , \end{aligned} \tag{7.4.24}$$

Thus, as the characteristic function $Z(\overline{\jmath}^*, \overline{\jmath})$ has been approximated by (7.4.4), the cumulant expansion of $\langle I^\nu \rangle$ given by

$$\langle I^\nu \rangle = \exp\left[\langle E \rangle \nu + 2^{-1} \{ \langle E^2 \rangle - \langle E \rangle^2 \} \nu^2 + O\{\nu^3\} \right] \tag{7.4.25a}$$

may have a good convergence in some cases; in fact, if E approximates to a nearly Gaussian distribution, then, the term of $O\{\nu^3\} \sim 0$, and (7.4.25a) lead to the expression,

$$\begin{aligned} m_\nu &= \exp\left[2^{-1} \nu (\nu - 1) \ln m_2 \right] \ , \\ \ln m_2 &= \langle E^2 \rangle - \langle E \rangle^2 \ . \end{aligned} \tag{7.4.25b}$$

in terms of the normalized intensity moments, m_ν, defined by (7.3.38), as can be directly confirmed by putting $\nu = 1$ and 2.

More generally, m_ν can be expressed by a cumulant series of the form

$$\begin{aligned} m_\nu = \exp\big[&(2!)^{-1} \nu (\nu - 1) \sigma_2^E + (3!)^{-1} \nu (\nu - 1)(\nu - 2) \sigma_3^E \\ &+ (4!)^{-1} \nu (\nu - 1)(\nu - 2)(\nu - 3) \sigma_4^E + \cdots \big] \ , \end{aligned} \tag{7.4.26}$$

which gives, for $\nu = 1, 2, 3, \dots$,

$$m_1 = 1, \quad m_2 = \exp(\sigma_2^E), \quad m_3 = \exp(3\sigma_2^E + \sigma_3^E), \quad m_4 = \exp(6\sigma_2^E + 4\sigma_3^E + \sigma_4^E), \quad \dots; \tag{7.4.27a}$$

hence,

$$\sigma_2^E = \ln m_2, \quad \sigma_3^E = \ln(m_3/m_2^3), \quad \sigma_4^E = \ln(m_4 m_2^6/m_3^4), \quad etc. \tag{7.4.27b}$$

Note that the coefficient, σ_n^E, is expressed in terms of m_n and the lower order moments. Thus, upon substitution of (7.4.27b) into (7.4.26), we obtain

$$m_\nu = \exp\left[\binom{\nu}{2}\ln m_2 + \binom{\nu}{3}\ln(m_3/m_2^3) + \binom{\nu}{4}\ln(m_4 m_2^6/m_3^4) + \cdots\right], \tag{7.4.28}$$

which is just the reproduction of series (7.3.49), but this time has been derived without any assumption except for its convergence.

Here, if the convergence is good enough to be approximated by the first two terms, the series can be expressed by (7.3.50a,b), which, in fact, showed a very good experimental agreement as demonstrated in Table 7.1, and which will also be presented in Fig. 7.7a,b. However, the approximated expression of m_ν is an entire function of ν, and also is not consistent with the required asymptotic condition (7.4.14), tending to $m_\nu \sim \exp(-2^{-1}K_1\Delta'\nu^3)$ for $\nu \sim +\infty$. Furthermore, since $\Delta' > 0$, as the experimental data suggest (Table 7.1), the expression (7.3.50a) leads to the unphysical result that $m_\nu \to 0$ as $\nu \to +\infty$.

On the other hand, in the range of $|(\nu - 2)\Delta'| \ll 1$, the same expression can also be written as

$$m_\nu = \exp\left[2^{-1}\nu(\nu-1)K_1\{1 + (\nu-2)\Delta\}^{-1}\right], \tag{7.4.29a}$$

with

$$\Delta = \Delta'(1-\Delta')^{-1} = 3\ln m_2/\ln m_3 - 1 \sim \Delta', \tag{7.4.29b}$$

which has been chosen so that the expression gives the right values, m_1, m_2, and m_3 for $\nu = 1$, 2, and 3, respectively, as the original (7.3.50a) does.

Here, pronounced features of the new expression (7.4.29a) are that, as $\nu \to \infty$, it tends to $\exp\left[2^{-1}(K_1/\Delta)\nu\right]$ and, therefore, satisfies the asymptotic condition (7.4.14); and further, that if $1 \gg \Delta > 0$, as the experiments show, it has a singular pole at $\nu = 2 - \Delta^{-1} < 0$ on the negative real axis, and is analytic everywhere else on the ν plane. Thus the modified expression of m_ν is found to satisfy all the necessary conditions to use the integral representation (7.4.20) of $P(E)$.

In Table 7.1, experimental values of m_ν, observed in [7.27], are shown along with theoretical values given by the modified expressions (7.4.29a,b) to demonstrate a good agreement. Here, however, the theoretical values are actually the same as those given by the original (7.3.50a,b) for the shown data of $\Delta \ll 1$ and $\nu = 2 \sim 4$.

7.4.4 Intensity Distribution for a Beam Wave Purely in the State of Wandering

Exact solutions of the moment equations can be obtained for all orders if the medium term $V(\vec{r})$ is given by (7.1.42a) over the entire $\vec{r}$ range; the intensity moments $\langle I^\nu \rangle$ are then given by (7.2.81) for a collimated beam wave of incidence, or more generally, by (7.2.76,78) which explicitly show that the moments are exactly those given when the undistorted original wave is displaced by the distance $\vec{\rho} - \vec{\rho}'$ on the plane perpendicular to the direction of wave propagation with the probability $P_S(\vec{\rho} - \vec{\rho}')$.

The moment $\langle I^\nu \rangle$ of (7.2.81) certainly satisfies the condition (7.4.14) to use the integral representation (7.4.20) of $P(E)$. Hence,

$$P(E) = \frac{1}{2\pi i} \int_{-i\infty - \varepsilon}^{i\infty - \varepsilon} d\nu \, (1 + \nu \sigma_E)^{-1} \times \exp\left[-\nu (1 + \nu \sigma_E)^{-1} E_0 - \nu E\right] , \qquad (7.4.30a)$$

where

$$E = \ln(I f^2 / I_0), \qquad E_0 = (\rho / \bar{b} f)^2 , \qquad (7.4.30b)$$

E being redefined with an additional constant. Here the integrand has a singular pole at $\nu = -\sigma_E^{-1}$ on the negative real axis, and is analytic everywhere else. Now, with the replacement of the variable of integration ν by it, the integral becomes exactly the same form as that given in (7.4.5) with $f(t)$ of (7.4.7), showing that the distribution function is the Rice–Nakagami distribution with respect to the log-intensity E, i.e., [7.5]

$$P(E) = \begin{cases} \sigma_E^{-1} \exp\left[-(E_0 - E)/\sigma_E\right] \times I_0\left[2(-E E_0)^{1/2}/\sigma_E\right], & E < 0 \\ 0, & E > 0 . \end{cases} \qquad (7.4.31)$$

The central moments of E can be obtained according to formula (7.4.24), hence,

$$\begin{aligned} \langle E \rangle &= -(E_0 + \sigma_E) < 0 , \\ \langle (E - \langle E \rangle)^2 \rangle &= (2E_0 + \sigma_E)\sigma_E , \\ \langle (E - \langle E \rangle)^3 \rangle &= -2(3E_0 + \sigma_E)\sigma_E^2 < 0 . \end{aligned} \qquad (7.4.32)$$

Here, since $\sigma_E \to \infty$ as $\phi \to \infty$ (7.2.40), the variance of E does not saturate for the increasing medium fluctuation.

7.4.5 Intensity Distribution Function in a Turbulent Medium

We have seen that the modified expression (7.4.29a,b) for m_ν shows a very good experimental agreement, and that it also satisfies the asymptotic conditions (7.4.14). Hence, with a log-intensity E defined by

$$E = \ln(I/\langle I \rangle) \ , \tag{7.4.33}$$

we obtain

$$P(E) = \frac{1}{2\pi \mathrm{i}} \int_{-\mathrm{i}\infty-\varepsilon}^{\mathrm{i}\infty-\varepsilon} d\nu \\ \times \exp\Big\{2^{-1}\nu(\nu-1)K_1\left[1+(\nu-2)\Delta\right]^{-1} - \nu E\Big\} \ , \tag{7.4.34}$$

which can bé written in the form

$$P(E) = \frac{1}{2\pi \mathrm{i}} \int_{-\mathrm{i}\infty-\varepsilon}^{\mathrm{i}\infty-\varepsilon} d\nu \\ \times \exp\Big[2^{-1}\left[b^2(\nu/\nu_0+1) + a^2(\nu/\nu_0+1)^{-1} - a^2 - b^2\right]\Big] \ , \tag{7.4.35}$$

with

$$\begin{aligned} \nu_0 &= \Delta^{-1} - 2 > 0, \qquad E_C = (2\Delta)^{-1}K_1 \ , \\ a &= \Delta^{-1}(1-\Delta)^{1/2}K_1^{1/2} \gg 1 \ , \\ b &= \Delta^{-1}(1-2\Delta)^{1/2}K_1^{1/2}(1-E/E_C)^{1/2} \ . \end{aligned} \tag{7.4.36}$$

Here, as $\nu \to \infty$, the integrand tends to

$$\exp\left[2^{-1}(b^2\nu/\nu_0 - a^2)\right] = \exp\left[-2^{-1}a^2 + (E_C - E)\nu\right] \ . \tag{7.4.37}$$

Therefore, $P(E) = 0$ for $E > E_C$, in view of the integrand which is analytic on the right half-plane of ν.

To evaluate the integral (7.4.35), we observe that the integral for the asymptotic expression (7.4.37) is

$$\exp(-2^{-1}a^2)\delta(E - E_C) \ ,$$

and therefore the integral representation for the new function,

$$P'(E) = P(E) - \exp(-2^{-1}a^2)\delta(E - E_C) \ , \tag{7.4.38}$$

has an integrand which is analytic on the entire ν plane except at the singular pole, $\nu = -\nu_0$, on the negative real axis, and has the asymptotic expression proportional to $\nu^{-1}\exp[(E_0 - E)\nu]$ for $\nu \sim \infty$. Hence, when $E > E_C$,

$P'(E) = 0$; whereas, when $E < E_C$, it can be given in terms of the residue value at the pole. Thus, with a new variable of integration,

$$t = (b/a)(\nu/\nu_0 + 1) \ , \tag{7.4.39}$$

and also with the aid of formula (7.4.9), we obtain

$$P'(E) = \begin{cases} \nu_0(a/b)\exp\left[-2^{-1}(a^2+b^2)\right] I_1(ab), & E < E_C \\ 0, & E > E_C \ , \end{cases} \tag{7.4.40}$$

Here, together with (7.4.38),

$$\int_{-\infty}^{\infty} dE\, P(E) = 1 \ . \tag{7.4.41}$$

In the practically important range of $E/E_C \ll 1$ and $a \sim b \gg 1$, (7.4.38,40) give the asymptotic expression

$$P(E) = P'(E) \sim (2\pi K_1)^{-1/2} \exp(-2^{-1}F^2) \ , \tag{7.4.42}$$

where, to the first order of $\Delta \ll 1$,

$$\begin{aligned} F = a - b \sim K_1^{1/2} &\left[E/K_1 + \tfrac{1}{2}\right. \\ &\left. + (\Delta/2)(E/K_1 - \tfrac{1}{2})(E/K_1 - \tfrac{3}{2})\right] \ . \end{aligned} \tag{7.4.43}$$

Hence the expression (7.4.42) tends to the log-normal distribution, $P_0(E)$, say, as $\Delta \to 0$. It may be remarked that the same expression as (7.4.42) can also be obtained by the formal substitution of expression (7.3.50a) into (7.4.20) and the followed integration, perturbatively in Δ', with the aid of the saddle point method of approximation.

In Fig. 7.4, the ratio $P(E)/P_0(E)$ is shown as a function of the variable

$$F\,\big|_{\Delta=0} = K_1^{1/2}(E/K_1 + \tfrac{1}{2})$$

for $K_1 = 0.7793$, and $\Delta = 0.05$, 0.0346, and 0.015. Here, for curve 2, the parameters have the same values as the experimental values in Table 7.1, and the expression (7.4.40) is used for the theoreticals.

On the other hand, in the vicinity of the threshold value, E_C, (7.4.40) gives

$$P(E) \sim 2^{-1}(\Delta^{-1} - 2)a^2 \exp\left[-2^{-1}(a^2+b^2)\right], \qquad E < E_C \ , \tag{7.4.44}$$

which is very small, however, in view of $a^2 \gg 1$.

The central moments of the log-intensity can be obtained according to formula (7.4.24); hence, from the original (7.3.50a),

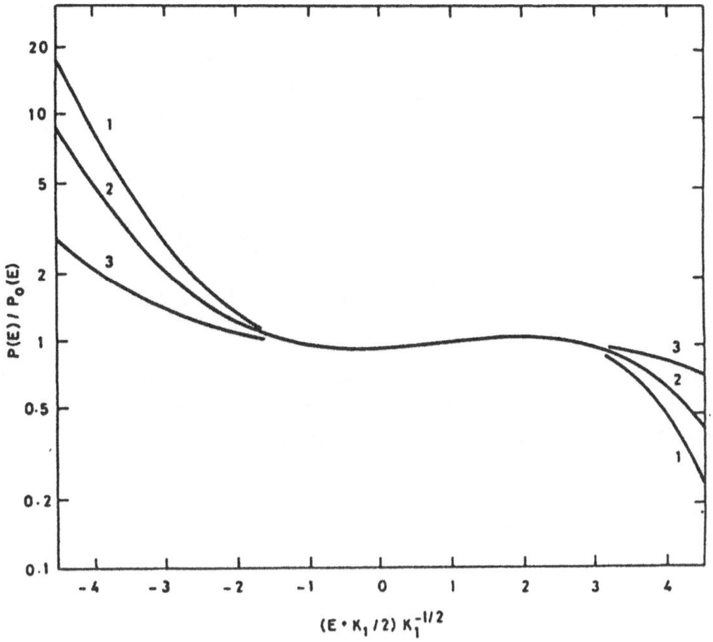

Fig. 7.4. Ratio of the distribution function of log-intensity $P(E)$ to the log-normal $P_0(E)$. $K_1 = 0.7793$ and (1) $\Delta = 0.05$; (2) $\Delta = 0.0346$; (3) $\Delta = 0.015$ (taken from [7.9])

$$\langle E\rangle = -2^{-1}(1+2\Delta')K_1 < 0 \ , \tag{7.4.45a}$$

$$\langle (E-\langle E\rangle)^2\rangle = (1+3\Delta')K_1 \ , \tag{7.4.45b}$$

$$\langle (E-\langle E\rangle)^3\rangle = -3\Delta' K_1 < 0 \ . \tag{7.4.45c}$$

while, from the modified (7.4.29a),

$$\langle E\rangle = -2^{-1}(1-2\Delta)^{-1}K_1 \ , \tag{7.4.46a}$$

$$\langle (E-\langle E\rangle)^2\rangle = (1-\Delta)(1-2\Delta)^{-2}K_1 \ , \tag{7.4.46b}$$

$$\langle (E-\langle E\rangle)^3\rangle = -3\Delta(1-\Delta)(1-2\Delta)^{-3}K_1 \ . \tag{7.4.46c}$$

which agree with (7.4.45a–c) to the first order of Δ'.

On the other hand, using (7.4.34), the cumulative probability distribution function, say $P_c(E)$, is expressed by the integral

$$P_c(E) = \int_{-\infty}^{E} dE'\, P(E') = \frac{1}{2\pi \mathrm{i}}\int_{-\mathrm{i}\infty-\varepsilon}^{\mathrm{i}\infty-\varepsilon} d\nu \\ \times (-\nu)^{-1}\exp\left\{2^{-1}\nu(\nu-1)K_1\left[1+(\nu-2)\Delta\right]^{-1} - \nu E\right\} \ , \tag{7.4.47}$$

which, upon changing the variable of integration to t of (7.4.39), becomes

$$
\begin{aligned}
P_c(E) &= \exp\left[-2^{-1}(a^2+b^2)\right]\frac{1}{2\pi \mathrm{i}}\int_{-\mathrm{i}\infty+\varepsilon}^{\mathrm{i}\infty+\varepsilon} dt \\
&\quad \times (b/a-t)^{-1}\exp\left[2^{-1}ab(t+t^{-1})\right] \\
&= \exp\left[-2^{-1}(a^2+b^2)\right]\sum_{n=0}^{\infty}\left(\frac{a}{b}\right)^{n+1} \\
&\quad \times \frac{1}{2\pi \mathrm{i}}\oint dt\, t^n \exp\left[2^{-1}ab(t+t^{-1})\right] , \qquad (7.4.48)
\end{aligned}
$$

where the last expression has been obtained by expansion of the factor $(b/a-t)^{-1}$ in the power series of t. Hence, with the aid of formula (7.4.9) and also with the replacement of $n+1$ by n, we obtain

$$
P_c(E) = \exp\left[-2^{-1}(a^2+b^2)\right]\sum_{n=1}^{\infty}(a/b)^n I_n(ab), \qquad E < E_C , \qquad (7.4.49)
$$

which is convenient for the range $a/b < 1$; while, using formula

$$
\sum_{n=-\infty}^{\infty}(a/b)^n I_n(ab) = \exp\left[2^{-1}(a^2+b^2)\right] ,
$$

equation (7.4.49) can also be written as

$$
P_c(E) = 1-\exp\left[-2^{-1}(a^2+b^2)\right]\sum_{n=0}^{\infty}(b/a)^n I_n(ab), , \qquad (7.4.50)
$$

which is convenient for $b/a < 1$. The average of the two expressions is also useful for the intermediate range $a/b \sim 1$.

As $E \to E_C$ or $b \to 0$ in (7.4.50),

$$
P_c(E) \to 1-\exp(-2^{-1}a^2) , \qquad (7.4.51)
$$

whose second term on the right-hand side is to be cancelled by the contribution from the $\delta(E-E_C)$ term in (7.4.38). In Fig 7.5, the cumulative probability function, $P_c(E)$, is illustrated against $K_1^{1/2}(E/K_1+\frac{1}{2})$, for the same values of the parameters as in Fig. 7.4.

It may be noticed that, besides the feature of being close to the log-normal, the probability density function has the threshold value E_C beyond which $P(E)$ becomes exactly zero, and where it has a very sharp distribution due to the $\delta(E-E_C)$ term, although the overall magnitude is very small. The last two points are direct mathematical consequences of the asymptotic

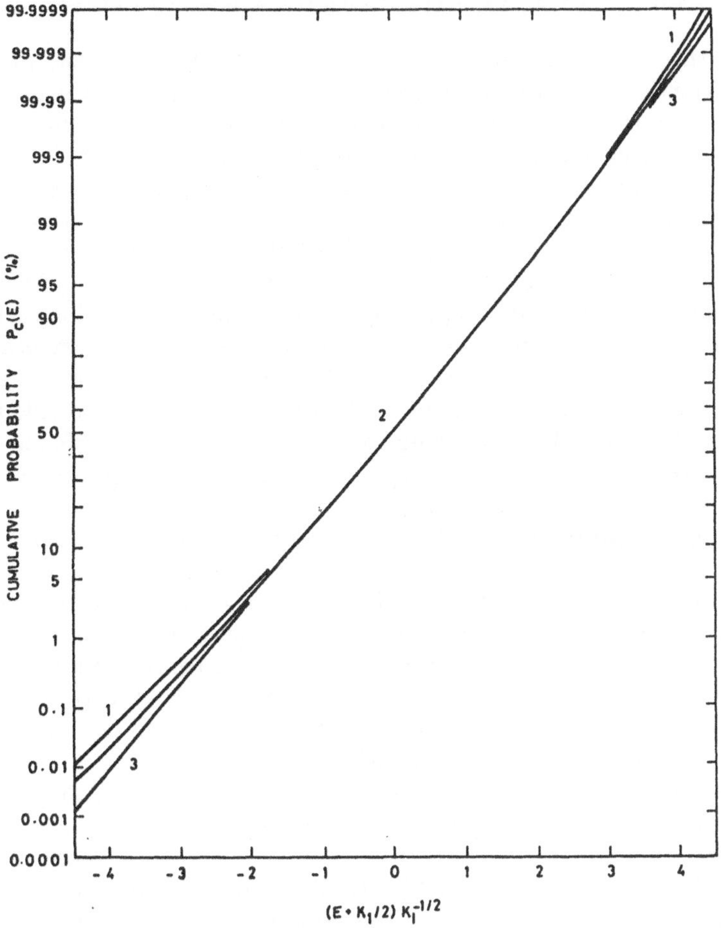

Fig. 7.5. Cumulative probability distribution function $P_c(E)$. The values of the parameters are the same as those in Fig. 7.4 (taken from [7.9])

expression of m_ν at $\nu \to \infty$ (7.4.14), while, the existence of a threshold value is what is physically required, being the condition that I be bounded no matter how large the medium fluctuation may be, and the other point is also physically quite reasonable, reflecting a contribution from the coherent wave, for example. The distribution is universal in the sense that the cluster expansion (7.3.49) of m_ν has been quite generally derived without any particular reference either to the medium spectrum or to the aperture size of the receiver, being involved only through the first two moments m_2 and m_3.

7.4.6 Comparison of Theories to Experimental Results

Besides the log-normal and the exponential distributions, many probability distributions of intensity have been proposed for wave propagation in a turbulent medium, mostly based on a model which assumes two independent parts of a wave subject to different statistics [7.32–35]. Here, a K-distribution [7.34] has received particular attention because of its simplicity and fairly good experimental agreement in the saturation region of intensity variance. A modified version of the exponential distribution was proposed [7.38] which is limited to the saturation region. Also a composite model of log-normal and exponential distributions (available also for the unsaturation region) was proposed [7.37], with an extensive comparison of it to experimental results using acoustic waves, along with several other distributions that have been proposed so far. Measurements of the high order intensity moments in a turbulent medium have been made up to the fifth order in [7.27, 35, 36, 37] (and others), over the range of $0.1 \lesssim \sigma_I^2/\langle I\rangle^2 \lesssim 6$, including the regions of both unsaturated and saturated scintillation.

Fitting the proposed distribution to experimental results seems to have all been good, as long as the comparison was made for the intensity moments within the saturation region, as a consequence of adjusting the parameter values properly. A remarkable experiment was performed [7.36], showing simultaneously observed values of both intensity moments and long-amplitude variance; the results were compared to those of available theories in [7.39, 40] to see how the theories could respond to the simultaneous fitting to both the intensity moments and the log-amplitude variance.

Shown in Fig. 7.6 are some of the experimental results of [7.34] in terms of the normalized intensity moments from m_3 to m_5 versus m_2, along with the theoretical values (of open circles and crosses) given by expressions (7.3.50a,b) and (7.4.29a,b), respectively [7.39]. Also shown in the same figure are the values of parameter Δ' deduced from the smoothed curve of m_4 versus m_2 with the aid of (7.3.50b,51). Shown in Fig. 7.7a are the corresponding moments observed by [7.36] over a wider range of m_2 up to about 7, along with several theoretical curves [7.39, 40]. Here, curves a and b are by expressions (7.3.50) and (7.4.29), say $m_\nu^{(a)}$ and $m_\nu^{(b)}$, respectively. The other three curves K, ME, and D are: K-distribution, say $m_\nu^{(K)}$, given by [7.34]

$$m_\nu^{(K)} = \nu!\Gamma(\nu+y)/y^\nu\Gamma(y) \ ,$$

where $y = 2/[\sigma_I^2/\langle I\rangle^2 - 1]$, and no distribution exists in the range $\sigma_I^2/\langle I\rangle^2 < 1$ or $m_2 < 2$; distribution given by [7.37], say $m_\nu^{(ME)}$, i.e.,

$$m_\nu^{(ME)} = (\nu!)^{c_1}\exp[2^{-1}\nu(\nu-1)c_2] \ ,$$

where c_1 and c_2 are constants, chosen such that $m_\nu^{(ME)}$ gives m_2 and m_3 for $\nu = 2$ and 3, hence,

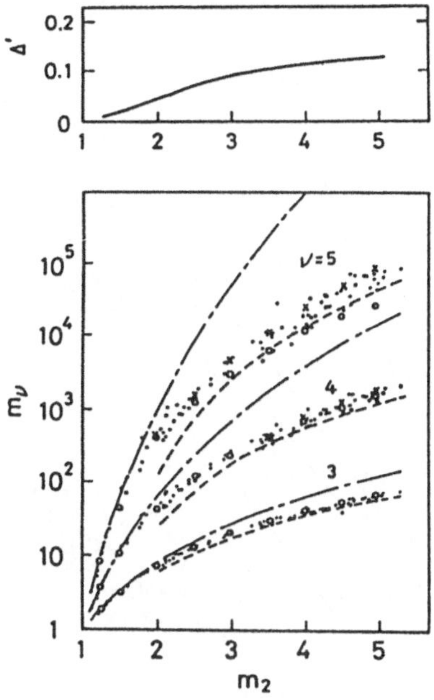

Fig. 7.6. Comparison of the experimental values of m_ν, $\nu = 2, 3, 4, 5$, observed by [7.36] with the theoretical values (*open circles* and *crosses*) given by expressions (7.3.50) and (7.4.29). The curve of Δ' shown in the upper panel was obtained according to expressions (7.3.50b,51) with the m_4 values given by the smoothed curve of the experimental m_4. The *dotted-dashed* and *broken curves* indicate the moments expected from the log-normal and the K distribution (taken from [7.39])

$$c_1 = \ln(m_3/m_2^3)/\ln(3!/2!^3) \ ,$$
$$c_2 = \ln m_2 - c_1 \ln 2! \ ;$$

and distribution by [7.38], say $m_\nu^{(D)}$, given by

$$m_\nu^{(D)} = \nu! \exp\left[\binom{\nu}{2} p_1 + \binom{\nu}{3} p_2\right] \ ,$$

with the constants

$$p_1 = \ln(m_2/2), \qquad p_2 = \ln(4m_3/3m_2^3) \ .$$

Here the log-amplitude variances resulting from $m_\nu^{(K)}$, $m_\nu^{(ME)}$, and $m_\nu^{(D)}$, say $\sigma_\chi^{(K)2}$, $\sigma_\chi^{(ME)2}$, and $\sigma_\chi^{(D)2}$, respectively, are given according to (7.4.24) by (see also Subsection C, following)

$$\sigma_\chi^{(K)2} = 4^{-1}\left[\psi'(1) + \psi'(y)\right], \qquad \psi'(1) = 1.64493 \ ,$$
$$\sigma_\chi^{(ME)2} = 4^{-1}\left[c_1\psi'(1) + c_2\right] \ ,$$
$$\sigma_\chi^{(D)2} = 4^{-1}\left[\psi'(1) + p_1 - p_2\right] \ ,$$

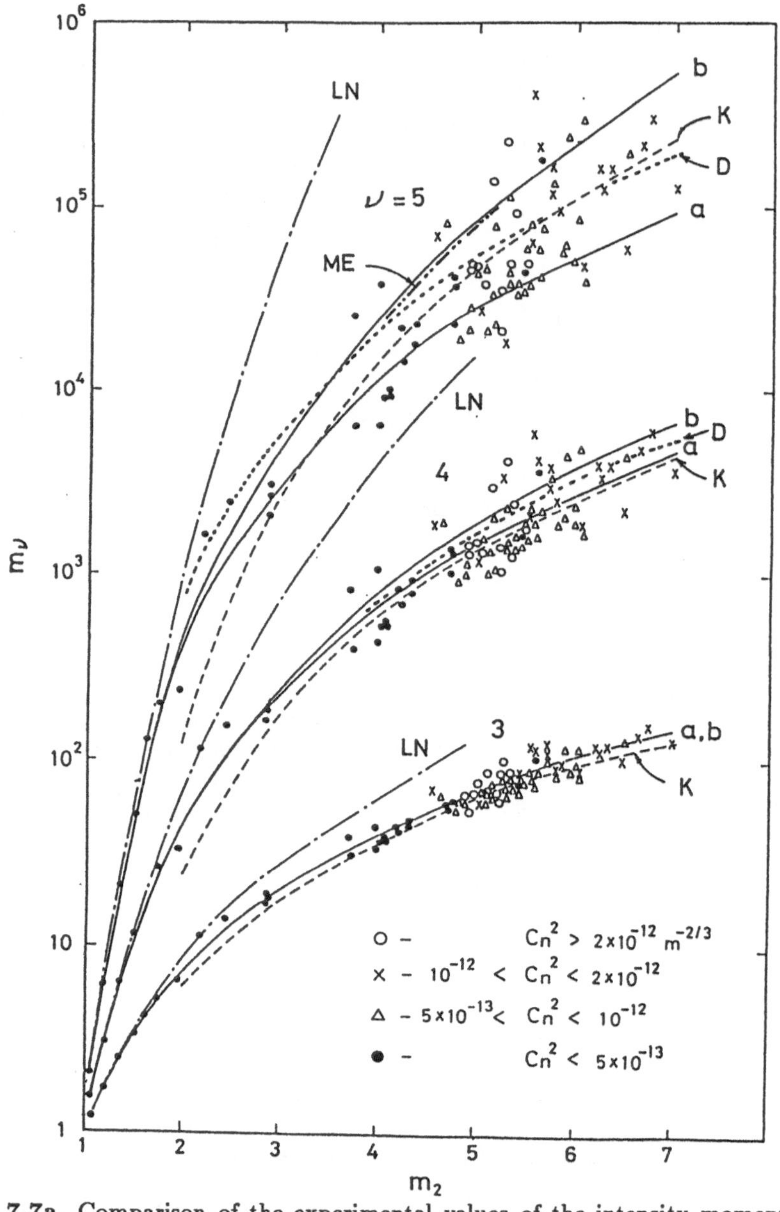

Fig. 7.7a. Comparison of the experimental values of the intensity moments m_ν, $\nu = 2, 3, 4, 5$, observed by [7.36] to the theoretical values given by $m_\nu^{(a)}$, $m_\nu^{(b)}$, m_ν^{ME}, and m_ν^{D}.

where $\psi(z) = \partial \ln \Gamma(z)/\partial z$ and $\psi'(z) = \partial\psi(z)/\partial z$. Note that, except for $m_\nu^{(b)}$ of (7.4.29a), all the other $m_\nu^{(\ )}$'s, including $m_\nu^{(a)}$ of (7.3.50a), do not tend to the asymptotic form (7.4.14) as $\mathrm{Re}\{\nu\} \to +\infty$.

a) Importance of Simultaneous Measurement of Both Intensity Moments and Log-Amplitude Variance

In Fig. 7.7a, all the curves within the range $7 \gtrsim m_2 \gtrsim 3$, except the log-normal (LN), are close to each other; this is particularly true for the curves b and ME over all the orders, as they were in the comparisons made in [7.37]. While, in Fig. 7.7b for the log-amplitude variance, a remarkable difference occures between those curves in the same range, i.e., $6 \gtrsim \sigma_I^2/\langle I\rangle^2 \gtrsim 2$, showing that the K and ME are clearly not good. In the range of $\sigma_I^2/\langle I\rangle^2 < 1$, on the other hand, the curve ME is not bad, while the distributions D and K are both not available.

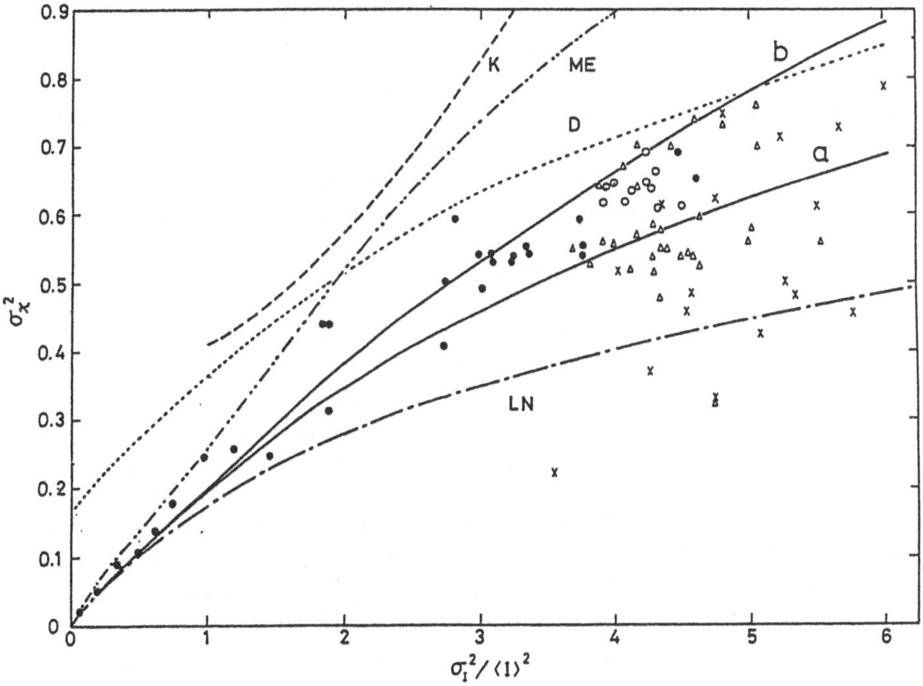

Fig. 7.7b Comparison of the log-amplitude variances $\sigma_\chi^{(\ )2}$ to the experimental values simultaneously measured with the intensity moments in Fig. 7.7a. Both figures are taken from [7.39] as supplemented by [7.40]

Here, because of a saturation effect of the detector used, the experimental values of the high order moments could be so underestimated that they might not have much quantitative meaning [7.41]. Still, this is not expected to be so large as to substantially change the pronounced feature of each of the curves of log-amplitude variance in Fig. 7.7b.

b) Discussion

A large discrepancy of the curves in Fig. 7.7b seems to mean that both $m_\nu^{(K)}$ and $m_\nu^{(ME)}$ are ultimately numerical approximations that express the true m_ν to a good accuracy within some finite range of ν; and therefore, in the cumulant expansion of m_ν, given by

$$m_\nu = \exp\left[2\langle\chi\rangle\nu + 2\sigma_\chi^2\nu^2 + \cdots\right] \qquad (7.4.52)$$

[which follows from (7.4.24)], the corresponding expansion coefficients $\sigma_\chi^{(\ )2}$ when using $m_\nu^{(\ )}$ instead of m_ν, have only a numerical meaning and not a physical meaning, after all. While, $m_\nu^{(a)}$ and $m_\nu^{(b)}$ are physical approximations that are derived based on the moment equations and are valid over some range of ν, m_2, and m_3, so that their expansion coefficients, $\langle\chi^{(\ )}\rangle$ and $\sigma_\chi^{(\ )2}$, should also keep the original meaning of (7.4.24).

It may be also emphasized that $m_\nu^{(a)}$ and $m_\nu^{(b)}$ are both universal in the sense of being independent of details of the medium spectrum, and also of the kind of wave, whether the incident wave is a plane, spherical, or beam wave; and this is true even of the aperture size of the receiver, so that the distributions remain unchanged with changed values of the parameters [Sect. (7.3.2)].

c) Moments of Truncated Intensity and Probability Distributions with No Finite Threshold Value

Since observed values of the high order intensity moments tend to be suppressed by the saturation effect of the detector used, direct comparison of theories to experimental results is a problem [7.41]. On the other hand, independent of whether the asymptotic condition (7.4.14) is satisfied or not, formal use of the integral representation (7.4.20) yields a probability distribution, as long as the integral is convergent, for any m_ν including those of both log-normal and exponential distributions which have no finite threshold value ($E_C = \infty$). On assuming m_ν of this general sort, we first obtain expressions for the moments of intensity truncated at some finite value, say E_t, so that $\langle E^n\rangle$, $E = \ln(I/\langle I\rangle)$, is replaced by [7.40]

$$\langle E^n\rangle_t \equiv \int_{-\infty}^{E_t} dE\, E^n P(E), \qquad n = 1, 2, 3, \ldots\ , \qquad (7.4.53)$$

which can be written using (7.4.20) and subsequent repeated partial integrations with respect to ν, as

$$\langle E^n\rangle_t = \frac{1}{2\pi \mathrm{i}} \int_{-\mathrm{i}\infty-\varepsilon}^{\mathrm{i}\infty-\varepsilon} d\nu\, (-\nu)^{-1} \exp(-E_t\nu) \left(\frac{\partial}{\partial\nu}\right)^n m_\nu\ . \qquad (7.4.54)$$

This can be rewritten, on changing the integration path infitnitesimally to the right side of the imaginary axis, as

$$\langle E^n\rangle_t = \left(\frac{\partial}{\partial\nu}\right)^n m_\nu \Big|_{\nu=0} + \frac{1}{2\pi \mathrm{i}}\int_{-\mathrm{i}\infty+\varepsilon}^{\mathrm{i}\infty+\varepsilon} d\nu\,(-\nu)^{-1}\exp(-E_t\nu)\left(\frac{\partial}{\partial\nu}\right)^n m_\nu \ . \quad (7.4.55)$$

Here, the first term is a contribution from the simple pole at $\nu = 0$, and the last integral tends to zero as $E_t \to +\infty$, in view of $\mathrm{Re}\{E_t\nu\} \to +\infty$. Hence, also, $\langle E^n\rangle_t \to \langle E^n\rangle$, as follows from (7.4.23a,b).

The situation is also the same for a truncated moment of $I/\langle I\rangle = \mathrm{e}^E$, say $m_n^{(t)}$, $n = 1, 2, 3, \ldots$, defined by

$$m_n^{(t)} = \langle (I/\langle I\rangle)^n\rangle_t = \int_{-\infty}^{E_t} dE\,\exp(nE)P(E) \quad (7.4.56)$$

$$= \frac{1}{2\pi \mathrm{i}}\int_{-\mathrm{i}\infty-\varepsilon}^{\mathrm{i}\infty-\varepsilon} d\nu\,(n-\nu)^{-1}\exp[(n-\nu)E_t]m_\nu \ . \quad (7.4.57)$$

Here, choosing a constant C such that $C > n$, we obtain another expression,

$$m_n^{(t)} = m_n + \frac{1}{2\pi \mathrm{i}}\int_{-\mathrm{i}\infty+C}^{\mathrm{i}\infty+C} d\nu\,(n-\nu)^{-1}\exp[(n-\nu)E_t]m_\nu \ , \quad (7.4.58)$$

where the first term is from the pole at $\nu = n$, and $\mathrm{Re}\{n-\nu\} < 0$. Hence, as $E_t \to +\infty$, the last integral tends to zero and, consequently, $m_n^{(t)} \to m_n$. Thus we find that the original moments are reproduced, independent of whether the asymptotic condition (7.4.14) is fulfilled or not. Nevertheless, the condition is necessary to derive the integral representation (7.4.20) for $P(E)$ from given $\langle I^n\rangle$, $n = 1, 2, 3, \ldots$, i.e., from those that solutions of the moment equations can provide, leading naturally to a finite threshold value of intensity for $P(E)$ being nonzero.

Thus the integral representation (7.4.20), as well as (7.4.55, 58) which result therefrom, hold true for a wide class of distributions; not only for physical distributions based on solutions of the moment equations of integral orders, but also for any unphysical distributions with no finite threshold value. And the asymptotic condition (7.4.14) rules the latter out.

7.5 Two-Scale Method

Recently, a two-scale method was proposed by [7.17, 18–20] and others to give an approximate integral representation of the two-point intensity correlation function, with the emphasis that its zero order term exactly agrees with the result obtained based on the multiple-phase screen method [7.15, 16], and others, for plane and spherical waves. The method was examined in some detail in [7.19], showing that the method is available not only for the strong medium fluctuation region but also for the weak fluctuation region, and that a particular first order correction introduced therein considerably improves the result. Besides, it may be mentioned that the first order correction so far proposed changes from one to another depending on the specific method used. The method was also used for the two frequency intensity correlation [7.42, 43].

The purpose of this section is as follows: to show that the method is essentially based on a variational principle (similar to the Lagrange variational principle in dynamics), and, therefore, it is naturally independent of whether the medium fluctuation is strong or weak, and also of the medium scales involved; and to specifically obtain an exact version of the two-scale integral representations for both the second and third order intensity correlations for a collimated beam wave, including spherical and plane waves as special cases. The two frequency intensity correlation is also obtained from a straightforward generalization of the procedure.

7.5.1 Exact Integral Representation of $M_{22}(1)$

Expression (7.2.27) of $M_{22}(1)$ can be written more explicitly as

$$M_{22}(1) = P\exp\left[\!\left[-\int_0^1 d\zeta\, V^{(2)}(\widehat{\boldsymbol{r}}_1(\zeta-1),\right.\right.$$

$$\left.\left.\widehat{\boldsymbol{r}}_2(\zeta-1), \widehat{\boldsymbol{\rho}}_{12}(\zeta-1))\right]\!\right] M_{22}^0(1) \qquad (7.5.1a)$$

with

$$V^{(2)}(\boldsymbol{r}_1, \boldsymbol{r}_2, \boldsymbol{\rho}_{12}) = \overline{V}(\boldsymbol{r}_1) + \overline{V}(\boldsymbol{r}_2) + \overline{V}_I(\boldsymbol{r}_1, \boldsymbol{r}_1, \boldsymbol{\rho}_{12}) \; . \qquad (7.5.1b)$$

In this section (7.5.1a) is represented by a functional integral perfectly free from any operator, meaning an exact multi-scale representation of $M_{22}(1)$ in the sense of using an infinite number of Fourier variables to describe the medium effect; and, in Sect. 7.6, the variable functions of integration involved therein are then determined so that the entire phase term becomes invariant against their arbitrary variation to the first order, so that the functions are uniquely determined once their end values are given, leading to a two-scale integral representation based on an unperturbative method [7.14].

For any functional of the Hermitian operator, $\widehat{\boldsymbol{\rho}}_{12}(\zeta)$, $1 \geq \zeta \geq 0$, say $f[\widehat{\boldsymbol{\rho}}_{12}]$, it is generally possible to represent it by a functional integral, as

$$f[\widehat{\rho}_{12}] = \int d[\rho']\,\delta[\widehat{\rho}_{12} - \rho']f[\rho']$$
$$= \int d[\rho']d[\boldsymbol{\lambda}'] \exp\left\{ \mathrm{i}\int_0^1 d\zeta\, \boldsymbol{\lambda}'(\zeta)\cdot [\rho'(\zeta) - \widehat{\rho}_{12}(\zeta)]\right\} f[\rho'] \ . \quad (7.5.2)$$

Here, $\boldsymbol{\lambda}'(\zeta)$, $1 \geq \zeta \geq 0$, is regarded as the (continuous) Fourier variable, and both $d[\rho']$ and $d[\boldsymbol{\lambda}']$ should be properly normalized with numerical factors so that the integral provides the right function. The representation according to (7.5.2) is similar to that employed in (7.2.43) in the sense that the operator, $\widehat{\rho}_{12}$, is involved only in the exponential factor, and, in fact, the latter can be regarded as a special case of the former.

Hence, applying (7.5.2) to (7.5.1a), we obtain a functional integral representation of $M_{22}(1)$ of the form

$$M_{22}(1) = \int d[\boldsymbol{\lambda}']d[\rho']\, U^{(2)}[\boldsymbol{\lambda}', \rho'] M^0_{22}(1) \ . \quad (7.5.3)$$

Here, $U^{(2)}[\boldsymbol{\lambda}', \rho']$ is an operator defined by

$$U^{(2)}[\boldsymbol{\lambda}', \rho'] = P \exp\left[\!\left[\int_0^1 d\zeta \left\{ \mathrm{i}\boldsymbol{\lambda}'(\zeta)\cdot[\rho'(\zeta) - \widehat{\rho}_{12}(\zeta - 1)] \right.\right.\right.$$
$$\left.\left.\left. - V^{(2)}[\widehat{\boldsymbol{r}}_1(\zeta - 1), \widehat{\boldsymbol{r}}_2(\zeta - 1), \rho'(\zeta)] \right\} \right]\!\right] \ , \quad (7.5.4)$$

(where the ordering P is still necessary) and can be reordered so that all the coordinates $(\boldsymbol{r}_1, \boldsymbol{r}_2, \boldsymbol{\rho}_{12})$ are placed to the left of all the operators ($\partial/\partial\boldsymbol{r}_j$ and $\partial/\partial\boldsymbol{\rho}_j$, $j = 1, 2$) involved (7.5.26–35). Hence, with the Fourier transformation function $\langle \boldsymbol{r}, \boldsymbol{\rho}|\boldsymbol{s}, \boldsymbol{\lambda}\rangle$ defined by

$$\langle \boldsymbol{r}, \boldsymbol{\rho}|\boldsymbol{s}, \boldsymbol{\lambda}\rangle = \exp\left[-\sum_{j=1}^{2} \mathrm{i}(\boldsymbol{s}_j\cdot\boldsymbol{r}_j + \boldsymbol{\lambda}_j\cdot\boldsymbol{\rho}_j)\right] \ , \quad (7.5.5)$$

we can explicitly define a functional $S^{(2)}[\boldsymbol{\lambda}', \rho']$ of $\boldsymbol{\lambda}'(\zeta)$ and $\rho'(\zeta)$ (free from any operator) according to

$$U^{(2)}[\boldsymbol{\lambda}', \rho']\langle \boldsymbol{r}, \boldsymbol{\rho}|\boldsymbol{s}, \boldsymbol{\lambda}\rangle = \exp\{S^{(2)}[\boldsymbol{\lambda}', \rho']\}\langle \boldsymbol{r}, \boldsymbol{\rho}|\boldsymbol{s}, \boldsymbol{\lambda}\rangle \quad (7.5.6a)$$
$$\equiv \langle \boldsymbol{r}, \boldsymbol{\rho}|U^{(2)}[\boldsymbol{\lambda}', \rho']|\boldsymbol{s}, \boldsymbol{\lambda}\rangle \ . \quad (7.5.6b)$$

Here, the last is a composite matrix expression of $U^{(2)}$ using the coordinates $\boldsymbol{r} = (\boldsymbol{r}_1, \boldsymbol{r}_2)$ and $\boldsymbol{\rho} = (\boldsymbol{\rho}_1, \boldsymbol{\rho}_2)$ for the column, and using the Fourier variables $\boldsymbol{s} = (\boldsymbol{s}_1, \boldsymbol{s}_2)$ and $\boldsymbol{\lambda} = (\boldsymbol{\lambda}_1, \boldsymbol{\lambda}_2)$ for the row, so that we also write $S^{(2)}[\boldsymbol{\lambda}', \rho']$ alternatively as $S^{(2)}(\boldsymbol{r}, \boldsymbol{\rho}|\boldsymbol{\lambda}', \rho'|\boldsymbol{s}, \boldsymbol{\lambda})$. Equation (7.5.6a) indicates that $\exp\{S^{(2)}[\boldsymbol{\lambda}', \rho']\}$ can be regarded as an eigenvalue of the operator

$U^{(2)}[\boldsymbol{\lambda}', \boldsymbol{\rho}']$. From (7.5.3), $M_{22}(1)$ is hence written in terms of the Fourier transform, $\widetilde{M}^0_{22}(\zeta = 1)$ of $M^0_{22}(1)$, by

$$M_{22}(\boldsymbol{r}, \boldsymbol{\rho}; \zeta = 1) = (2\pi)^{-8} \int d\boldsymbol{s} d\boldsymbol{\lambda} \exp\left[-\sum_{j=1}^{2} \mathrm{i}(\boldsymbol{s}_j \cdot \boldsymbol{r}_j + \boldsymbol{\lambda}_j \cdot \boldsymbol{\rho}_j)\right]$$

$$\times \int d[\boldsymbol{\lambda}'] d[\boldsymbol{\rho}'] \exp\left[S^{(2)}(\boldsymbol{r}, \boldsymbol{\rho} | \boldsymbol{\lambda}', \boldsymbol{\rho}' | \boldsymbol{s}, \boldsymbol{\lambda})\right] \widetilde{M}^0_{22}(\boldsymbol{s}, \boldsymbol{\lambda}; \zeta = 1) \ , \quad (7.5.7)$$

where $d\boldsymbol{s} = d\boldsymbol{s}_1 d\boldsymbol{s}_2$ and $d\boldsymbol{\lambda} = d\boldsymbol{\lambda}_1 d\boldsymbol{\lambda}_2$.

$S^{(2)}[\boldsymbol{\lambda}', \boldsymbol{\rho}']$ is evaluated in detail in (7.5.26–35), and is obtained in the form

$$S^{(2)}[\boldsymbol{\lambda}', \boldsymbol{\rho}'] = \int_0^1 d\zeta \left\{ \mathrm{i}\boldsymbol{\lambda}'(\zeta) \cdot [\boldsymbol{\rho}'(\zeta) - \overline{\boldsymbol{\rho}}_{12}(\zeta)] - \langle \widetilde{V}^{(2)}(\zeta) \rangle \right\} \ , \quad (7.5.8)$$

$$\langle \widetilde{V}^{(2)}(\zeta) = V^{(2)}[\langle \widetilde{\boldsymbol{r}}_1(\zeta) \rangle, \langle \widetilde{\boldsymbol{r}}_2(\zeta) \rangle, \boldsymbol{\rho}'(\zeta)] \ . \quad (7.5.9)$$

Here,

$$\langle \widetilde{\boldsymbol{r}}_j(\zeta) \rangle = \overline{\boldsymbol{r}}_j(\zeta) \pm \boldsymbol{\xi}(\zeta), \qquad j = 1, 2 \ , \quad (7.5.10)$$

where

$$\overline{\boldsymbol{r}}_j(\zeta) = \boldsymbol{r}_j + (\zeta - 1)\boldsymbol{\lambda}_j \ , \quad (7.5.11)$$

$$\overline{\boldsymbol{\rho}}_{12}(\zeta) = \boldsymbol{\rho}_{12} + (\zeta - 1)\boldsymbol{v}, \qquad \boldsymbol{v} = \boldsymbol{s}_1 - \boldsymbol{s}_2 \quad (7.5.12)$$

[$\overline{\boldsymbol{r}}_j(\zeta)$ should not be confused with the function $\overline{\boldsymbol{r}}'_j(\zeta)$ of (7.2.51a,b), although $\overline{\boldsymbol{\rho}}_{12}(\zeta) = \overline{\boldsymbol{\rho}}'_{12}(\zeta)$ of (7.2.51c)]. In (7.5.10) the sign is $\pm$ depending on whether $j = 1$ or 2, and

$$\boldsymbol{\xi}(\zeta) = \int_0^1 d\zeta' \, G(\zeta|\zeta') \boldsymbol{\lambda}'(\zeta') \quad (7.5.13)$$

is expressed in terms of a Green's function, $G(\zeta|\zeta')$, defined by

$$G(\zeta|\zeta') = \begin{cases} \zeta - 1, & \zeta > \zeta' \\ \zeta' - 1, & \zeta < \zeta' \ , \end{cases} \quad (7.5.14)$$

which is the solution of

$$\left(\frac{\partial}{\partial \zeta}\right)^2 G(\zeta|\zeta') = \delta(\zeta - \zeta') \ , \quad (7.5.15a)$$

subjected to the boundary conditions

$$G(\zeta = 1|\zeta') = \frac{\partial}{\partial \zeta} G(\zeta|\zeta') \Big|_{\zeta=0} = 0 \ . \quad (7.5.15b)$$

Thus, when $r_1 = r_2 = 0$, the expression (7.5.7) for $M_{22}(1)$ is explicitly written, in terms of $\widetilde{M}_{22}(\zeta = 0)$ and the relative variables of (7.2.44–48), as

$$M_{22}(\boldsymbol{r}, \boldsymbol{\rho}; \zeta = 1) = (2\pi)^{-8} \int d\boldsymbol{s} d\boldsymbol{\lambda} \exp\{\mathrm{i}[\boldsymbol{\nu} \cdot (\boldsymbol{v} - \boldsymbol{\rho}_{12}) + \overline{\boldsymbol{\nu}} \cdot (\overline{\boldsymbol{v}} - \overline{\boldsymbol{\rho}})]\}$$

$$\times \int d[\boldsymbol{\lambda}']d[\boldsymbol{\rho}'] \exp[S^{(2)}(\boldsymbol{r}, \boldsymbol{\rho}|\boldsymbol{\lambda}', \boldsymbol{\rho}'|\boldsymbol{s}, \boldsymbol{\lambda})] \widetilde{M}_{22}(\boldsymbol{s}, \boldsymbol{\lambda}, \zeta = 0) \ . \quad (7.5.16)$$

This gives the exact integral representation, and is a multi-scale expression in the sense of using an infinite number of the Fourier variables, $\boldsymbol{\lambda}'(\zeta)$, $1 \geq \zeta \geq 0$; that is, the right-hand side of (7.5.16) depends on the coordinates $\boldsymbol{\rho}_1$ and $\boldsymbol{\rho}_2$ only through the factor

$$\exp\left\{-\mathrm{i}\left[\int_0^1 d\zeta\, \boldsymbol{\lambda}'(\zeta) + \boldsymbol{\nu}\right] \cdot \boldsymbol{\rho}_{12} - \mathrm{i}\overline{\boldsymbol{\nu}} \cdot \overline{\boldsymbol{\rho}}\right\} , \quad (7.5.17)$$

in view of (7.5.8), showing that the Fourier variable of the coordinates $\boldsymbol{\rho}_{12}$, say $\boldsymbol{\lambda}''$, is given by [compare to (7.2.50)]

$$\boldsymbol{\lambda}'' = \int_0^1 d\zeta\, \boldsymbol{\lambda}'(\zeta) + \boldsymbol{\nu} \ . \quad (7.5.18)$$

Also, we note that in the case of a phase screen, the exact (two-scale) expression (7.2.53–55) is reproduced by setting $\boldsymbol{\lambda}'(\zeta') = \delta(\zeta' - \zeta)\boldsymbol{\lambda}'$, with $d[\boldsymbol{\lambda}']d[\boldsymbol{\rho}'] = (2\pi)^{-2} d\boldsymbol{\lambda}' d\boldsymbol{\rho}'$, and changing the variables of integration to $\boldsymbol{\lambda}''$ and $\boldsymbol{\rho}''$ of (7.2.50) [(7.2.52)].

Expression (7.5.16) can be written in a more symmetrical form by introducing a new function, $\boldsymbol{\eta}(\zeta)$, defined by

$$\boldsymbol{\eta}(\zeta) = \boldsymbol{\rho}'(\zeta) - \overline{\boldsymbol{\rho}}_{12}(\zeta) \quad (7.5.19)$$

and using the relation

$$\ddot{\boldsymbol{\xi}}(\zeta) \equiv \left(\frac{\partial}{\partial \zeta}\right)^2 \boldsymbol{\xi}(\zeta) = \boldsymbol{\lambda}'(\zeta) \quad (7.5.20)$$

from (7.5.13,15a). Hence, from (7.5.8),

$$S^{(2)}[\boldsymbol{\lambda}', \boldsymbol{\rho}'] = \int_0^1 d\zeta\, [\mathrm{i}\ddot{\boldsymbol{\xi}} \cdot \boldsymbol{\eta} - \langle \widetilde{V}^{(2)}(\zeta)\rangle] \quad (7.5.21)$$

$$= \mathrm{i}\dot{\boldsymbol{\xi}} \cdot \boldsymbol{\eta} \Big|_{\zeta=0}^{1} + \int_0^1 d\zeta\, \mathcal{L}(\dot{\boldsymbol{\xi}}, \dot{\boldsymbol{\eta}}, \boldsymbol{\xi}, \boldsymbol{\eta}, \zeta) \ . \quad (7.5.22)$$

Here, the last expression is obtained by partial integration, with $\dot{\boldsymbol{\xi}} = \partial \boldsymbol{\xi}/\partial \zeta$, $\dot{\boldsymbol{\eta}} = \partial \boldsymbol{\eta}/\partial \zeta$, and

$$\mathcal{L} = -i\dot{\boldsymbol{\xi}} \cdot \dot{\boldsymbol{\eta}} - \langle \widetilde{V}^{(2)}(\boldsymbol{\xi}, \boldsymbol{\eta}, \zeta) \rangle \ , \tag{7.5.23}$$

where, using (7.5.19) in (7.5.9),

$$\langle \widetilde{V}^{(2)}(\boldsymbol{\xi}, \boldsymbol{\eta}, \zeta) \rangle = V^{(2)}[\overline{\boldsymbol{r}}_1(\zeta) + \boldsymbol{\xi}(\zeta), \overline{\boldsymbol{r}}_2(\zeta) - \boldsymbol{\xi}(\zeta), \overline{\boldsymbol{\rho}}_{12}(\zeta) + \boldsymbol{\eta}(\zeta)] \ ; \tag{7.5.24}$$

and from (7.5.13,15b), $\boldsymbol{\xi}(\zeta)$ is subject to the relations

$$\dot{\boldsymbol{\xi}}(\zeta) = \int_0^{\zeta} d\zeta' \, \boldsymbol{\lambda}'(\zeta'), \qquad \dot{\boldsymbol{\xi}}(\zeta = 0) = \boldsymbol{\xi}(\zeta = 1) = 0 \ . \tag{7.5.25}$$

We note that $\dot{\boldsymbol{\xi}}(\zeta = 1)$ provides the integrated value of integration variable function, $\boldsymbol{\lambda}'(\zeta)$, over the whole range $1 \geq \zeta \geq 0$.

a) Evaluation of $S^{(2)}[\boldsymbol{\lambda}', \boldsymbol{\rho}']$

To obtain the reordered expression of $U^{(2)}[\boldsymbol{\lambda}', \boldsymbol{\rho}']$ as described below (7.5.4), we first observe that the $\widehat{\boldsymbol{\rho}}_{12}$ operators placed on the left-hand side of $\widehat{\boldsymbol{r}}_j(\zeta - 1)$ at given ζ, are $\widehat{\boldsymbol{\rho}}_{12}(\zeta' - 1)$ over the range $1 \geq \zeta' \geq \zeta$, and are involved only through the factor

$$\mathcal{T}(\zeta) = \exp\left[-i \int_{\zeta}^{1} d\zeta' \, \boldsymbol{\lambda}'(\zeta') \cdot \widehat{\boldsymbol{\rho}}_{12}(\zeta' - 1)\right] \ . \tag{7.5.26}$$

Also, if we introduce new operators, $\widetilde{\boldsymbol{r}}_j(\zeta)$, defined by

$$\widetilde{\boldsymbol{r}}_j(\zeta) = \mathcal{T}(\zeta) \widehat{\boldsymbol{r}}_j(\zeta - 1) \mathcal{T}^{-1}(\zeta), \qquad j = 1, 2 \ , \tag{7.5.27}$$

we obtain, as in (7.2.11),

$$\begin{aligned} \widetilde{\boldsymbol{r}}_j(\zeta) &= \widehat{\boldsymbol{r}}_j(\zeta - 1) - i \int_{\zeta}^{1} d\zeta' \, [\boldsymbol{\lambda}'(\zeta') \cdot \widehat{\boldsymbol{\rho}}_{12}(\zeta' - 1), \widehat{\boldsymbol{r}}_j(\zeta - 1)] \\ &= \widehat{\boldsymbol{r}}_j(\zeta - 1) \pm \int_{\zeta}^{1} d\zeta' \, \boldsymbol{\lambda}'(\zeta')(\zeta' - \zeta), \qquad j = 1, 2 \ , \end{aligned} \tag{7.5.28}$$

as a consequence of the commutation relation (7.2.23), showing that $\widetilde{\boldsymbol{r}}_j(\zeta)$ and $\widetilde{\boldsymbol{r}}_j(\zeta')$ are mutually commutable, being independent of the $\boldsymbol{\rho}_j$ coordinates, as are $\widehat{\boldsymbol{r}}_j(\zeta)$ and $\widehat{\boldsymbol{r}}_j(\zeta')$ (7.2.24). Thus, all the $\widehat{\boldsymbol{\rho}}_{12}$ operators involved in (7.5.4) can be transferred to the right-hand side of all the $\widehat{\boldsymbol{r}}_j$ operators, with the result that

$$\begin{aligned} U^{(2)}[\boldsymbol{\lambda}', \boldsymbol{\rho}'] = \exp&\left\{-\int_0^1 d\zeta \, V^{(2)}[\widetilde{\boldsymbol{r}}_1(\zeta), \widetilde{\boldsymbol{r}}_2(\zeta), \boldsymbol{\rho}'(\zeta)]\right\} \\ &\times \exp\left\{\int_0^1 d\zeta \, i\boldsymbol{\lambda}'(\zeta) \cdot [\boldsymbol{\rho}'(\zeta) - \widehat{\boldsymbol{\rho}}_{12}(\zeta - 1)]\right\} \ , \end{aligned} \tag{7.5.29}$$

as a consequence of

$$\mathcal{T}(\zeta)V^{(2)}[\widehat{\boldsymbol{r}}_1(\zeta-1),\widehat{\boldsymbol{r}}_2(\zeta-1),\boldsymbol{\rho}'(\zeta)]\mathcal{T}^{-1}(\zeta) \\ = V^{(2)}[\widetilde{\boldsymbol{r}}_1(\zeta),\widetilde{\boldsymbol{r}}_2(\zeta),\boldsymbol{\rho}'(\zeta))] \ , \qquad (7.5.30)$$

resulting from

$$\mathcal{T}(\zeta)[\widehat{\boldsymbol{r}}_j(\zeta-1)]^n\mathcal{T}^{-1}(\zeta)=\widetilde{\boldsymbol{r}}_j^n(\zeta), \qquad n=1,2,3,\dots \ .$$

Here, the ordering symbol P is no longer necessary, and, with the Fourier transformation function of (7.5.5), and $\overline{\boldsymbol{\rho}}_{12}$ of (7.5.12), it holds the relation

$$\exp\left[-\mathrm{i}\int_0^1 d\zeta\,\boldsymbol{\lambda}'(\zeta)\cdot\widehat{\boldsymbol{\rho}}_{12}(\zeta-1)\right]\langle\boldsymbol{r},\boldsymbol{\rho}|\boldsymbol{s},\boldsymbol{\lambda}\rangle \\ =\exp\left[-\mathrm{i}\int_0^1 d\zeta\,\boldsymbol{\lambda}'(\zeta)\cdot\overline{\boldsymbol{\rho}}_{12}(\zeta)\right]\langle\boldsymbol{r},\boldsymbol{\rho}|\boldsymbol{s},\boldsymbol{\lambda}\rangle \ . \qquad (7.5.31)$$

Hence, using (7.5.29,31), we can write

$$U^{(2)}[\boldsymbol{\lambda}',\boldsymbol{\rho}']\langle\boldsymbol{r},\boldsymbol{\rho}|\boldsymbol{s},\boldsymbol{\lambda}\rangle \\ =\exp\left\{-\int_0^1 d\zeta\,V^{(2)}[\widetilde{\boldsymbol{r}}_1(\zeta),\widetilde{\boldsymbol{r}}_2(\zeta),\boldsymbol{\rho}'(\zeta)]\right\}\exp(\varphi) \ , \qquad (7.5.32)$$

where

$$\varphi=\int_0^1 d\zeta\,\mathrm{i}\boldsymbol{\lambda}'(\zeta)\cdot[\boldsymbol{\rho}'(\zeta)-\overline{\boldsymbol{\rho}}_{12}(\zeta)]-\mathrm{i}\sum_{j=1}^{2}(\boldsymbol{s}_j\cdot\boldsymbol{r}_j+\boldsymbol{\lambda}_j\cdot\boldsymbol{\rho}_j) \ . \qquad (7.5.33)$$

Here, with $\langle\widetilde{\boldsymbol{r}}_j(\zeta)\rangle$ of (7.5.10), it holds the relation

$$\widetilde{\boldsymbol{r}}_j(\zeta)\mathrm{e}^{\varphi}=\langle\widetilde{\boldsymbol{r}}_j(\zeta)\rangle\mathrm{e}^{\varphi} \ , \qquad (7.5.34)$$

in view of (7.5.28,12), and hence, more generally,

$$[\widetilde{\boldsymbol{r}}_j(\zeta)]^n\mathrm{e}^{\varphi}=\langle\widetilde{\boldsymbol{r}}_j(\zeta)\rangle^n\mathrm{e}^{\varphi} \ , \qquad (7.5.35)$$

in view of $\langle\widetilde{\boldsymbol{r}}_j(\zeta)\rangle$, which is independent of $\boldsymbol{\rho}_1$ and $\boldsymbol{\rho}_2$. Thus all the $\widetilde{\boldsymbol{r}}_j(\zeta)$'s involved in the right-hand side of (7.5.32) can be replaced by $\langle\widetilde{\boldsymbol{r}}_j(\zeta)\rangle$, leading to the expression (7.5.8,9) for $S^{(2)}[\boldsymbol{\lambda}',\boldsymbol{\rho}']$ according to the definition (7.5.6a).

7.5.2 Two-Scale Representation as an Effective Approximation

An important range of contribution of the Fourier variables $\boldsymbol{\lambda}'(\zeta)$ and the associated $\boldsymbol{\rho}'(\zeta)$, involved in (7.5.16) and distributed over the range $1 \geq \zeta \geq 0$, are expected to be very limited to such a set of functions for which the phase of the integrand, $S^{(2)}[\boldsymbol{\lambda}', \boldsymbol{\rho}']$, is stationary against their arbitrary variations so that $\boldsymbol{\lambda}'(\zeta)$ and $\boldsymbol{\rho}'(\zeta)$ are uniquely determined once their boundary values at $\zeta = 0$ and/or $\zeta = 1$ are given. This observation is reasonable, particularly because the original wave equations (7.1.2,3) were derived under the assumption that the medium change be negligibly small within a distance of the wave length.

Here, to find the "equations of motion" for those $\boldsymbol{\xi}(\zeta)$ and $\boldsymbol{\eta}(\zeta)$, we can directly utilize the Lagrange variational principle in dynamics: That is, we first introduce a Hamilton integral $S^L(\zeta)$ defined by

$$S^L(\zeta') = \int_0^{\zeta'} d\zeta\, \mathcal{L}(\dot{\boldsymbol{\xi}}, \dot{\boldsymbol{\eta}}, \boldsymbol{\xi}, \boldsymbol{\eta}, \zeta) \ , \tag{7.5.36}$$

to write, from (7.5.22),

$$S^{(2)}[\boldsymbol{\lambda}', \boldsymbol{\rho}'] = \mathrm{i}\dot{\boldsymbol{\xi}} \cdot \boldsymbol{\eta} \,\Big|_{\zeta=0}^{1} + S^L(\zeta = 1) \ . \tag{7.5.37}$$

We then derive the equations of motion for $\boldsymbol{\xi}(\zeta)$ and $\boldsymbol{\eta}(\zeta)$ according to the variational principle $\delta S^L = 0$ [Subsection A]. Hence,

$$\mathrm{i}\ddot{\boldsymbol{\xi}} = \langle \widetilde{V}_{\eta}^{(2)} \rangle, \qquad \mathrm{i}\ddot{\boldsymbol{\eta}} = \langle \widetilde{V}_{\xi}^{(2)} \rangle \ . \tag{7.5.38}$$

Here,

$$\langle \widetilde{V}_{\eta}^{(2)} \rangle = \frac{\partial}{\partial \boldsymbol{\eta}} \langle \widetilde{V}^{(2)} \rangle, \qquad \langle \widetilde{V}_{\xi}^{(2)} \rangle = \frac{\partial}{\partial \boldsymbol{\xi}} \langle \widetilde{V}^{(2)} \rangle \ , \tag{7.5.39}$$

where $\langle \widetilde{V}^{(2)} \rangle$ is given by (7.5.24) with $\overline{\boldsymbol{r}}_j(\zeta)$ and $\overline{\boldsymbol{\rho}}_{12}(\zeta)$ of (7.5.11,12).

When $\boldsymbol{\xi}(\zeta)$ and $\boldsymbol{\eta}(\zeta)$ are solutions of (7.5.38), equation (7.5.21) can be written as

$$S^{(2)}[\boldsymbol{\lambda}', \boldsymbol{\rho}'] = +\mathrm{i}\int_{-0}^{+0} d\zeta\, \ddot{\boldsymbol{\xi}} \cdot \boldsymbol{\eta} + \int_{+0}^{1} d\zeta\, [\langle \widetilde{V}_{\eta}^{(2)} \rangle \cdot \boldsymbol{\eta} - \langle \widetilde{V}^{(2)} \rangle] \ . \tag{7.5.40}$$

Here, as for the first term, we assume, on considering (7.5.25), that $\dot{\boldsymbol{\xi}}(\zeta < 0) = 0$ and $\dot{\boldsymbol{\xi}}(\zeta = +0) \neq 0$; whereas $\dot{\boldsymbol{\xi}}(\zeta)$ is continuous at $\zeta = 1$ $[\boldsymbol{\lambda}'(\zeta) = 0, \zeta > 1]$, so that

$$S_0 = \mathrm{i}\int_{-0}^{+0} d\zeta\, \ddot{\boldsymbol{\xi}} \cdot \boldsymbol{\eta} = \mathrm{i}\dot{\boldsymbol{\xi}} \cdot \boldsymbol{\eta}(\zeta = +0) \ . \tag{7.5.41}$$

While, in the other range, (7.5.38) is assumed to be subjected to

$$\boldsymbol{\xi}(\varsigma = 1) = 0, \qquad \dot{\boldsymbol{\xi}}(\varsigma = 1) = \boldsymbol{\lambda}' \ , \tag{7.5.42}$$

$$\boldsymbol{\eta}(\varsigma = 0) = \boldsymbol{\rho}', \qquad \dot{\boldsymbol{\eta}}(\varsigma = 0) = 0 \ , \tag{7.5.43}$$

with arbitrary boundary values $\boldsymbol{\lambda}'$ and $\boldsymbol{\rho}'$ [which should not be confused with the functions $\boldsymbol{\lambda}'(\varsigma)$ and $\boldsymbol{\rho}'(\varsigma)$]. Here, the first condition of (7.5.42) is required from (7.5.25), and the second condition of (7.5.43) means, in view of (7.5.19,12), that

$$\dot{\boldsymbol{\rho}}(\varsigma = 0) = \boldsymbol{v} \ . \tag{7.5.44}$$

This results from the fact that

$$\boldsymbol{\rho}'(\varsigma) = \overline{\boldsymbol{\rho}}_{12}(\varsigma)$$

over any ς range where the medium is free from fluctuation, in view of the factor $\delta[\boldsymbol{\rho}' - \overline{\boldsymbol{\rho}}_{12}]$ which comes from (7.5.7,8) by the functional integration with respect to $\boldsymbol{\lambda}'(\varsigma)$ over that ς range. Thus, even in free space $S^{(2)}$ from (7.5.40,41) is nonzero.

To express the solutions of (7.5.38) subject to (7.5.42,43), we introduce two new Green's functions, $G_\xi(\varsigma|\varsigma')$ and $G_\eta(\varsigma|\varsigma')$, defined by

$$\left(\frac{\partial}{\partial\varsigma}\right)^2 G_\xi(\varsigma|\varsigma') = \left(\frac{\partial}{\partial\varsigma}\right)^2 G_\eta(\varsigma|\varsigma') = \delta(\varsigma - \varsigma') \tag{7.5.45}$$

subjected to the boundary conditions

$$G_\xi(\varsigma = 1|\varsigma') = \left(\frac{\partial}{\partial\varsigma}\right) G_\xi(\varsigma = 1|\varsigma') = 0 \ , \tag{7.5.46}$$

$$G_\eta(\varsigma = 0|\varsigma') = \left(\frac{\partial}{\partial\varsigma}\right) G_\eta(\varsigma = 0|\varsigma') = 0 \ ; \tag{7.5.47}$$

hence,

$$G_\xi(\varsigma|\varsigma') = \begin{cases} 0, & \varsigma > \varsigma' \\ \varsigma' - \varsigma, & \varsigma < \varsigma' \ , \end{cases} \tag{7.5.48}$$

$$G_\eta(\varsigma|\varsigma') = \begin{cases} \varsigma - \varsigma', & \varsigma > \varsigma' \\ 0, & \varsigma < \varsigma' \ . \end{cases} \tag{7.5.49}$$

Thus,

$$\boldsymbol{\xi}(\varsigma) = -\mathrm{i} \int d\varsigma' \, G_\xi(\varsigma|\varsigma') \langle \widetilde{\boldsymbol{V}}_\eta^{(2)}(\varsigma') \rangle + (\varsigma - 1)\boldsymbol{\lambda}' \ , \tag{7.5.50}$$

$$\boldsymbol{\eta}(\varsigma) = -\mathrm{i} \int d\varsigma' \, G_\eta(\varsigma|\varsigma') \langle \widetilde{\boldsymbol{V}}_\xi^{(2)}(\varsigma') \rangle + \boldsymbol{\rho}' \ . \tag{7.5.51}$$

Here, (7.5.50) leads to the relation

$$\dot{\boldsymbol{\xi}}(\zeta=+0)=\mathrm{i}\int_0^1 d\zeta'\,\langle\widetilde{\boldsymbol{V}}_\eta^{(2)}(\zeta')\rangle+\boldsymbol{\lambda}' \ . \tag{7.5.52}$$

Thus $S^{(2)}[\boldsymbol{\lambda}',\boldsymbol{\rho}']$ of (7.5.40) becomes a function of only the boundary values $\boldsymbol{\lambda}'$ and $\boldsymbol{\rho}'$, say $S^{(2)}(\boldsymbol{\lambda}',\boldsymbol{\rho}')$, and is reduced to [see also (7.5.81)]

$$\begin{aligned}S^{(2)}(\boldsymbol{\lambda}',\boldsymbol{\rho}')=\mathrm{i}\boldsymbol{\lambda}'\cdot\boldsymbol{\rho}'-\int_0^1 d\zeta\,\langle\widetilde{V}^{(2)}(\zeta)\rangle\\ -\mathrm{i}\int_0^1 d\zeta d\zeta'\,\langle\widetilde{\boldsymbol{V}}_\eta^{(2)}(\zeta)\rangle\cdot G_\eta(\zeta|\zeta')\langle\widetilde{\boldsymbol{V}}_\xi^{(2)}(\zeta')\rangle \ ,\end{aligned} \tag{7.5.53}$$

as a consequence of (7.5.41,51,52). Here, $\langle\widetilde{V}^{(2)}(\zeta)\rangle$ is given by (7.5.24), where $\boldsymbol{\xi}(\zeta)$ and $\boldsymbol{\eta}(\zeta)$ are from (7.5.50,51).

In the special case of $V^{(2)}=0$, (7.5.53) gives $S^{(2)}=\mathrm{i}\boldsymbol{\lambda}'\cdot\boldsymbol{\rho}'$ and, by setting

$$d[\boldsymbol{\lambda}']d[\boldsymbol{\rho}']=(2\pi)^{-2}d\boldsymbol{\lambda}'d\boldsymbol{\rho}' \ , \tag{7.5.54}$$

the functional integral in (7.5.16) is reduced to

$$\int d[\boldsymbol{\lambda}']d[\boldsymbol{\rho}']\exp[S^{(2)}]=1 \ ,$$

yielding $M_{22}(1)=M_{22}^0(1)$, as it should be.

Thus, with the replacement of the functional $S^{(2)}[\boldsymbol{\lambda}',\boldsymbol{\rho}']$ by the end value function $S^{(2)}(\boldsymbol{\lambda}',\boldsymbol{\rho}')$ of (7.5.53,54), expression (7.5.16) is converted to an exact two-scale integral representation of M_{22} as: Firstly, the functional integration in (7.5.7), with respect to $\boldsymbol{\lambda}'(\zeta)$ and $\boldsymbol{\rho}'(\zeta)$ within the intermediate range $1>\zeta>0$ (excluding those at the end points $\zeta=0$ and $\zeta=1$), yields a factor of unity to the approximation of (7.5.40), in view of a particular structure of the integrand as given by (7.5.8); i.e., on writing $\boldsymbol{\lambda}'(\zeta)=\boldsymbol{\lambda}_0'(\zeta)+\Delta\boldsymbol{\lambda}'(\zeta)$ and $\boldsymbol{\rho}'(\zeta)=\boldsymbol{\rho}_0'(\zeta)+\Delta\boldsymbol{\rho}'(\zeta)$ with the functions $\boldsymbol{\lambda}_0'(\zeta)$ and $\boldsymbol{\rho}_0'(\zeta)$ along the phase stationary points subject to (7.5.38), use of (7.5.8) yields an additional factor given by the integral

$$\int d[\boldsymbol{\lambda}']d[\boldsymbol{\rho}']\exp\left[\mathrm{i}\int_{1>\zeta>0} d\zeta\,\Delta\boldsymbol{\lambda}'(\zeta)\cdot\Delta\boldsymbol{\rho}'(\zeta)\right] \ ,$$

which is unity by virtue of the definition according to (7.5.2). Here, in the integrand, the exponential term is the second order in $\Delta\boldsymbol{\lambda}'(\zeta)$ and $\Delta\boldsymbol{\rho}'(\zeta)$, $\zeta\neq 0,\,1$, including those from $\langle\widetilde{V}^{(2)}(\zeta)\rangle$ that have been neglected. Secondly, (7.5.25,42) indicate that

$$\dot{\boldsymbol{\xi}}(\zeta=1)=\int_{-0}^1 d\zeta\,\boldsymbol{\lambda}'(\zeta)=\boldsymbol{\lambda}' \ , \tag{7.5.55}$$

so that (7.5.18) is simply written by

$$\boldsymbol{\lambda}'' = \boldsymbol{\lambda}' + \boldsymbol{\nu} \ , \tag{7.5.56}$$

which is the same equation as (7.2.50), and, therefore, showing that the spectrum variable of $M_{22}(1)$ with respect to the coordinates $\boldsymbol{\rho}_{12}$ (i.e., $\boldsymbol{\lambda}''$), is given just by a sum of two kinds of variables: $\boldsymbol{\lambda}'$ from the medium and $\boldsymbol{\nu}$ from the incident wave.

More precisely, we observe that the introduction of a new variable, $\boldsymbol{\rho}''$, defined by

$$\boldsymbol{\rho}'' = \boldsymbol{\rho}' - \boldsymbol{v} + \boldsymbol{\rho}_{12} \tag{5.5.57}$$

(where $\boldsymbol{\rho}''$ is the same as that in (7.2.50), but $\boldsymbol{\rho}'$ is not), and also two new functions, $\Delta\boldsymbol{\xi}(\zeta)$ and $\Delta\boldsymbol{\eta}(\zeta)$, given by

$$\Delta\boldsymbol{\xi}(\zeta) = -\mathrm{i}\int d\zeta'\, G_\xi(\zeta|\zeta')\langle\widetilde{\boldsymbol{V}}_\eta^{(2)}(\zeta')\rangle \ , \tag{7.5.58}$$

$$\Delta\boldsymbol{\eta}(\zeta) = -\mathrm{i}\int d\zeta'\, G_\eta(\zeta|\zeta')\langle\widetilde{\boldsymbol{V}}_\xi^{(2)}(\zeta')\rangle \ , \tag{7.5.59}$$

enables $\langle\widetilde{V}^{(2)}(\zeta)\rangle$ of (7.5.24) to be written, when $\boldsymbol{r}_1 = \boldsymbol{r}_2 = 0$ and with (7.2.46b) and (7.5.56), by

$$\begin{aligned}\langle\widetilde{V}^{(2)}(\zeta)\rangle = V^{(2)}[&(\zeta-1)(\boldsymbol{\lambda}''+\overline{\boldsymbol{\nu}}/2)+\Delta\boldsymbol{\xi}(\zeta),\\ &(\zeta-1)(-\boldsymbol{\lambda}''+\overline{\boldsymbol{\nu}}/2)-\Delta\boldsymbol{\xi}(\zeta), \boldsymbol{\rho}''+\zeta\boldsymbol{v}+\Delta\boldsymbol{\eta}(\zeta)] \ .\end{aligned} \tag{7.5.60}$$

This is exactly the same function as $V^{(2)}[\ ]$ involved in (7.2.54), except that it contains the additional terms $\Delta\boldsymbol{\xi}(\zeta)$ and $\Delta\boldsymbol{\eta}(\zeta)$; and furthermore, to the first order of $V^{(2)}$, the last integral in (7.5.53), as well as $\Delta\boldsymbol{\xi}(\zeta)$ and $\Delta\boldsymbol{\eta}(\zeta)$ involved, is negligible so that the result from substituting $S^{(2)}(\boldsymbol{\lambda}', \boldsymbol{\rho}')$ of (7.5.53,54) into (7.5.16) becomes exactly the same as that which would be given by the two-scale expression (7.2.53) with the spectrum function $\widetilde{M}_{22}(\boldsymbol{\lambda}'', \overline{\boldsymbol{\rho}})$ of (7.2.54) and the replacement of $\Delta\zeta \to \int_0^1 d\zeta$ (7.2.56). This is true including the resultant phase term which is given by

$$\begin{aligned}&\boldsymbol{\lambda}'\cdot\boldsymbol{\rho}' + \boldsymbol{\nu}\cdot(\boldsymbol{v}-\boldsymbol{\rho}_{12}) + \overline{\boldsymbol{\nu}}\cdot(\overline{\boldsymbol{v}}-\overline{\boldsymbol{\rho}})\\ &\quad = \boldsymbol{\lambda}''\cdot(\boldsymbol{\rho}''+\boldsymbol{v}-\boldsymbol{\rho}_{12}) - \boldsymbol{\nu}\cdot\boldsymbol{\rho}'' + \overline{\boldsymbol{\nu}}\cdot(\overline{\boldsymbol{v}}-\overline{\boldsymbol{\rho}}) \ .\end{aligned} \tag{7.5.61}$$

(7.5.61) is the same as that given (7.2.52), in terms of $\boldsymbol{\lambda}''$ and $\boldsymbol{\rho}''$ of (7.2.50), and the relative coordinates and variables of (7.2.44–48); that is, the exact version of the spectrum function is

$$\begin{aligned}\widetilde{M}_{22}(\boldsymbol{\lambda}'', \overline{\boldsymbol{\rho}}) &= (2\pi)^{-4}\int d\boldsymbol{\rho}''d\boldsymbol{v}d\overline{\boldsymbol{\nu}}\, \exp\{\mathrm{i}[\boldsymbol{\lambda}''\cdot(\boldsymbol{\rho}''+\boldsymbol{v}) - \overline{\boldsymbol{\nu}}\cdot\overline{\boldsymbol{\rho}}]\}\\ &\times \exp\left[-\int_0^1 d\zeta\,\langle\widetilde{V}^{(2)}(\zeta)\rangle - \mathrm{i}\int_0^1 d\zeta d\zeta'\,\langle\widetilde{\boldsymbol{V}}_\eta^{(2)}(\zeta)\rangle\cdot G_\eta(\zeta|\zeta')\right.\\ &\times \left.\langle\widetilde{\boldsymbol{V}}_\xi^{(2)}(\zeta')\rangle\right]\widetilde{M}_{22}^0(\boldsymbol{\rho}''; \boldsymbol{v}, \overline{\boldsymbol{\nu}}) \ .\end{aligned} \tag{7.5.62}$$

Here, from (7.5.39, 24), $\langle \widetilde{V}_\xi^{(2)} \rangle$ and $\langle \widetilde{V}_\eta^{(2)} \rangle$ are given in terms of $\langle \widetilde{V}^{(2)} \rangle$ of (7.5.60) by

$$\langle \widetilde{\boldsymbol{V}}_\xi^{(2)} \rangle = \frac{\partial}{\partial \Delta \boldsymbol{\xi}} \langle \widetilde{V}^{(2)} \rangle , \qquad \langle \widetilde{\boldsymbol{V}}_\eta^{(2)} \rangle = \frac{\partial}{\partial \Delta \boldsymbol{\eta}} \langle \widetilde{V}^{(2)} \rangle \ ; \tag{7.5.63}$$

and $\widetilde{M_{22}^0}(\boldsymbol{\rho}''; \boldsymbol{v}, \overline{\boldsymbol{v}})$ is given by (7.2.55) for the beam wave, and is reduced to (7.2.57) in the cases of spherical and plane waves.

In the case of a phase screen, $\Delta\boldsymbol{\xi} = \Delta\boldsymbol{\eta} = 0$, according to (7.5.48,49), and (7.5.62) agrees with the exact expression (7.2.54). In the case of a turbulent continuum, on the other hand, $\Delta\boldsymbol{\xi}$ and $\Delta\boldsymbol{\eta}$ may be treated perturbatively (without justification); if so, the last exponential factor in (7.5.62) becomes, in terms of the notation $\langle \cdots \rangle_0$ for $\langle \cdots \rangle$ when $\Delta\boldsymbol{\xi} = \Delta\boldsymbol{\eta} = 0$:

$$\begin{aligned} \exp[\,] \sim \exp\Bigg[&- \int_0^1 d\zeta \, \langle \widetilde{V}^{(2)}(\zeta) \rangle_0 \\ &+ \mathrm{i} \int_0^1 d\zeta d\zeta' \, \langle \widetilde{\boldsymbol{V}}_\xi^{(2)}(\zeta) \rangle_0 \cdot G_\xi(\zeta|\zeta') \langle \widetilde{\boldsymbol{V}}_\eta^{(2)}(\zeta') \rangle_0 \Bigg] \ . \end{aligned} \tag{7.5.64}$$

Here, the G_η terms have been cancelled out, and the first term is exactly what was given by (7.2.54,56) as the "zero order" expression. The second term gives the first order correction, and the resulting asymptotic expressions are briefly summarized for both spherical and plane waves in Subsection A.

Generally, equation (7.5.62) can be evaluated by iteratively solving the closed form equations (7.5.58–60) for $\Delta\boldsymbol{\xi}$, $\Delta\boldsymbol{\eta}$, and $\langle \widetilde{V}^{(2)} \rangle$; finally, $M_{22}(1)$ is given by the spectrum integral (7.2.53).

(i) Case of a Spherical Wave.

With (7.2.57) and (7.5.62), expression (7.2.53) is reduced simply to

$$\begin{aligned} M_{22}(\boldsymbol{\rho}_{12} = 0) = {} & (4\pi L)^{-4} (2\pi)^{-2} \int d\boldsymbol{\lambda}'' d\boldsymbol{v} \, \exp(\mathrm{i}\boldsymbol{\lambda}'' \cdot \boldsymbol{v}) \\ & \times \exp\Bigg[- \int_0^1 d\zeta \, \langle \widetilde{V}^{(2)}(\zeta) \rangle \\ & \quad - \mathrm{i} \int_{\zeta > \zeta' > 0}^1 d\zeta d\zeta' \, (\zeta - \zeta') \langle \widetilde{\boldsymbol{V}}_\eta^{(2)}(\zeta) \rangle \cdot \langle \widetilde{\boldsymbol{V}}_\xi^{(2)}(\zeta') \rangle \Bigg] \ . \end{aligned} \tag{7.5.65}$$

Here, $\langle \widetilde{\boldsymbol{V}}_\xi^{(2)} \rangle$ and $\langle \widetilde{\boldsymbol{V}}_\eta^{(2)} \rangle$ are given by (7.5.63), with

$$\begin{aligned} \langle \widetilde{V}^{(2)}(\zeta) \rangle = V^{(2)}[& (\zeta - 1)\boldsymbol{\lambda}'' + \Delta\boldsymbol{\xi}(\zeta), \\ & - (\zeta - 1)\boldsymbol{\lambda}'' - \Delta\boldsymbol{\xi}(\zeta), \zeta\boldsymbol{v} + \Delta\boldsymbol{\eta}(\zeta)] \ , \end{aligned} \tag{7.5.66}$$

wherein $\Delta\boldsymbol{\xi}$ and $\Delta\boldsymbol{\eta}$ are given by (7.5.58,59) in terms of $\langle \widetilde{\boldsymbol{V}}_\eta^{(2)} \rangle$ and $\langle \widetilde{\boldsymbol{V}}_\xi^{(2)} \rangle$, so that these three equations constitute a complete set of equations to obtain the unknown $\langle \widetilde{V}^{(2)}(\zeta) \rangle$.

(ii) Case of a Plane Wave.

Substitution of (7.2.57) in (7.5.62) leads to the expression

$$M_{22}(\boldsymbol{\rho}_{12}=0) = I_0^2(2\pi)^{-2}\int d\boldsymbol{\lambda}''d\boldsymbol{\rho}''\exp(\mathrm{i}\boldsymbol{\lambda}''\cdot\boldsymbol{\rho}'')$$
$$\times\exp\Big[-\int_0^1 d\zeta\,\langle\widetilde{V}^{(2)}(\zeta)\rangle$$
$$-\mathrm{i}\int_{\zeta>\zeta'>0}^1 d\zeta d\zeta'\,(\zeta-\zeta')\langle\widetilde{\boldsymbol{V}}_{\eta}^{(2)}(\zeta)\rangle\cdot\langle\widetilde{\boldsymbol{V}}_{\xi}^{(2)}(\zeta')\rangle\Big]\ . \quad (7.5.67)$$

Here,

$$\langle\widetilde{V}^{(2)}(\zeta)\rangle = V^{(2)}[(\zeta-1)\boldsymbol{\lambda}''+\Delta\boldsymbol{\xi}(\zeta),$$
$$-(\zeta-1)\boldsymbol{\lambda}''-\Delta\boldsymbol{\xi}(\zeta),\boldsymbol{\rho}''+\Delta\boldsymbol{\eta}(\zeta)]\ , \quad (7.5.68)$$

and the other functions are defined in the same way as in (7.5.65,66).

Hence the zero order expressions (7.2.58,59) are reproduced by (7.5.65,67), respectively, to the approximation of setting $\Delta\boldsymbol{\xi}=\Delta\boldsymbol{\eta}=0$ and neglecting the last exponential terms; and, to the first order approximation of $\Delta\boldsymbol{\xi}$ and $\Delta\boldsymbol{\eta}$, the exponential term is given, according to (7.5.64), by

$$-\int_0^1 d\zeta\,\langle\widetilde{V}^{(2)}(\zeta)\rangle_0$$
$$+\mathrm{i}\int_{\zeta'>\zeta>0}^1 d\zeta d\zeta'\,(\zeta'-\zeta)\langle\widetilde{\boldsymbol{V}}_{\eta}^{(2)}(\zeta')\rangle_0\cdot\langle\widetilde{\boldsymbol{V}}_{\xi}^{(2)}(\zeta)\rangle_0\ . \quad (7.5.69)$$

The sign of the second term is opposite to that in the original (7.5.65,67). To this approximation a similar expression of M_{22} for a plane wave was obtained in [7.19].

a) Asymptotic Expressions

The first order correction to the exponential terms of (7.5.65,67) is given by the second term of (7.5.69), and its effect on the asymptotic expression of $M_{22}(1)$ for $\phi\gg 1$ in two-dimensional space was investigated for the pure Kolmogorov medium, with two results [7.14]:

(i) Case of a Spherical Wave.

$$M_{22}(1)\sim(4\pi L)^{-4}(2+1.032\,\phi^{-2/5})\ ; \quad (7.5.70a)$$

(ii) Case of a Plane Wave.

$$M_{22}(1)\sim I_0^2(2+0.4942\,\phi^{-2/5})\ . \quad (7.5.70b)$$

It was suggested, however, that the second and higher order terms could also yield additional $\phi^{-2/5}$ terms with numerical factors of the same order of magnitude, so that determining numerical factors of the $\phi^{-2/5}$ terms in (7.5.70a,b) is still an unsolved problem.

b) $S^L(\lambda', \rho')$ as a Solution of the Hamilton-Jacobi Equation

On rewriting (7.5.36) as

$$S^L(\zeta_1|\zeta_2) = \int_{\zeta_2}^{\zeta_1} d\zeta\, \mathcal{L}(\dot{\boldsymbol{\xi}}, \dot{\boldsymbol{\eta}}, \boldsymbol{\xi}, \boldsymbol{\eta}, \zeta) \;, \tag{7.5.71}$$

we first observe that, when $\boldsymbol{\xi}(\zeta)$ and $\boldsymbol{\eta}(\zeta)$ are solutions of (7.5.38), the integral $S^L(\zeta_1|\zeta_2)$ is invariant against arbitrary variations, $\delta\boldsymbol{\xi}(\zeta)$ and $\delta\boldsymbol{\eta}(\zeta)$, $\zeta_1 > \zeta > \zeta_2$, to the first order, depending only on the end values of $\delta\boldsymbol{\xi}$ and $\delta\boldsymbol{\eta}$ at $\zeta = \zeta_1$ and ζ_2. That is, with $\mathcal{L}$ of (7.5.23) and partial integration, we obtain

$$\begin{aligned} \delta S^L(\zeta_1|\zeta_2) &= \int_{\zeta_2}^{\zeta_1} d\zeta\, \big[-\mathrm{i}(\delta\dot{\boldsymbol{\xi}} \cdot \dot{\boldsymbol{\eta}} + \dot{\boldsymbol{\xi}} \cdot \delta\dot{\boldsymbol{\eta}}) \\ &\quad - \langle \widetilde{\boldsymbol{V}}_{\xi}^{(2)} \rangle \cdot \delta\boldsymbol{\xi} - \langle \widetilde{\boldsymbol{V}}_{\eta}^{(2)} \rangle \cdot \delta\boldsymbol{\eta}\big] \\ &= -\mathrm{i}(\dot{\boldsymbol{\eta}} \cdot \delta\boldsymbol{\xi} + \dot{\boldsymbol{\xi}} \cdot \delta\boldsymbol{\eta})\, \big|_{\zeta=\zeta_2}^{\zeta_1} \\ &\quad + \int_{\zeta_2}^{\zeta_1} d\zeta\, \big[(\mathrm{i}\ddot{\boldsymbol{\eta}} - \langle \widetilde{\boldsymbol{V}}_{\xi}^{(2)} \rangle) \cdot \delta\boldsymbol{\xi} + (\mathrm{i}\ddot{\boldsymbol{\xi}} - \langle \widetilde{\boldsymbol{V}}_{\eta}^{(2)} \rangle) \cdot \delta\boldsymbol{\eta}\big] \;, \end{aligned} \tag{7.5.72}$$

where the last integral is zero in view of (7.5.38), showing that $\delta S^L(\zeta_1|\zeta_2)$ depends only on the end values of $\delta\boldsymbol{\xi}$ and $\delta\boldsymbol{\eta}$ at ζ_1 and ζ_2. Equation (7.5.72) can be generalized to also include a response for arbitrary displacement $\delta\zeta$ at each point ζ on keeping the $\boldsymbol{\xi}(\zeta)$ and $\boldsymbol{\eta}(\zeta)$ unchanged; the result is obtained in the conventional form:

$$\delta S^L(\zeta_1|\zeta_2) = \big[\boldsymbol{\alpha} \cdot \delta\boldsymbol{\xi} + \boldsymbol{\beta} \cdot \delta\boldsymbol{\eta} - H\, d\zeta\big]\, \big|_{\zeta=\zeta_2}^{\zeta_1} \;. \tag{7.5.73}$$

Here, $\boldsymbol{\alpha}$ and $\boldsymbol{\beta}$ are the canonical momenta of $\boldsymbol{\xi}$ and $\boldsymbol{\eta}$, respectively, defined by

$$\boldsymbol{\alpha} = \frac{\partial}{\partial \dot{\boldsymbol{\xi}}} \mathcal{L} = -\mathrm{i}\dot{\boldsymbol{\eta}}, \qquad \boldsymbol{\beta} = \frac{\partial}{\partial \dot{\boldsymbol{\eta}}} \mathcal{L} = -\mathrm{i}\dot{\boldsymbol{\xi}} \;, \tag{7.5.74}$$

and

$$H = \boldsymbol{\alpha} \cdot \dot{\boldsymbol{\xi}} + \boldsymbol{\beta} \cdot \dot{\boldsymbol{\eta}} - \mathcal{L} = \mathrm{i}\boldsymbol{\alpha} \cdot \boldsymbol{\beta} + \langle \widetilde{V}^{(2)}(\boldsymbol{\xi}, \boldsymbol{\eta}, \zeta) \rangle \;, \tag{7.5.75}$$

which is regarded as a function of $\boldsymbol{\alpha}$, $\boldsymbol{\beta}$, $\boldsymbol{\xi}$, $\boldsymbol{\eta}$, and ζ. The canonical equations in dynamics derived from (7.5.75) are

$$\begin{aligned} \dot{\boldsymbol{\xi}} &= \frac{\partial H}{\partial \boldsymbol{\alpha}}, & \dot{\boldsymbol{\alpha}} &= -\frac{\partial H}{\partial \boldsymbol{\xi}} \;, \\ \dot{\boldsymbol{\eta}} &= \frac{\partial H}{\partial \boldsymbol{\beta}}, & \dot{\boldsymbol{\beta}} &= -\frac{\partial H}{\partial \boldsymbol{\eta}} \;, \end{aligned} \tag{7.5.76}$$

which reproduce the original equations (7.5.38).

Equation (7.5.73) indicates that $S^L(\zeta_1|\zeta_2)$ depends only on the end values of $\boldsymbol{\xi}$ and $\boldsymbol{\eta}$ at ζ_1 and ζ_2, say $\boldsymbol{\xi}_1$, $\boldsymbol{\eta}_1$, and $\boldsymbol{\xi}_2$, $\boldsymbol{\eta}_2$, respectively; and that

$$\boldsymbol{\alpha}_1 = \frac{\partial}{\partial \boldsymbol{\xi}_1} S^L(\zeta_1|\zeta_2), \qquad \boldsymbol{\beta}_1 = \frac{\partial}{\partial \boldsymbol{\eta}_1} S^L(\zeta_1|\zeta_2) \ , \tag{7.5.77}$$

$$-\frac{\partial}{\partial \zeta_1} S^L(\zeta_1|\zeta_2) = H(\boldsymbol{\alpha}_1, \boldsymbol{\beta}_1, \boldsymbol{\xi}_1, \boldsymbol{\eta}_1, \zeta_1) \ . \tag{7.5.78}$$

The same set of equations with respect to $\boldsymbol{\xi}_2$, $\boldsymbol{\eta}_2$, and ζ_2 also holds true with the opposite sign of the right-hand sides. These equations constitute the Hamilton-Jacobi equation, and the complete solution can be obtained in the form

$$S^L(\zeta_1|\zeta_2) = S^L(\boldsymbol{\xi}_1, \boldsymbol{\eta}_1, \zeta_1|\boldsymbol{\xi}_2, \boldsymbol{\eta}_2, \zeta_2) \ ,$$

with $\boldsymbol{\xi}_2$, $\boldsymbol{\eta}_2$, and ζ_2 as the initial values. Here, the boundary conditions (7.5.42,43) enable us to determine all the end values completely, i.e., in view of relation (7.5.74), the conditions are reduced to

$$\boldsymbol{\xi}_1 = 0, \qquad \frac{\partial}{\partial \boldsymbol{\eta}_1} S^L(\zeta_1 = 1|\zeta_2 = 0) = -\mathrm{i}\boldsymbol{\lambda}' \ , \tag{7.5.79}$$

$$\boldsymbol{\eta}_2 = \boldsymbol{\rho}', \qquad \frac{\partial}{\partial \boldsymbol{\xi}_2} S^L(\zeta_1 = 1|\zeta_2 = 0) = 0 \ , \tag{7.5.80}$$

which determine $\boldsymbol{\eta}_1$ and $\boldsymbol{\xi}_2$, and hence, also $S^L(\zeta_1 = 1|\zeta_2 = 0)$ as a function of $\boldsymbol{\lambda}'$ and $\boldsymbol{\rho}'$. Finally, $S^{(2)}(\boldsymbol{\lambda}', \boldsymbol{\rho}')$ of (7.5.53) is obtained from (7.5.37) with the additional term, S_0, of (7.5.41), as

$$S^{(2)}(\boldsymbol{\lambda}', \boldsymbol{\rho}') = \mathrm{i}\boldsymbol{\lambda}' \cdot \boldsymbol{\eta}_1 + S^L(\zeta_1 = 1|\zeta_2 = 0) \ , \tag{7.5.81}$$

where the first term is from $\mathrm{i}\dot{\boldsymbol{\xi}}_1 \cdot \boldsymbol{\eta}(\zeta = 1)$.

In free space where $\langle \widetilde{V}^{(2)}(\zeta) \rangle = 0$, for example, we obtain, as the solution of (7.5.77, 78),

$$S^L(\zeta_1|\zeta_2) = -\mathrm{i}(\zeta_1 - \zeta_2)^{-1}(\boldsymbol{\xi}_1 - \boldsymbol{\xi}_2) \cdot (\boldsymbol{\eta}_1 - \boldsymbol{\eta}_2) \ ;$$

hence, from (7.5.79,80), $\boldsymbol{\xi}_1 = 0$, $\boldsymbol{\xi}_2 = -\boldsymbol{\lambda}'$, $\boldsymbol{\eta}_1 = \boldsymbol{\eta}_2 = \boldsymbol{\rho}'$, which from (7.5.81), reproduces $S^{(2)}(\boldsymbol{\lambda}', \boldsymbol{\rho}')$ of (7.5.53) in view of $S^L = 0$.

7.5.3 Two-Frequency Intensity Correlation

It is straightforward to generalize the previous results to include the intensity correlation between two waves of different frequencies, say, with the intensities I_1 and I_2 of wave numbers k_1 and k_2, respectively. Here, the basic equations to be modified are as follows: We redefine the dimensionless coordinates $\boldsymbol{r}_j$

and $\boldsymbol{\rho}_j$ according to (7.2.1) by using the mean wave number $k = (k_1 + k_2)/2$; and we write $k_j = k\omega_j$, $j = 1, 2$, with

$$\omega_1 = 1 - \Delta\omega/2, \quad \omega_2 = 1 + \Delta\omega/2, \quad \Delta\omega = (k_2 - k_1)/k \ . \tag{7.5.82}$$

Hence, (7.2.6,16) are replaced by [see the original moment equations (7.1.38,39)]

$$T_j = -\mathrm{i}\omega_j^{-1}\frac{\partial}{\partial \boldsymbol{r}_j}\cdot\frac{\partial}{\partial \boldsymbol{\rho}_j}, \qquad V_j = \omega_j^2\overline{V}(\boldsymbol{r}_j)\ , \tag{7.5.83}$$

$$V_{12} = \omega_1\omega_2\overline{V}_I(\boldsymbol{r}_1, \boldsymbol{r}_2, \boldsymbol{\rho}_{12})\ ; \tag{7.5.84}$$

and, from (7.2.11,22), $\widehat{\boldsymbol{r}}_j(\zeta)$ and $\widehat{\boldsymbol{\rho}}_j(\zeta)$ are replaced by

$$\widehat{\boldsymbol{r}}_j(\zeta) = \boldsymbol{r}_j + \mathrm{i}\omega_j^{-1}\zeta\frac{\partial}{\partial \boldsymbol{\rho}_j}\ , \tag{7.5.85a}$$

$$\widehat{\boldsymbol{\rho}}_j(\zeta) = \boldsymbol{\rho}_j + \mathrm{i}\omega_j^{-1}\zeta\frac{\partial}{\partial \boldsymbol{r}_j}\ , \tag{7.5.85b}$$

with the commutation relations

$$[\widehat{\boldsymbol{r}}_i(\zeta), \widehat{\boldsymbol{\rho}}_j(\zeta')] = \mathrm{i}(\zeta - \zeta')\omega_j^{-1}\delta_{ij}\ , \tag{7.5.86a}$$

$$[\widehat{\boldsymbol{r}}_i(\zeta), \widehat{\boldsymbol{r}}_j(\zeta')] = [\widehat{\boldsymbol{\rho}}_i(\zeta), \widehat{\boldsymbol{\rho}}_j(\zeta')] = 0\ . \tag{7.5.86b}$$

Hence expression (7.2.27) for $M_{22}(1)$ remains formally unchanged; more generally, explicit expression of the νth order intensity correlation function between the intensities I_1, I_2, ..., I_ν, with the wave numbers $k_1 = k\omega_1$, $k_2 = k\omega_2$, ..., $k_\nu = k\omega_\nu$ (where $k = \nu^{-1}\sum_{j=1}^{\nu} k_j$), respectively, is taken from (7.2.29, 30),

$$M_{\nu\nu}(1) = P\exp\left[\!\left[-\int_0^1 d\zeta\left\{\sum_{j=1}^{\nu}\omega_j^2\overline{V}[\widehat{\boldsymbol{r}}_j(\zeta-1)] + \sum_{i>j=1}^{\nu}\omega_i\omega_j\overline{V}_I[\widehat{\boldsymbol{r}}_i(\zeta-1), \widehat{\boldsymbol{r}}_j(\zeta-1), \widehat{\boldsymbol{\rho}}_{ij}(\zeta-1)]\right\}\right]\!\right] M^0_{\nu\nu}(1)\ . \tag{7.5.87}$$

Hence, with a new medium term, $V^{(2)}$, redefined by

$$V^{(2)}(\boldsymbol{r}_1, \boldsymbol{r}_2, \boldsymbol{\rho}_{12}) = \omega_1^2\overline{V}(\boldsymbol{r}_1) + \omega_2^2\overline{V}(\boldsymbol{r}_2) + \omega_1\omega_2\overline{V}_I(\boldsymbol{r}_1, \boldsymbol{r}_2, \boldsymbol{\rho}_{12})\ , \tag{7.5.88}$$

the spectrum function with respect to $\boldsymbol{\rho}_{12}$, (7.5.62), can be written by the same equation [Subsection **a)**] as

$$\widetilde{M}_{22}(\boldsymbol{\lambda}'', \overline{\boldsymbol{\rho}}) = (2\pi)^{-4} \int d\boldsymbol{\rho}'' d\boldsymbol{v}' d\overline{\boldsymbol{\nu}} \exp\{\mathrm{i}[\boldsymbol{\lambda}'' \cdot (\boldsymbol{\rho}'' + \boldsymbol{v}') - \overline{\boldsymbol{\nu}} \cdot \overline{\boldsymbol{\rho}}]\}$$
$$\times \exp\left[-\int_0^1 d\zeta \, \langle \widetilde{\boldsymbol{V}}^{(2)}(\zeta)\rangle - \mathrm{i} \int_0^1 d\zeta d\zeta' \, \langle \widetilde{\boldsymbol{V}}_\eta^{(2)}(\zeta)\rangle \cdot G_\eta(\zeta|\zeta') \langle \widetilde{\boldsymbol{V}}_\xi^{(2)}(\zeta')\rangle\right]$$
$$\times \widetilde{M}_{22}^0(\boldsymbol{\rho}''; \boldsymbol{v}', \overline{\boldsymbol{\nu}}) \ , \tag{7.5.89}$$

except for the replacement of $\boldsymbol{v} \to \boldsymbol{v}'$ of (7.5.95b). Here, from (7.5.100),

$$\langle \widetilde{V}^{(2)}(\zeta)\rangle = V^{(2)}[(\zeta - 1)\omega_1^{-1}(\boldsymbol{\lambda}'' + \overline{\boldsymbol{\nu}}/2) + \omega_1^{-1}\Delta\boldsymbol{\xi}(\zeta), \\ (\zeta - 1)\omega_2^{-1}(-\boldsymbol{\lambda}'' + \overline{\boldsymbol{\nu}}/2) - \omega_2^{-1}\Delta\boldsymbol{\xi}(\zeta), \boldsymbol{\rho}'' + \zeta\boldsymbol{v}' + \Delta\boldsymbol{\eta}(\zeta))] \ , \tag{7.5.90}$$

in which $\Delta\boldsymbol{\xi}(\zeta)$ and $\Delta\boldsymbol{\eta}(\zeta)$ are still given by (7.5.58,59) with $\langle \widetilde{\boldsymbol{V}}_\xi^{(2)}\rangle$ and $\langle \widetilde{\boldsymbol{V}}_\eta^{(2)}\rangle$ given, according to (7.5.63), by the new $\langle \widetilde{V}^{(2)}(\zeta)\rangle$; and, from (7.5.102),

$$\widetilde{M}_{22}^0(\boldsymbol{\rho}''; \boldsymbol{v}', \overline{\boldsymbol{\nu}}) = I_0^2(\pi\overline{b}^2)^2\omega^2\overline{\omega}^{-1} \exp\left[-(2\overline{b}^2)^{-1}\boldsymbol{\rho}''^2 - \overline{b}^2\left(\tfrac{1}{2}\overline{\omega}\boldsymbol{v}'^2 + \tfrac{1}{8}\overline{\boldsymbol{\nu}}^2\right) - \frac{1}{8\overline{b}^2\overline{\omega}}\left(\overline{\boldsymbol{\nu}} - \mathrm{i}2\Delta\omega\overline{b}^2\boldsymbol{v}'\right)^2\right] \ , \tag{7.5.91}$$

where

$$\overline{\omega} = 1 + (\Delta\omega/2)^2 = 2 - \omega \ , \tag{7.5.92a}$$
$$\omega = 1 - (\Delta\omega/2)^2 = \omega_1\omega_2 \ . \tag{7.5.92b}$$

Equation (7.5.91) is reduced to (7.2.55b) for $\Delta\omega \to 0$, and to:

(i) Case of a Spherical Wave ($b = 0$).

$$\widetilde{M}_{22}^0(\boldsymbol{\rho}''; \boldsymbol{v}', \overline{\boldsymbol{\nu}}) = (4\pi L)^{-4}\omega^2(2\pi)^4\delta(\overline{\boldsymbol{\nu}})\delta(\boldsymbol{\rho}'') \ ; \tag{7.5.93}$$

(ii) Case of a Plane Wave ($b = \infty$).

$$\widetilde{M}_{22}^0(\boldsymbol{\rho}''; \boldsymbol{v}', \overline{\boldsymbol{\nu}}) = I_0^2(\omega/\overline{\omega})^2(2\pi)^4\delta(\overline{\boldsymbol{\nu}})\delta(\boldsymbol{v}') \ . \tag{7.5.94}$$

Thus the spectrum function of (7.5.89) becomes a single integral in both cases, leading to expressions of $M_{22}(1)$ similar to (7.5.65,67), with $\langle \widetilde{V}^{(2)}(\zeta)\rangle$ from (7.5.88,90) or, to the zero order approximation, with $\langle \widetilde{V}^{(2)}(\zeta)\rangle_0$ of (7.5.104–106).

a) Basic Equations for the Two-Frequency Intensity Correlation Function

By following the procedure of Sect. 7.5.2a, it is straightforward to find that expressions (7.5.8,9) for the functional $S^{(2)}[\boldsymbol{\lambda}', \boldsymbol{\rho}']$ remain unchanged with the new $\overline{\boldsymbol{\rho}}_{12}(\zeta)$ redefined by

$$\overline{\boldsymbol{\rho}}_{12}(\zeta) = \boldsymbol{\rho}_{12} + (\zeta - 1)\boldsymbol{v}' \ , \tag{7.5.95a}$$

$$\boldsymbol{v}' = \omega_1^{-1}\boldsymbol{s}_1 - \omega_2^{-1}\boldsymbol{s}_2 = \omega^{-1}(\boldsymbol{v} + \Delta\omega\overline{\boldsymbol{v}}) \ ; \tag{7.5.95b}$$

and the new $\langle\widetilde{\boldsymbol{r}}_j(\zeta)\rangle$ redefined by

$$\langle\widetilde{\boldsymbol{r}}_j(\zeta)\rangle = \overline{\boldsymbol{r}}_j(\zeta) \pm \omega_j^{-1}\boldsymbol{\xi}(\zeta), \qquad j = 1,2 \ , \tag{7.5.96}$$

$$\overline{\boldsymbol{r}}_j(\zeta) = \boldsymbol{r}_j + (\zeta - 1)\omega_j^{-1}\boldsymbol{\lambda}_j \ , \tag{7.5.97}$$

where $\boldsymbol{\xi}(\zeta)$ is the same as defined by (7.5.13). Here, $\overline{\boldsymbol{\rho}}_{12}(\zeta)$ is still an eigenvalue of the (new) operator, $\widehat{\boldsymbol{\rho}}_{12}(\zeta - 1)$, as in (7.5.31), and is also the same for $\langle\widetilde{\boldsymbol{r}}_j(\zeta)\rangle$, defined by (7.5.34), being an eigenvalue of the operator $\widetilde{\boldsymbol{r}}_j(\zeta)$ of (7.5.27), which is presently given by

$$\widetilde{\boldsymbol{r}}_j(\zeta) = \widehat{\boldsymbol{r}}_j(\zeta - 1) \pm \omega_j^{-1} \int_\zeta^1 d\zeta' \, \boldsymbol{\lambda}'(\zeta')(\zeta' - \zeta) \ , \qquad j = 1,2 \ , \tag{7.5.98}$$

similar to (7.5.28).

The variable $\boldsymbol{v}'$, and another variable $\overline{\boldsymbol{v}}'$, defined by

$$\overline{\boldsymbol{v}}' = \tfrac{1}{2}(\omega_1^{-1}\boldsymbol{s}_1 + \omega_2^{-1}\boldsymbol{s}_2) = \omega^{-1}(\overline{\boldsymbol{v}} + \tfrac{1}{4}\Delta\omega\boldsymbol{v}) \ , \tag{7.5.99}$$

constitute a new set of variables to substitute for $\boldsymbol{v}$ and $\overline{\boldsymbol{v}}$, with the element $d\boldsymbol{v}'d\overline{\boldsymbol{v}}' = \omega^{-2}\, d\boldsymbol{v}d\overline{\boldsymbol{v}}$.

Hence the functional integral representation (7.5.16) for M_{22}, as well as the expressions (7.5.8,21–23) for $S^{(2)}[\boldsymbol{\lambda}', \boldsymbol{\rho}']$, hold true except that $\langle\widetilde{V}^{(2)}(\zeta)\rangle$ (7.5.24) is changed to

$$\langle\widetilde{V}^{(2)}(\zeta)\rangle = V^{(2)}\left[\overline{\boldsymbol{r}}_1(\zeta) + \omega_1^{-1}\boldsymbol{\xi}(\zeta), \overline{\boldsymbol{r}}_2(\zeta) - \omega_2^{-1}\boldsymbol{\xi}(\zeta), \overline{\boldsymbol{\rho}}_{12} + \boldsymbol{\eta}(\zeta)\right] \ , \tag{7.5.100}$$

in terms of the function $\boldsymbol{\eta}(\zeta)$ by the same definition (7.5.19). Consequently, the "equations of motion" (7.5.38,39) for $\boldsymbol{\xi}(\zeta)$ and $\boldsymbol{\eta}(\zeta)$ also remain the same, as well as the resulting expressions (7.5.53) for $S^{(2)}(\boldsymbol{\lambda}', \boldsymbol{\rho}')$; hence, when $\boldsymbol{r}_1 = \boldsymbol{r}_2 = 0$, the sum of the phase terms in (7.5.16) is written, using (7.5.56,57) with $\boldsymbol{v} \to \boldsymbol{v}'$, by

$$\begin{aligned}\boldsymbol{\lambda}' \cdot \boldsymbol{\rho}' &+ \sum_{j=1}^{2} \boldsymbol{\lambda}_j \cdot (\omega_j^{-1}\boldsymbol{s}_j - \boldsymbol{\rho}_j) \\ &= \boldsymbol{\lambda}'' \cdot (\boldsymbol{\rho}'' + \boldsymbol{v}' - \boldsymbol{\rho}_{12}) - \boldsymbol{\nu} \cdot \boldsymbol{\rho}'' + \overline{\boldsymbol{\nu}} \cdot (\overline{\boldsymbol{v}}' - \overline{\boldsymbol{\rho}}) \ ,\end{aligned} \tag{7.5.101}$$

which is the same equation as (7.5.61) except for the replacement $\boldsymbol{v}, \overline{\boldsymbol{v}} \to \boldsymbol{v}', \overline{\boldsymbol{v}}'$.

Thus the spectrum function (7.5.62) formally remains the same except for the above replacement, and $\widetilde{M}^0_{22}(\boldsymbol{\rho}''; \boldsymbol{v}, \overline{\boldsymbol{\nu}}) \to \widetilde{M}^0_{22}(\boldsymbol{\rho}''; \boldsymbol{v}', \overline{\boldsymbol{\nu}})$, defined by (7.2.55a,b)

$$\widetilde{M}^0_{22}(\boldsymbol{\rho}''; \boldsymbol{v}', \overline{\boldsymbol{\nu}}) = (2\pi)^{-4}\omega^2 \int d\boldsymbol{\nu}\, d\overline{\boldsymbol{v}}' \\ \times \exp\{-\mathrm{i}[\boldsymbol{\nu} \cdot \boldsymbol{\rho}'' - \overline{\boldsymbol{\nu}} \cdot \overline{\boldsymbol{v}}']\} \widetilde{M}^0_{22}(\boldsymbol{s}, \boldsymbol{\lambda}, \zeta = 0) \ , \quad (7.5.102)$$

and given by (7.5.91) for the collimated beam wave of (7.2.32) where

$$\boldsymbol{s}_1 = \omega_1(\overline{\boldsymbol{v}}' + \tfrac{1}{2}\boldsymbol{v}'), \qquad \boldsymbol{s}_2 = \omega_2(\overline{\boldsymbol{v}}' - \tfrac{1}{2}\boldsymbol{v}') \ . \quad (7.5.103)$$

For example, expression (7.5.65) for a spherical wave holds true, as well as the first order expression (7.5.64), with $\boldsymbol{v} \to \boldsymbol{v}'$, and an additional factor ω^2 (7.5.92b). Here,

$$\langle \widetilde{V}^{(2)}(\zeta) \rangle_0 = \langle \widetilde{V}_1(\zeta) \rangle_0 + \langle \widetilde{V}_2(\zeta) \rangle_0 + \langle \widetilde{V}_{12}(\zeta) \rangle_0 \ , \quad (7.5.104)$$

where, from (7.5.83,84,90),

$$\langle \widetilde{V}_j(\zeta) \rangle_0 = \omega_j^2 \overline{V}[(\zeta - 1)\omega_j^{-1}(\pm\boldsymbol{\lambda}'' + \overline{\boldsymbol{\nu}}/2)], \quad j = 1, 2 \ , \quad (7.5.105)$$

$$\langle \widetilde{V}_{12}(\zeta) \rangle_0 = \sum_{\pm} \omega\{\overline{V}[\boldsymbol{\rho}'' + \zeta\boldsymbol{v}' \pm (\zeta - 1)(2\omega)^{-1}(\overline{\boldsymbol{\nu}} + \Delta\omega\boldsymbol{\lambda}'')] \\ - \overline{V}[\boldsymbol{\rho}'' + \zeta\boldsymbol{v}' \pm (\zeta - 1)\omega^{-1}(\boldsymbol{\lambda}'' + 4^{-1}\Delta\omega\overline{\boldsymbol{\nu}})]\} \ , \quad (7.5.106)$$

with the signs $\pm$ in (7.5.105), depending on whether $j = 1$ or 2.

7.5.4 Third Order Intensity Correlation Function

It is straightforward to generalize the previous method to an exact two-scale integral representation of the third and higher order intensity correlation functions. We begin by obtaining the (exact) integral representation of M_{33} for a phase screen, and the formally generalized version of it for a turbulent continuum, following the procedure of Sect. 7.5.2 and also limiting ourselves to the case of a plane wave.

From (7.2.29,30), $M_{33}(1)$ can be written by

$$M_{33}(1) = P \exp\Big\{ - \int_0^1 d\zeta\, V^{(3)}[\widehat{\boldsymbol{r}}_1(\zeta - 1), \widehat{\boldsymbol{r}}_2(\zeta - 1), \widehat{\boldsymbol{r}}_3(\zeta - 1); \\ \widehat{\boldsymbol{\rho}}_{12}(\zeta - 1), \widehat{\boldsymbol{\rho}}_{23}(\zeta - 1), \widehat{\boldsymbol{\rho}}_{31}(\zeta - 1)] \Big\} M^0_{33}(1) \quad (7.5.107)$$

with

$$V^{(3)}(\boldsymbol{r}_1, \boldsymbol{r}_2, \boldsymbol{r}_3; \boldsymbol{\rho}_{12}, \boldsymbol{\rho}_{23}, \boldsymbol{\rho}_{31}) = \sum_{j=1}^{3} \overline{V}(\boldsymbol{r}_j) + \sum_{i>j=1}^{3} \overline{V}_I(\boldsymbol{r}_i, \boldsymbol{r}_j, \boldsymbol{\rho}_{ij}) \ . \tag{7.5.108}$$

Hence, when a phase screen with a width $\Delta\zeta$ is located at ζ, we obtain an integral representation of $M_{33}(1)$ similar to (7.2.43), as

$$\begin{aligned} M_{33}(1) = (2\pi)^{-2\times 3} \int d\boldsymbol{\lambda}'_{12} d\boldsymbol{\lambda}'_{23} d\boldsymbol{\lambda}'_{31} \int d\boldsymbol{\rho}'_{12} d\boldsymbol{\rho}'_{23} d\boldsymbol{\rho}'_{31} \\ \times \exp\{-\Delta\zeta V^{(3)}[\widehat{\boldsymbol{r}}_1(\zeta-1), \widehat{\boldsymbol{r}}_2(\zeta-1), \widehat{\boldsymbol{r}}_3(\zeta-1); \boldsymbol{\rho}'_{12}, \boldsymbol{\rho}'_{23}, \boldsymbol{\rho}'_{31}]\} \\ \times \exp\left\{\sum_{i>j=1}^{3} \mathrm{i}\boldsymbol{\lambda}'_{ij} \cdot [\boldsymbol{\rho}'_{ij} - \widehat{\boldsymbol{\rho}}_{ij}(\zeta-1)]\right\} M^0_{33}(1) \ . \end{aligned} \tag{7.5.109}$$

Here, $\boldsymbol{\lambda}'_{ij} = -\boldsymbol{\lambda}'_{ji}$ and $\boldsymbol{\rho}'_{ij} = -\boldsymbol{\rho}'_{ji}$; and, in the present case of a plane wave where $M^0_{33} = I^3_0$ is a constant, the operators $\widehat{\boldsymbol{r}}_j(\zeta-1)$ involved can be replaced by their eigenvalues, $\overline{\boldsymbol{r}}'_j(\zeta)$, given by

$$\overline{\boldsymbol{r}}'_j(\zeta) = \boldsymbol{r}_j + (\zeta-1)\boldsymbol{\lambda}'_j, \qquad j = 1, 2, 3 \ , \tag{7.5.110}$$

similar to (7.2.51,50), where

$$\boldsymbol{\lambda}'_1 = \boldsymbol{\lambda}'_{12} - \boldsymbol{\lambda}'_{31}, \quad \boldsymbol{\lambda}'_2 = \boldsymbol{\lambda}'_{23} - \boldsymbol{\lambda}'_{12}, \quad \boldsymbol{\lambda}'_3 = \boldsymbol{\lambda}'_{31} - \boldsymbol{\lambda}'_{23} \ , \tag{7.5.111}$$

$$\boldsymbol{\lambda}'_1 + \boldsymbol{\lambda}'_2 + \boldsymbol{\lambda}'_3 = 0 \tag{7.5.112}$$

[which should not be confused with the unprimed $\boldsymbol{\lambda}_j$'s to be used exclusively for the Fourier variables of $\widetilde{M^0_{33}}$ (7.5.165)].

Here, instead of $\boldsymbol{\lambda}'_{12}$ and $\boldsymbol{\lambda}'_{23}$, we can conveniently use $\boldsymbol{\lambda}'_1$ and $\boldsymbol{\lambda}'_2$ as alternative integration variables in (7.5.109), so that the first exponential factor of $V^{(3)}$ becomes perfectly free from the variable $\boldsymbol{\lambda}'_{31}$, in consequence of (7.5.112); thus the $\boldsymbol{\lambda}'_{31}$ integration yields a factor from the second exponential as

$$\begin{aligned} \delta(\boldsymbol{\rho}'_{12} + \boldsymbol{\rho}'_{23} + \boldsymbol{\rho}'_{31}) \\ \times \exp\{\mathrm{i}[(\boldsymbol{\rho}'_{12} - \boldsymbol{\rho}_{12}) \cdot \boldsymbol{\lambda}'_1 + (\boldsymbol{\rho}'_{23} - \boldsymbol{\rho}_{23}) \cdot (\boldsymbol{\lambda}'_1 + \boldsymbol{\lambda}'_2)]\} \ . \end{aligned} \tag{7.5.113}$$

Hence (7.5.109) is reduced, when $\boldsymbol{r}_1 = \boldsymbol{r}_2 = \boldsymbol{r}_3 = 0$, to

$$\begin{aligned} M_{33}(1) = I^3_0 (2\pi)^{-4} \int d\boldsymbol{\lambda}'_1 d\boldsymbol{\lambda}'_2 \int \boldsymbol{\rho}'_{12}\, d\boldsymbol{\rho}'_{23} \\ \times \exp\{-\mathrm{i}[\boldsymbol{\lambda}'_1 \cdot (\boldsymbol{\rho}'_{31} - \boldsymbol{\rho}_{31}) - \boldsymbol{\lambda}'_2 \cdot (\boldsymbol{\rho}'_{23} - \boldsymbol{\rho}_{23})]\} \\ \times \exp\{-\Delta\zeta V^{(3)}[(\zeta-1)\boldsymbol{\lambda}'_1, (\zeta-1)\boldsymbol{\lambda}'_2, -(\zeta-1)(\boldsymbol{\lambda}'_1 + \boldsymbol{\lambda}'_2); \\ \boldsymbol{\rho}'_{12}, \boldsymbol{\rho}'_{23}, -\boldsymbol{\rho}'_{12} - \boldsymbol{\rho}'_{23}]\} \ , \end{aligned} \tag{7.5.114}$$

corresponding to (7.2.53, 59). And, for a turbulent continuum, the zero order expression is obtained therefrom with the replacement (7.2.56).

An exact functional integral representation of M_{33} can also be constructed (7.5.120–130) in the same way as (7.5.16) was for M_{22}; the two-scale expression is derived therefrom, based on the variational principle, and given by (7.5.161–164) with (7.5.165) for the collimated beam wave. In the case of a spherical wave, the exact (7.5.161) is reduced to

$$
\begin{aligned}
M_{33}(1) = {} & (4\pi L)^{-6}(2\pi)^{-4} \int \boldsymbol{\lambda}_1'' d\boldsymbol{\lambda}_2'' \int d\boldsymbol{v}_{12} d\boldsymbol{v}_{23} \\
& \times \exp\{-\mathrm{i}[\boldsymbol{\lambda}_1'' \cdot (\boldsymbol{v}_{31} - \boldsymbol{\rho}_{31}) - \boldsymbol{\lambda}_2'' \cdot (\boldsymbol{v}_{23} - \boldsymbol{\rho}_{23})]\} \\
& \times \exp\Bigg\{ -\int_0^1 d\zeta \, \langle \widetilde{V}^{(3)}(\zeta) \rangle - \sum_{i>j=1}^{3} \mathrm{i} \int_{\zeta>\zeta'>0}^{1} d\zeta d\zeta' \\
& \times (\zeta - \zeta') \langle \widetilde{\boldsymbol{V}}^{(3)}_{\eta ij}(\zeta) \rangle \cdot \langle \widetilde{\boldsymbol{V}}^{(3)}_{\xi ij}(\zeta) \rangle \Bigg\} ,
\end{aligned}
\tag{7.5.115}
$$

being an expression corresponding to (7.5.65) for M_{22}. Here,

$$
\begin{aligned}
\langle \widetilde{V}^{(3)}(\zeta) \rangle = V^{(3)} [& (\zeta - 1)\boldsymbol{\lambda}_1'' + \Delta\boldsymbol{\xi}_1(\zeta), (\zeta - 1)\boldsymbol{\lambda}_2'' + \Delta\boldsymbol{\xi}_2(\zeta), \\
& (\zeta - 1)\boldsymbol{\lambda}_3'' + \Delta\boldsymbol{\xi}_3(\zeta); \zeta \boldsymbol{v}_{12} + \Delta\boldsymbol{\eta}_{12}(\zeta), \\
& \zeta \boldsymbol{v}_{23} + \Delta\boldsymbol{\eta}_{23}(\zeta), \zeta \boldsymbol{v}_{31} + \Delta\boldsymbol{\eta}_{31}(\zeta)] ,
\end{aligned}
\tag{7.5.116}
$$

where

$$
\begin{aligned}
\boldsymbol{\lambda}_3'' &= -\boldsymbol{\lambda}_1'' - \boldsymbol{\lambda}_2'', & \boldsymbol{v}_{31} &= -\boldsymbol{v}_{12} - \boldsymbol{v}_{23} , \\
\Delta\boldsymbol{\xi}_3 &= -\Delta\boldsymbol{\xi}_1 - \Delta\boldsymbol{\xi}_2, & \Delta\boldsymbol{\eta}_{31} &= -\Delta\boldsymbol{\eta}_{12} - \Delta\boldsymbol{\eta}_{23} ;
\end{aligned}
\tag{7.5.117}
$$

and $\Delta\boldsymbol{\xi}_j$ and $\Delta\boldsymbol{\eta}_{ij}$ are taken from (7.5.142, 138, 139), as

$$
\Delta\boldsymbol{\eta}_{ij}(\zeta) = -\mathrm{i} \int d\zeta' \, G_\eta(\zeta|\zeta') \left[\frac{\partial}{\partial \Delta\boldsymbol{\xi}_i} - \frac{\partial}{\partial \Delta\boldsymbol{\xi}_j} \right] \langle \widetilde{V}^{(3)}(\zeta') \rangle ,
\tag{7.5.118}
$$

$$
\begin{aligned}
\Delta\boldsymbol{\xi}_1(\zeta) &= -\mathrm{i} \int d\zeta' \, G_\xi(\zeta|\zeta') \left[\frac{\partial}{\partial \Delta\boldsymbol{\eta}_{12}} - \frac{\partial}{\partial \Delta\boldsymbol{\eta}_{31}} \right] \langle \widetilde{V}^{(3)}(\zeta') \rangle , \\
\Delta\boldsymbol{\xi}_2(\zeta) &= -\mathrm{i} \int d\zeta' \, G_\xi(\zeta|\zeta') \left[\frac{\partial}{\partial \Delta\boldsymbol{\eta}_{23}} - \frac{\partial}{\partial \Delta\boldsymbol{\eta}_{12}} \right] \langle \widetilde{V}^{(3)}(\zeta') \rangle, \dots .
\end{aligned}
\tag{7.5.119}
$$

To the zero order approximation of $\Delta\boldsymbol{\xi}_j = \Delta\boldsymbol{\eta}_{ij} = 0$, (7.5.115) becomes the same as (7.5.114) for a plane wave with $\zeta \boldsymbol{v}_{ij} \to \boldsymbol{v}_{ij}$ in $V^{(3)}[\,]$ of (7.5.116), besides $\Delta\zeta \to \int_0^1 d\zeta$; and the expression resulting therefrom is exact when including the terms of $\Delta\boldsymbol{\xi}_j$ and $\Delta\boldsymbol{\eta}_{ij}$.

a) Derivation of Two-Scale Integral Representation for $M_{33}(1)$

By following the procedure of Sect. 7.5.1, we can obtain an exact functional integral representation of M_{33}, corresponding to (7.5.16) for M_{22} as

$$\begin{aligned} M_{33}(\boldsymbol{r},\boldsymbol{\rho};\zeta=1) &= (2\pi)^{-2\times 6}\int d\boldsymbol{s}d\boldsymbol{\lambda} \\ &\times \exp\left[\sum_{j=1}^{3} \mathrm{i}(\boldsymbol{s}_j\cdot\boldsymbol{\lambda}_j - \boldsymbol{s}_j\cdot\boldsymbol{r}_j - \boldsymbol{\lambda}_j\cdot\boldsymbol{\rho}_j)\right] \\ &\times \prod_{i>j=1}^{3}\int d[\boldsymbol{\lambda}'_{ij}]d[\boldsymbol{\rho}'_{ij}]\exp\left[S^{(3)}(\boldsymbol{r},\boldsymbol{\rho}|\boldsymbol{\lambda}',\boldsymbol{\rho}'|\boldsymbol{s},\boldsymbol{\lambda})\right] \\ &\times \widetilde{M}^{0}_{33}(\boldsymbol{s},\boldsymbol{\lambda};\zeta=0)\ . \end{aligned} \tag{7.5.120}$$

Here, $S^{(3)}(\boldsymbol{r},\boldsymbol{\rho}|\boldsymbol{\lambda}',\boldsymbol{\rho}'|\boldsymbol{s},\boldsymbol{\lambda}) \equiv S^{(3)}[\boldsymbol{\lambda}',\boldsymbol{\rho}']$ is a function of the coordinates $\boldsymbol{r}=(\boldsymbol{r}_1,\boldsymbol{r}_2,\boldsymbol{r}_3)$, $\boldsymbol{\rho}=(\boldsymbol{\rho}_1,\boldsymbol{\rho}_2,\boldsymbol{\rho}_3)$, and the Fourier variables $\boldsymbol{s}=(\boldsymbol{s}_1,\boldsymbol{s}_2,\boldsymbol{s}_3)$, $\boldsymbol{\lambda}=(\boldsymbol{\lambda}_1,\boldsymbol{\lambda}_2,\boldsymbol{\lambda}_3)$, as well as a functional of $\boldsymbol{\lambda}'_{ij}(\zeta)=-\boldsymbol{\lambda}'_{ji}(\zeta)$ and $\boldsymbol{\rho}'_{ij}(\zeta)=-\boldsymbol{\rho}'_{ji}(\zeta)$, given by [see (7.5.8–13)]

$$S^{(3)}[\boldsymbol{\lambda}',\boldsymbol{\rho}'] = \int_0^1 d\zeta\left\{\sum_{i>j>1}^{3}\mathrm{i}\boldsymbol{\lambda}'_{ij}(\zeta)\cdot[\boldsymbol{\rho}'_{ij}(\zeta)-\overline{\boldsymbol{\rho}}_{ij}(\zeta)]-\langle\widetilde{V}^{(3)}(\zeta)\rangle\right\}\ . \tag{7.5.121}$$

Expression (7.5.121) can be written in a more symmetrical form in terms of the functions $\boldsymbol{\xi}_{ij}(\zeta)$ and $\boldsymbol{\eta}_{ij}(\zeta)$ defined by

$$\boldsymbol{\xi}_{ij}(\zeta) = \int d\zeta'\, G(\zeta|\zeta')\boldsymbol{\lambda}'_{ij}(\zeta')\ , \tag{7.5.122}$$

$$\boldsymbol{\eta}_{ij}(\zeta) = \boldsymbol{\rho}'_{ij}(\zeta) - \overline{\boldsymbol{\rho}}_{ij}(\zeta)\ , \tag{7.5.123}$$

similar to (7.5.13,19), and also the associated $\boldsymbol{\xi}_j(\zeta)$, $j=1$, 2, 3, defined by

$$\boldsymbol{\xi}_1=\boldsymbol{\xi}_{12}-\boldsymbol{\xi}_{31},\quad \boldsymbol{\xi}_2=\boldsymbol{\xi}_{23}-\boldsymbol{\xi}_{12},\quad \boldsymbol{\xi}_3=\boldsymbol{\xi}_{31}-\boldsymbol{\xi}_{23}\ , \tag{7.5.124}$$

to write the $\langle\widetilde{V}^{(3)}(\zeta)\rangle$ by [compare to (7.5.24)]

$$\begin{aligned} \langle\widetilde{V}^{(3)}\rangle = V^{(3)}(&\overline{\boldsymbol{r}}_1+\boldsymbol{\xi}_1,\overline{\boldsymbol{r}}_2+\boldsymbol{\xi}_2,\overline{\boldsymbol{r}}_3+\boldsymbol{\xi}_3; \\ &\overline{\boldsymbol{\rho}}_{12}+\boldsymbol{\eta}_{12},\overline{\boldsymbol{\rho}}_{23}+\boldsymbol{\eta}_{23},\overline{\boldsymbol{\rho}}_{31}+\boldsymbol{\eta}_{31})\ . \end{aligned} \tag{7.5.125}$$

Here,

$$\overline{\boldsymbol{r}}_j(\zeta) = \boldsymbol{r}_j + (\zeta-1)\boldsymbol{\lambda}_j,\qquad j=1,2,3\ , \tag{7.5.126}$$

being the same as defined by (7.5.11), and $\overline{\boldsymbol{\rho}}_{ij}(\zeta)$ is now defined by

$$\overline{\boldsymbol{\rho}}_{ij}(\zeta) = \boldsymbol{\rho}_{ij} + (\zeta-1)\boldsymbol{v}_{ij},\qquad \boldsymbol{v}_{ij}=\boldsymbol{s}_i-\boldsymbol{s}_j\ . \tag{7.5.127}$$

Hence we obtain

$$S^{(3)}[\boldsymbol{\lambda}', \boldsymbol{\rho}'] = \int_0^1 d\zeta \left[\sum_{i>j} \mathrm{i}\ddot{\boldsymbol{\xi}}_{ij} \cdot \boldsymbol{\eta}_{ij} - \langle \widetilde{V}^{(3)} \rangle \right] \tag{7.5.128}$$

$$= \sum_{i>j} \mathrm{i}\dot{\boldsymbol{\xi}}_{ij} \cdot \boldsymbol{\eta}_{ij} \Big|_{\zeta=0}^{1} + \int_0^1 d\zeta\, \mathcal{L}^{(3)}(\dot{\boldsymbol{\xi}}, \dot{\boldsymbol{\eta}}, \boldsymbol{\xi}, \boldsymbol{\eta}, \zeta) \ , \tag{7.5.129}$$

where $\boldsymbol{\xi} = \{\boldsymbol{\xi}_{ij}\}$, $\boldsymbol{\eta} = \{\boldsymbol{\eta}_{ij}\}$, $i, j = 1, 2, 3$, and

$$\mathcal{L}^{(3)} = -\sum_{i>j} \mathrm{i}\dot{\boldsymbol{\xi}}_{ij} \cdot \dot{\boldsymbol{\eta}}_{ij} - \langle \widetilde{V}^{(3)}(\boldsymbol{\xi}, \boldsymbol{\eta}, \zeta) \rangle \ , \tag{7.5.130}$$

being written in the same form as (7.5.21,22). Here, in view of (7.5.122), $\boldsymbol{\xi}_{ij}(\zeta)$ is subject to the boundary conditions

$$\boldsymbol{\xi}_{ij}(\zeta = 1) = \dot{\boldsymbol{\xi}}_{ij}(\zeta = 0) = 0 \ , \tag{7.5.131}$$

with the relation

$$\dot{\boldsymbol{\xi}}_{ij}(\zeta = 1) = \int_0^1 d\zeta\, \boldsymbol{\lambda}'_{ij}(\zeta) \ . \tag{7.5.132}$$

Hence the "equations of motion" to find the $\boldsymbol{\xi}_{ij}(\zeta)$ and $\boldsymbol{\eta}_{ij}(\zeta)$ of letting $S^{(3)}[\boldsymbol{\lambda}', \boldsymbol{\rho}']$ be stationary, are also obtained based on the variational principle in the same form as

$$\mathrm{i}\ddot{\boldsymbol{\xi}}_{ij} = \langle \widetilde{\boldsymbol{V}}^{(3)}_{\eta ij} \rangle, \qquad \mathrm{i}\ddot{\boldsymbol{\eta}}_{ij} = \langle \widetilde{\boldsymbol{V}}^{(3)}_{\xi ij} \rangle \ , \tag{7.5.133}$$

with

$$\langle \widetilde{\boldsymbol{V}}^{(3)}_{\eta ij} \rangle = \frac{\partial}{\partial \boldsymbol{\eta}_{ij}} \langle \widetilde{V}^{(3)} \rangle, \qquad \langle \widetilde{\boldsymbol{V}}^{(3)}_{\xi ij} \rangle = \frac{\partial}{\partial \boldsymbol{\xi}_{ij}} \langle \widetilde{V}^{(3)} \rangle \ ; \tag{7.5.134}$$

and they are subject to boundary conditions

$$\boldsymbol{\xi}_{ij}(\zeta = 1) = 0, \qquad \dot{\boldsymbol{\xi}}_{ij}(\zeta = 1) = \boldsymbol{\lambda}'_{ij} \ , \tag{7.5.135}$$

$$\boldsymbol{\eta}_{ij}(\zeta = 0) = \boldsymbol{\rho}'_{ij}, \qquad \dot{\boldsymbol{\eta}}_{ij}(\zeta = 0) = 0 \ , \tag{7.5.136}$$

with arbitrary constants $\boldsymbol{\lambda}'_{ij}$ and $\boldsymbol{\rho}'_{ij}$ (not to be confused with the functions $\boldsymbol{\lambda}'_{ij}(\zeta)$ and $\boldsymbol{\rho}'_{ij}(\zeta)$), in the same form as (7.5.42,43). Hence the solutions are also obtained in the form of (7.5.50,51), in terms of the Green's functions G_ξ and G_η.

Thus $S^{(3)}[\boldsymbol{\lambda}', \boldsymbol{\rho}']$ of (7.5.128) becomes a function only of the boundary values $\boldsymbol{\lambda}'_{ij}$ and $\boldsymbol{\rho}'_{ij}$, say $S^{(3)}(\boldsymbol{\lambda}', \boldsymbol{\rho}')$, and is given, with the additional term,

$$S_0^{(3)} \equiv \mathrm{i}\int_{-0}^{+0} d\zeta \sum_{i>j} \ddot{\boldsymbol{\xi}}_{ij}\cdot\boldsymbol{\eta}_{ij} = \mathrm{i}\sum_{i>j}\dot{\boldsymbol{\xi}}_{ij}\cdot\boldsymbol{\eta}_{ij}(\zeta=+0) \ , \quad (7.5.137)$$

$$\dot{\boldsymbol{\xi}}_{ij}(\zeta=+0) = \mathrm{i}\int_0^1 d\zeta' \,\langle \widetilde{\boldsymbol{V}}^{(3)}_{\eta ij}(\zeta')\rangle + \boldsymbol{\lambda}'_{ij} \ ,$$

similar to (7.5.41,52), and in terms of the functions $\Delta\boldsymbol{\xi}_{ij}(\zeta)$ and $\Delta\boldsymbol{\eta}_{ij}(\zeta)$ defined by

$$\Delta\boldsymbol{\xi}_{ij}(\zeta) = -\mathrm{i}\int d\zeta'\, G_\xi(\zeta|\zeta')\langle \widetilde{\boldsymbol{V}}^{(3)}_{\eta ij}(\zeta')\rangle \ , \quad (7.5.138)$$

$$\Delta\boldsymbol{\eta}_{ij}(\zeta) = -\mathrm{i}\int d\zeta'\, G_\eta(\zeta|\zeta')\langle \widetilde{\boldsymbol{V}}^{(3)}_{\xi ij}(\zeta')\rangle \ , \quad (7.5.139)$$

by

$$S^{(3)}(\boldsymbol{\lambda}',\boldsymbol{\rho}') = \sum_{i>j}\mathrm{i}\boldsymbol{\lambda}'_{ij}\cdot\boldsymbol{\rho}'_{ij} + \int_0^1 d\zeta \left[\sum_{i>j}\langle \widetilde{\boldsymbol{V}}^{(3)}_{\eta ij}\rangle\cdot\Delta\boldsymbol{\eta}_{ij} - \langle \widetilde{V}^{(3)}\rangle\right] \ . \quad (7.5.140)$$

Here, when $\boldsymbol{r}_1 = \boldsymbol{r}_2 = \boldsymbol{r}_3 = 0$, $\langle \widetilde{V}^{(3)}\rangle$ from (7.5.125) can be written as [cf. (7.5.60)]

$$\begin{aligned}\langle \widetilde{V}^{(3)}(\zeta)\rangle = V^{(3)}[&(\zeta-1)(\boldsymbol{\lambda}''_1+\overline{\boldsymbol{\nu}}/3)+\Delta\boldsymbol{\xi}_1(\zeta), (\zeta-1)(\boldsymbol{\lambda}''_2+\overline{\boldsymbol{\nu}}/3)+\Delta\boldsymbol{\xi}_2(\zeta),\\ &(\zeta-1)(\boldsymbol{\lambda}''_3+\overline{\boldsymbol{\nu}}/3)+\Delta\boldsymbol{\xi}_3(\zeta); \boldsymbol{\rho}''_{12}+\zeta\boldsymbol{v}_{12}+\Delta\boldsymbol{\eta}_{12}(\zeta),\\ &\boldsymbol{\rho}''_{23}+\zeta\boldsymbol{v}_{23}+\Delta\boldsymbol{\eta}_{23}(\zeta), \boldsymbol{\rho}''_{31}+\zeta\boldsymbol{v}_{31}+\Delta\boldsymbol{\eta}_{31}(\zeta)] \ , \end{aligned} \quad (7.5.141)$$

where

$$\begin{aligned}\Delta\boldsymbol{\xi}_1 &= \Delta\boldsymbol{\xi}_{12} - \Delta\boldsymbol{\xi}_{31} \ ,\\ \Delta\boldsymbol{\xi}_2 &= \Delta\boldsymbol{\xi}_{23} - \Delta\boldsymbol{\xi}_{12} \ ,\\ \Delta\boldsymbol{\xi}_3 &= \Delta\boldsymbol{\xi}_{31} - \Delta\boldsymbol{\xi}_{23} \ ;\end{aligned} \quad (7.5.142)$$

and the other new variables introduced are defined as follows:

$$\boldsymbol{\rho}''_{ij} = \boldsymbol{\rho}'_{ij} - \boldsymbol{v}_{ij} + \boldsymbol{\rho}_{ij}, \qquad \boldsymbol{v}_{ij} = \boldsymbol{s}_i - \boldsymbol{s}_j \ , \quad (7.5.143)$$

$$\boldsymbol{\lambda}''_{ij} = \boldsymbol{\lambda}'_{ij} + \boldsymbol{\nu}_{ij}, \qquad \boldsymbol{\lambda}''_j = \boldsymbol{\lambda}'_j + \boldsymbol{\nu}_j \ . \quad (7.5.144)$$

Here, $\boldsymbol{\rho}''_{ij}$ corresponds to $\boldsymbol{\rho}''$ of (7.5.57), and $\boldsymbol{\lambda}''_j$ can be defined by (7.5.111) with $\boldsymbol{\lambda}'_{ij} \to \boldsymbol{\lambda}''_{ij}$, while $\boldsymbol{\nu}_{ij}$, $\boldsymbol{\nu}_j$, and $\overline{\boldsymbol{\nu}}$ are connected to the Fourier variables $\boldsymbol{\lambda}_j$, $j = 1, 2, 3$, according to

$$\boldsymbol{\lambda}_j = \boldsymbol{\nu}_j + \overline{\boldsymbol{\nu}}/3, \qquad \boldsymbol{\nu}_1 + \boldsymbol{\nu}_2 + \boldsymbol{\nu}_3 = 0 \ . \quad (7.5.145)$$

Here,

$$\overline{\boldsymbol{\nu}} = \boldsymbol{\lambda}_1 + \boldsymbol{\lambda}_2 + \boldsymbol{\lambda}_3 \ , \tag{7.5.146}$$

$$\boldsymbol{\nu}_1 = \boldsymbol{\nu}_{12} - \boldsymbol{\nu}_{31}, \quad \boldsymbol{\nu}_2 = \boldsymbol{\nu}_{23} - \boldsymbol{\nu}_{12}, \quad \boldsymbol{\nu}_3 = \boldsymbol{\nu}_{31} - \boldsymbol{\nu}_{23} \ , \tag{7.5.147}$$

where

$$\boldsymbol{\nu}_{ij} = \tfrac{1}{3}(\boldsymbol{\lambda}_i - \boldsymbol{\lambda}_j) = \tfrac{1}{3}(\boldsymbol{\nu}_i - \boldsymbol{\nu}_j), \qquad \boldsymbol{\nu}_{12} + \boldsymbol{\nu}_{23} + \boldsymbol{\nu}_{31} = 0 \ . \tag{7.5.148}$$

Also, we use a similar expression for the coordinates $\boldsymbol{\rho}_j$,

$$\boldsymbol{\rho}_j = \overline{\boldsymbol{\rho}}_j + \overline{\boldsymbol{\rho}}, \quad \overline{\boldsymbol{\rho}} = \tfrac{1}{3}(\boldsymbol{\rho}_1 + \boldsymbol{\rho}_2 + \boldsymbol{\rho}_3), \quad \overline{\boldsymbol{\rho}}_1 + \overline{\boldsymbol{\rho}}_2 + \overline{\boldsymbol{\rho}}_3 = 0 \ , \tag{7.5.149}$$

$$\overline{\boldsymbol{\rho}}_1 = \tfrac{1}{3}(\boldsymbol{\rho}_{12} - \boldsymbol{\rho}_{31}), \quad \overline{\boldsymbol{\rho}}_2 = \tfrac{1}{3}(\boldsymbol{\rho}_{23} - \boldsymbol{\rho}_{12}), \quad \overline{\boldsymbol{\rho}}_3 = \tfrac{1}{3}(\boldsymbol{\rho}_{31} - \boldsymbol{\rho}_{23}) \ , \tag{7.5.150}$$

and the corresponding expression of $\boldsymbol{s}_j$,

$$\boldsymbol{s}_j = \overline{\boldsymbol{v}}_j + \overline{\boldsymbol{v}}, \quad \overline{\boldsymbol{v}} = \tfrac{1}{3}(\boldsymbol{s}_1 + \boldsymbol{s}_2 + \boldsymbol{s}_3), \quad \overline{\boldsymbol{v}}_1 + \overline{\boldsymbol{v}}_2 + \overline{\boldsymbol{v}}_3 = 0 \ , \tag{7.5.151}$$

$$\overline{\boldsymbol{v}}_1 = \tfrac{1}{3}(\boldsymbol{v}_{12} - \boldsymbol{v}_{31}), \quad \overline{\boldsymbol{v}}_2 = \tfrac{1}{3}(\boldsymbol{v}_{23} - \boldsymbol{v}_{12}), \quad \overline{\boldsymbol{v}}_3 = \tfrac{1}{3}(\boldsymbol{v}_{31} - \boldsymbol{v}_{23}) \ , \tag{7.5.152}$$

so that we can write

$$\sum_{j=1}^{3} \boldsymbol{\lambda}_j \cdot \boldsymbol{\rho}_j = \sum_{j=1}^{3} \boldsymbol{\nu}_j \cdot \overline{\boldsymbol{\rho}}_j + \overline{\boldsymbol{\nu}} \cdot \overline{\boldsymbol{\rho}} \ , \tag{7.5.153}$$

$$\sum_{j=1}^{3} \boldsymbol{\lambda}_j \cdot \boldsymbol{s}_j = \sum_{j=1}^{3} \boldsymbol{\nu}_j \cdot \overline{\boldsymbol{v}}_j + \overline{\boldsymbol{\nu}} \cdot \overline{\boldsymbol{v}} \ , \tag{7.5.154}$$

Hence the whole phase term involved in (7.5.120) with $S^{(3)}$ of (7.5.140) can be written, when $\boldsymbol{r}_j = 0$, as

$$\sum_{i>j} \boldsymbol{\lambda}'_{ij} \cdot \boldsymbol{\rho}'_{ji} + \sum_{j} \boldsymbol{\lambda}_j \cdot (\boldsymbol{s}_j - \boldsymbol{\rho}_j)$$
$$= \sum_{i>j} [\boldsymbol{\lambda}''_{ij} \cdot (\boldsymbol{\rho}''_{ij} + \boldsymbol{v}_{ij} - \boldsymbol{\rho}_{ij}) - \boldsymbol{\nu}_{ij} \cdot \boldsymbol{\rho}''_{ji}] + \overline{\boldsymbol{\nu}} \cdot (\overline{\boldsymbol{v}} - \overline{\boldsymbol{\rho}}) \ , \tag{7.5.155}$$

in terms of $\boldsymbol{\rho}''_{ij}$ and $\boldsymbol{\lambda}''_{ij}$ of (7.5.143,144), and in the same form as (7.5.61). While, from (7.5.54),

$$\prod_{i>j=1}^{3} d[\boldsymbol{\lambda}'_{ij}] d[\boldsymbol{\rho}'_{ij}] = (2\pi)^{-2\times 3} \prod_{i>j=1}^{3} d\boldsymbol{\lambda}'_{ij} d\boldsymbol{\rho}'_{ij} \ , \tag{7.5.156}$$

and, since $\langle \widetilde{V}^{(3)} \rangle$ of (7.5.141) is dependent on $\boldsymbol{\lambda}'_{ij}$ only through $\boldsymbol{\lambda}''_{ij}$, we can conveniently change the variables $\boldsymbol{\lambda}'_{ij} \to \boldsymbol{\lambda}''_{ij} \to \boldsymbol{\lambda}''_j$ with the element of variables,

$$d\boldsymbol{\lambda}'_{12}d\boldsymbol{\lambda}'_{23}d\boldsymbol{\lambda}'_{31} = \delta(\boldsymbol{\lambda}''_1 + \boldsymbol{\lambda}''_2 + \boldsymbol{\lambda}''_3)\, d\boldsymbol{\lambda}''_1 d\boldsymbol{\lambda}''_2 d\boldsymbol{\lambda}''_3 d\overline{\boldsymbol{\lambda}}'' \ , \tag{7.5.157}$$

where the additional variable $\overline{\boldsymbol{\lambda}}''$ is defined by

$$\overline{\boldsymbol{\lambda}}'' = \tfrac{1}{3}(\boldsymbol{\lambda}''_{12} + \boldsymbol{\lambda}''_{23} + \boldsymbol{\lambda}''_{31}), \qquad \boldsymbol{\lambda}''_{ij} = \tfrac{1}{3}(\boldsymbol{\lambda}''_i - \boldsymbol{\lambda}''_j) + \overline{\boldsymbol{\lambda}}'' \ . \tag{7.5.158}$$

The right-hand side of (7.5.155) can be written in terms of $\boldsymbol{\lambda}''_j$ by using the relation

$$\sum_{i>j} \boldsymbol{\lambda}''_{ij} \cdot \boldsymbol{\rho}''_{ij} = \sum_j \boldsymbol{\lambda}''_j \cdot \overline{\boldsymbol{\rho}}''_j + \overline{\boldsymbol{\lambda}}'' \cdot (\boldsymbol{\rho}''_{12} + \boldsymbol{\rho}''_{23} + \boldsymbol{\rho}''_{31}) \ , \tag{7.5.159}$$

where $\overline{\boldsymbol{\rho}}''_j$ is the same function of $\boldsymbol{\rho}''_{ij}$ as $\overline{\boldsymbol{\rho}}_j$ of (7.5.150) is of $\boldsymbol{\rho}_{ij}$.

Hence, since $\langle \widetilde{V}^{(3)} \rangle$ is perfectly free from $\overline{\boldsymbol{\lambda}}''$, (7.5.141), and the variable $\overline{\boldsymbol{\lambda}}''$ is involved in the integrand only through the exponential term $\mathrm{i}\overline{\boldsymbol{\lambda}}'' \cdot (\boldsymbol{\rho}''_{12} + \boldsymbol{\rho}''_{23} + \boldsymbol{\rho}''_{31})$, in view of (7.5.159), the $\overline{\boldsymbol{\lambda}}''$ integration yields the factor $\delta(\boldsymbol{\rho}''_{12} + \boldsymbol{\rho}''_{23} + \boldsymbol{\rho}''_{31})$, leading to the relation

$$\sum_{i>j} \boldsymbol{\lambda}''_{ij} \cdot (\boldsymbol{\rho}''_{ij} + \boldsymbol{v}_{ij} - \boldsymbol{\rho}_{ij}) = \sum_j \boldsymbol{\lambda}''_j \cdot (\overline{\boldsymbol{\rho}}''_j + \overline{\boldsymbol{v}}_j - \overline{\boldsymbol{\rho}}_j) \ , \tag{7.5.160}$$

in the right-hand side of (7.5.155).

Thus the two-scale expression derived from (7.5.120) can be written by a spectrum integral with respect to the relative coordinates $\overline{\boldsymbol{\rho}}_j$, $j = 1, 2, 3$, of (7.5.149), as

$$M_{33}(1) = (2\pi)^{-4} \int d\boldsymbol{\lambda}''_1 d\boldsymbol{\lambda}''_2 d\boldsymbol{\lambda}''_3\, \delta(\boldsymbol{\lambda}''_1 + \boldsymbol{\lambda}''_2 + \boldsymbol{\lambda}''_3) \times \exp\left[-\sum_{j=1}^{3} \mathrm{i}\boldsymbol{\lambda}''_j \cdot \overline{\boldsymbol{\rho}}_j \right] \widetilde{M}_{33}(\boldsymbol{\lambda}'', \overline{\boldsymbol{\rho}}) \ . \tag{7.5.161}$$

Here, $\widetilde{M}_{33}(\boldsymbol{\lambda}'', \overline{\boldsymbol{\rho}})$ means the spectrum function at the center of coordinates $\overline{\boldsymbol{\rho}}$, and, when $\boldsymbol{r}_1 = \boldsymbol{r}_2 = \boldsymbol{r}_3 = 0$, changing the variables of integration from $\boldsymbol{s} = (\boldsymbol{s}_1, \boldsymbol{s}_2, \boldsymbol{s}_3)$ and $\boldsymbol{\lambda} = (\boldsymbol{\lambda}_1, \boldsymbol{\lambda}_2, \boldsymbol{\lambda}_3)$ to $(\boldsymbol{v}_{ij}, \overline{\boldsymbol{v}})$ and $(\boldsymbol{\nu}_{ij}, \overline{\boldsymbol{\nu}})$, with the elements from (7.5.151, 148)

$$\begin{aligned} d\boldsymbol{s}_1 d\boldsymbol{s}_2 d\boldsymbol{s}_3 &= \delta(\boldsymbol{v}_{12} + \boldsymbol{v}_{23} + \boldsymbol{v}_{31})\, d\boldsymbol{v}_{12} d\boldsymbol{v}_{23} d\boldsymbol{v}_{31} d\overline{\boldsymbol{v}} \ , \\ d\boldsymbol{\lambda}_1 d\boldsymbol{\lambda}_2 d\boldsymbol{\lambda}_3 &= 3^2 \delta(\boldsymbol{\nu}_{12} + \boldsymbol{\nu}_{23} + \boldsymbol{\nu}_{31})\, d\boldsymbol{\nu}_{12} d\boldsymbol{\nu}_{23} d\boldsymbol{\nu}_{31} d\overline{\boldsymbol{\nu}} \ , \end{aligned} \tag{7.5.162}$$

leads to the expression [compare to (7.5.62)]

$$
\begin{aligned}
\widetilde{M}_{33}(\boldsymbol{\lambda}'',\overline{\boldsymbol{\rho}}) = \int d\boldsymbol{\rho}''_{12}d\boldsymbol{\rho}''_{23}d\boldsymbol{\rho}''_{31}\,\delta(\boldsymbol{\rho}''_{12}+\boldsymbol{\rho}''_{23}+\boldsymbol{\rho}''_{31}) \\
\times (2\pi)^{-6}\int d\boldsymbol{v}_{12}d\boldsymbol{v}_{23}d\boldsymbol{v}_{31}\,\delta(\boldsymbol{v}_{12}+\boldsymbol{v}_{23}+\boldsymbol{v}_{31}) \\
\times \int d\overline{\boldsymbol{\nu}}\exp\Bigg\{-\mathrm{i}\Bigg[\overline{\boldsymbol{\nu}}\cdot\overline{\boldsymbol{\rho}}-\sum_{j=1}^{3}\boldsymbol{\lambda}''_j\cdot(\overline{\boldsymbol{\rho}}''_j+\overline{\boldsymbol{v}}_j)\Bigg] \\
+\int_0^1 d\zeta\Bigg[\sum_{i>j=1}^{3}\langle\widetilde{\boldsymbol{V}}^{(3)}_{\eta ij}\rangle\cdot\Delta\boldsymbol{\eta}_{ij}-\langle\widetilde{V}^{(3)}\rangle\Bigg]\Bigg\} \\
\times \widetilde{M}^0_{33}(\boldsymbol{\rho}'';\boldsymbol{v},\overline{\boldsymbol{\nu}}) \ , \qquad (7.5.163)
\end{aligned}
$$

where $\langle\widetilde{V}^{(3)}\rangle$ is given by (7.5.141,142) in terms of $\Delta\boldsymbol{\xi}_{ij}$ and $\Delta\boldsymbol{\eta}_{ij}$ of (7.5.138, 139); $\overline{\boldsymbol{\rho}}''_j$ is the same function of $\boldsymbol{\rho}''_{ij}$ as $\overline{\boldsymbol{\rho}}_j$ is of $\boldsymbol{\rho}_{ij}$ [cf. (7.5.150)], and (7.2.55)

$$
\begin{aligned}
\widetilde{M}^0_{33}(\boldsymbol{\rho}'';\boldsymbol{v},\overline{\boldsymbol{\nu}}) = (2\pi)^{-6}\int d\boldsymbol{\nu}_{12}d\boldsymbol{\nu}_{23}d\boldsymbol{\nu}_{31}\ 3^2\delta(\boldsymbol{\nu}_{12}+\boldsymbol{\nu}_{23}+\boldsymbol{\nu}_{31}) \\
\times \int d\overline{\boldsymbol{v}}\exp\Bigg\{-\mathrm{i}\Bigg[\sum_{i>j=1}^{3}\boldsymbol{\nu}_{ij}\cdot\boldsymbol{\rho}''_{ij}-\overline{\boldsymbol{\nu}}\cdot\overline{\boldsymbol{v}}\Bigg]\Bigg\}\widetilde{M}^0_{33}(\boldsymbol{s},\boldsymbol{\lambda};\zeta=0)\ . \qquad (7.5.164)
\end{aligned}
$$

Here, in the case of a collimated beam wave of (7.2.32),

$$
\widetilde{M}^0_{33}(\boldsymbol{s},\boldsymbol{\lambda};\zeta=0) = I_0^3(2\pi\overline{b}^2)^6\exp\Bigg[-\sum_{j=1}^{3}\overline{b}^2\left(\boldsymbol{s}_j^2+\tfrac{1}{4}\boldsymbol{\lambda}_j^2\right)\Bigg]\ , \qquad (7.5.165)
$$

where the variables $\boldsymbol{s}_j$ and $\boldsymbol{\lambda}_j$ are given in terms of $(\boldsymbol{v}_{ij},\overline{\boldsymbol{v}})$ and $(\boldsymbol{\nu}_{ij},\overline{\boldsymbol{\nu}})$ by (7.5.151,152) and (7.5.145,147), respectively.

In the case of a spherical wave where $\widetilde{M}^0_{33}(\zeta=0)=(4\pi L)^{-6}(2\pi)^6$, (7.5.164) is reduced to

$$
\widetilde{M}^0_{33}(\boldsymbol{\rho}'';\boldsymbol{v},\overline{\boldsymbol{\nu}}) = (4\pi L)^{-6}(2\pi)^6\delta(\boldsymbol{\rho}''_{12}-\boldsymbol{\rho}''_{31})\delta(\boldsymbol{\rho}''_{23}-\boldsymbol{\rho}''_{31})\delta(\overline{\boldsymbol{\nu}})\ , \qquad (7.5.166)
$$

and expression (7.5.115) is obtained by the substitution in (7.5.163, 161).

7.5.5 Summary and Discussion

An exact functional integral representation for the two-point intensity correlation function was first obtained by solving the moment equation and was regarded as a multiscale representation (7.5.16). The variable functions of integration involved therein can be effectively limited to a set of functions so that the entire phase term of the integrand becomes stationary against their arbitrary variation, exactly according to the Lagrange variational principle in dynamics (7.5.36–39). Here, solutions of the "equations of motion" thus obtained are completely determined once their end values are given at $\zeta = 0$ and/or 1. This means an approximation of the multiscale by a two-scale; an exact version of the integral representations by the conventional two-scale method [7.17–20] is thereby obtained specifically for a collimated beam wave, including both spherical and plane waves as special cases. Here, the "two-scale" is manifested by the fact that the Fourier variable of the relative coordinates $\boldsymbol{\rho}_{12}$, i.e., $\boldsymbol{\lambda}''$, is given by (7.5.56) as the sum of $\boldsymbol{\lambda}'$ from the medium and $\boldsymbol{\nu}$ from the incident wave; while the counterpart equation is given by (7.5.18) when using the exact functional integral (multiscale) representation (7.5.16) for M_{22}. Therefore, it has nothing to do with the scales of the medium nor with the medium fluctuation intensity. Nevertheless, it leads to the same zero order expression as that obtained by the conventional, and also to a similar first order expression when making a formal expansion. The result is free from any expansion, and is presented with a set of unperturbative equations of closed form which can be solved iteratively (7.5.58–60). With exactly the same procedure, the three-point intensity correlation and the two-frequency intensity correlation were also obtained for the same incident beam wave.

The two-scale representations thus obtained are expected to be accurate solutions of the moment equations, particularly because the validity of using the variational principle seems to be almost ensured to the same approximation as when using the parabolic wave equations (7.1.2,3); and, therefore, to provide a basic means of investigating, for a practical beam wave, tough problems such as very large values of the scintillation index observed up to 6 [7.34, 36, 39], finding a specific expression of the fundamental parameter Δ of the intensity probability distribution function, derived based on the cluster expansion, and described consistently in terms of the normalized second and third order moments of the intensity (Sect. 7.4.5).

References

Chapter 1

1.1 For operator algebra, see, e.g., W.H.Louisel: *Quantum Statistical Properties of Radiation* (Wiley, New York 1973), §3.1
1.2 K.Furutsu: J. Math. Phys. **21**, 2764 (1980); Radio Sci. **10**, 29 (1975); J. Opt. Soc. Am.**62**, 240 (1972)

Chapter 2

2.1 K.Furutsu: J. Math. Phys. **21**, 2764 (1980); Radio Sci. **10**, 29 (1975)
2.2 K.Furutsu: J. Opt. Soc. Am. **A2**, 913, 932 (1985)
2.3 S.Chandrasekhar: *Radiative Transfer* (Dover, New York 1960)
2.4 S.Twomey, H.Jacobowitz, H.B.Howell: J. Atmos. Sci. **23**, 289 (1966)
2.5 H.B.Howell, H.Jacobowitz: J. Atmos. Sci., **27**, 1195 (1970)
2.6 K.Furutsu: J. Math. Phys. **21**, 765 (1980)

Chapter 3

3.1 K.Furutsu: Proc. IEE (London) **130**, Part F, 601 (1983)
3.2 F.G.Bass, I.M.Fuks: *Wave Scattering from Statistically Rough Surface*, ed. by C.B.Vesecky, J.F.Vesecky, International Series in Natural Philosophy, Vol.93 (Pergamon, Oxford 1979)
3.3 K.A.Norton: Proc. IRE **29**, 623 (1941)
3.4 M.A.Biot: J. Appl. Phys. **28**, 1455 (1957)
3.5 P.Beckmann, A.Spizzichino: *The Scattering of Electromagnetic Waves from Rough Surfaces* (Macmillan, New York 1963) pp.80–89
3.6 B.G.Smith: IEEE Trans. **AP–15**, 668 (1967)
3.7 M.I.Sancer: IEEE Trans. **AP–17**, 577 (1969)
3.8 G.S.Brown: IEEE Trans. **AP–26**, 472 (1978)
3.9 D.E.Barrick, E.Bahar: IEEE Trans. **AP-29**, 798 (1981)
3.10 A.K.Fung, H.J.Eom: IEEE Trans. **AP–29**, 463 (1981)
3.11 K.Furutsu: J. Math. Phys. **26**, 2352 (1985)
3.12 For electromagnetic waves, see K.Furutsu: J. Opt. Soc. Am. **A2**, 2244, 2260 (1985)
3.13 D.E.Barrick, W.H.Peake: Radio Sci. **3**, 865 (1968)
3.14 G.R.Valenzuela: IEEE Trans. **AP–15**, 552 (1967)
3.15 J.W.Wright: IEEE Trans. **AP–14**, 749 (1966)

Chapter 4

4.1 K.Furutsu: J. Opt. Soc. Am. **A2**, 913, 932 (1985); Phys. Rev. **A39**, 1386 (1989)
4.2 K.Furutsu: Phys. Rev. **A36**, 2080 (1987)

4.3 A.K.Fung, H.J.Eom: IEEE Trans. **AP–29**, 899 (1981)

Chapter 5

5.1 K.Furutsu: Phys. Rev. **A43**, 2741 (1991); **A39**, 1386 (1989)
5.2 E.Akkermans, P.E.Wolf, R.Maynard: Phys. Rev. Lett. **56**, 1471 (1986)
5.3 M.B. Van der Mark, M.P. Van Albada, A.Lagendijk: Phys. Rev. **B37**, 3575 (1988)
5.4 A.Ishimaru, L.Tsang: J. Opt. Soc. Am. **A5**, 228 (1988)

Chapter 6

6.1 K.Furutsu: J. Opt. Soc. Am. **A2**, 932 (1985); Phys. Rev. **A36**, 2080 (1987)

Chapter 7

7.1 For literature of the moment equations in the USSR before 1970, see, e.g., A.M.Prokhorov, F.V.Bunkin, K.S.Gochelashvili, V.I.Shishov: Proc. IEEE **63**, 790 (1975)
7.2 M.J.Beran, T.L.Ho: J. Opt. Soc. Am. **59**, 1134 (1969)
7.3 J.E.Molyneux: J. Opt. Soc. Am. **61**, 248 (1971)
7.4 W.P.Brown, Jr.: J. Opt. Soc. Am. **62**, 45 (1972)
7.5 K.Furutsu: J. Opt. Soc. Am. **62**, 240 (1972)
7.6 V.I.Tatarskii: *The Effects of the Turbulent Atmosphere on Wave Propagation*, available from the US Dept. of Commerce, NTIS, Springfield, VA 22151
7.7 J.W.Strohbehn: Proc. IEEE **56**, 1301 (1968)
7.8 K.S.Gochelashvili, V.I.Shishov: Sov. Phys. - JETP **39**, 605 (1974)
7.9 K.Furutsu: J. Math. Phys. **17**, 1252 (1976)
7.10 A.M.Whitman, M.J.Beran: J. Opt. Soc. Am. **A5**, 735 (1988)
7.11 F.V.Bunkin, K.S.Gochelashvili: Radiophys. Quantum Electron. (USSR) **13**, 811 (1970)
7.12 A.M.Whitman, M.J.Beran: J. Opt. Soc. Am. **60**, 1595 (1970)
7.13 V.I.Klyatskin, V.I.Tatarskii: Radiophys. Quantum Electron. (USSR) **13**, 828 (1970)
7.14 K.Furutsu: Appl. Opt. **27**, 2127 (1988)
7.15 B.J.Uscinski: Proc. R. Soc. Lond. **A380**, 137 (1982); J. Opt. Soc. Am. **A2**, 2077 (1985)
7.16 B.J.Uscinski, C.Macaskill, J.E.Ewart: J. Acoust. Soc. Am. **74**, 1474 (1983)
7.17 C.Macaskill: Proc. R. Soc. Lond. **A386**, 461 (1983)
7.18 S.Frankenthal, A.M.Whitman, M.J.Beran: J. Opt. Soc. Am. **A1**, 585 (1984)
7.19 A.M.Whitman, M.J.Beran: J. Opt. Soc. Am. **A2**, 2133 (1985)
7.20 J.Gozani: J. Opt. Soc. Am. **B3**, 721 (1987)
7.21 R.L.Fante: J. Opt. Soc. Am. **73**, 277 (1983)
7.22 M.Tateiba: IEEE Trans. **AP–23**, 493 (1975)
7.23 J.Strohbehn, T.-I.Wang, J.Speck: Radio Sci. **10**, 59, (1975)
7.24 K.Furutsu, Y.Furuhama: Opt. Acta **20**, 707 (1973)
7.25 I.M.Besieris: J. Math. Phys. **19**, 2533 (1976)
7.26 R.Kubo: J. Phys. Soc. Jpn. **17**, 1100 (1962)
7.27 M.E.Gracheva, A.S.Gurvich, S.S.Kashkarov, V.V.Pokasov: Trans. LRG–73–T–28 (Aerospace Corporation, Los Angeles, CA 1973); M.E.Gracheva, A.S.Gurvich, S.O.Lomadze, V.V.Pokasov, A.S.Khrupin: Radiophys. Quantum Elctron. **17**, 83 (1974)
7.28 G.R.Ochs, R.S.Lawrence: J. Opt. Soc. Am. **59**, 226 (1969)
7.29 J.Schwinger: J. Math. Phys. **2**, 407 (1961)
7.30 M.Nakagami: J. Res. Natl. Bur. Stand. **68D**, 995 (1964)
7.31 D.A.DeWolf: J. Opt. Soc. Am. **58**, 461 (1968); ibid. **59**, 1455 (1969)

7.32 T.-I.Wang, J.W.Strohbehn: J. Opt. Soc. Am. **64**, 994 (1974)
7.33 K.S.Gochelashvili, V.I.Shishov: Sov. Phys. – JETP **47**, 1028 (1978)
7.34 G.Parry, P.N.Pusey: J. Opt. Soc. Am. **69**, 796 (1979); G.Parry: Opt. Acta **28**, 715 (1981)
7.35 R.L.Phillips, L.C.Andrews: J. Opt. Soc. Am. **71**, 1140 (1981); L.C.Andrews, R.L.Phillips: J. Opt. Soc. Am. **A2**, 160 (1985)
7.36 G.R.Ochs, T.-I.Wang: NOAA/ERL, US Dept. of Commerce, Boulder, CO 80303 (private communication)
7.37 C.Macaskill, T.E.Ewart: J. Acoust. Soc. Am. **76**, 1466 (1984)
7.38 R.Dashen: Opt. Lett. **10**, 110 (1984)
7.39 S.Ito, K.Furutsu: J. Opt. Soc. Am. **72**, 760 (1982)
7.40 K.Furutsu: J. Phys. Soc. Jpn. **57**, 1167 (1988)
7.41 A.Consortini, E.Briccolani, G.Conforti: J. Opt. Soc. Am. **A3**, 101 (1986); A.Consortini, G.Conforti: J. Opt. Soc. Am. **A1**, 1075 (1984)
7.42 R.Mazar, J.Gozani, M.Tur: J. Opt. Soc. Am. **A2**, 2152 (1985)
7.43 B.J.Uscinski, C.Macaskill: Opt. Acta **32**, 71 (1985)

Bibliography

Chapter 1

J.Schwinger: *Brownian Motion of a Quantum Oscillator*. J. Math. Phys. **2**, 407 (1961)

Chapter 5

A.Ishimaru: *Wave Propagation and Scattering in Random Media*, Vols. 1, 2 (Academic Press, New York 1978)

Subject Index

Addition formula of incoherent scattering matrices 25
– – of scattering matrices 139
– of scattering matrices 23
Asymptotic form 151, 152, 233
– condition 223, 225, 235
– expression 77, 79, 108, 153–155, 168, 182, 225
– region 77
Attenuation coefficient 68, 105, 109
– factor 25, 106
Average symbol 209

Barnes-type integral 220
Bessel function 199, 219
Bethe-Salpeter (BS) equation 14–16, 101, 113, 119, 173, 174
Boundary equation 46, 50, 159, 160
– of no reflection 165
– scattering matrix 126, 135
– space 86, 90, 114–116

Canonical equations 248
– momenta 248
Cauchy's integral formula 37
Characteristic function 1, 4, 5, 217, 219, 221, 222
– functional 1, 6–9, 11
Cluster approximation 207, 212, 216
– expansion 229, 262
Coherent propagator 16, 24, 121, *see also* propagator for the coherent wave 174
Collimated beam wave 198, 201, 205, 224, 236, 253, 255, 262
Commutation relation 3, 7, 205, 250
Commutator 3, 4
Composite expression 19
Condition of no reflection 125, 147, *see also* no boundary scattering 150, 154
Contour path 37
Coordinate matrix 10, 13, 16, *see also* transposed version 25 and transposition 26
Correlation distance 8, 9
Cross section 27, 34, 58, 60, 107
– – per unit area 61
Cumulant 2, 5
– expansion 234
– series 3, 210, 216, 217, 222
Cumulative probability distribution 227, 228

Deterministic Green's function 68, 75, 111, 115, 171
Diffusion constant 165
– approximation 147, 155, 167
– eigenfunction 32
– equation 35, 127, 147, 156, 157, 160, 162, 168, 170
– expression 168
– mode wave 40
– region 29, 158
– term 34, 41, 156, 158, 169
Dispersive medium 119, 121

Effective medium 13, 15, 116, 117, 171
– scatterer 171
– (boundary) scattering matrix 127, 130, 133, 149

Eigenfunction expansion 27
- series 156
Eigenvalue 27, 29, 30, 32, 33, 35, 43, 201, 237, 252, 254
- equation 27
- function 27, 30, 35, 156
Electro-magnetic wave 97
Enhanced backscattering 174, 175
Equation of continuity 10, 22, 46
- of motion 242, 252, 257, 261
Exponential distribution 219, 234

Fixed scatterer 113
Forward scattering approximation 185, 192
- (and backward) reflection 64, 66
Fourier representation 1
- inversion 20, 21, 23, 28, 37, 38, 54, 189
- transform 15, 19, 20, 43, 50
- transformation 18
- variable 18
Free boundary 22
- surface 47
- - Green's function 47, 49, 51
Functional 1, 6, 7, 13, 196, 237, 256
- derivative, differentiation 6
- differential operator 7
- displacement operator 12
- integral 236, 252, 255, 256, 261
- integration 243, 244

Gaussian 4, 5
- distribution 222
- statistics 7, 118, 185, 186
Green's function 46, 238, 243, 257
Green's theorem 47, 95, 116
Ground wave propagation 47

Hamilton integral 242
Hamilton-Jacobi equation 248, 249
Hermitian 47, 51, 115, 122, 197
see also anti-Hermitian 82
- condition 53, 90, 98
- conjugation 90, 95
- matrix 88
- operator 53, 94, 236

Incoherent factor 15
- cross section 79, 109
- scattering matrix 16, 69, 101
- wave 21, 22
Inner product 96, 209
Integrated (optical) relation 18, 103
Intensity distribution 212, 216
- moments 219
Interference term 60, 75, 79, 80, 106, 124, 135, 178, 182
Intrinsically dispersive 19, 36

Jacobian 59, 63, 80

K-distribution 230
Kolmogorov spectrum 192, 194, 199
- medium 194, 195, 204
- theory of turbulence 193, 194

Ladder approximation 67, 69, 172
Lagrange variational principle 205, 236, 242, 261
Laplace transformation 221, 222
Large-scale rough surface 58, 64, 172
Legendre functions 34
Local ensemble averaging 62
- optical relation 70, 72, 102, 180
log-normal 212, 226–228, 233

Mode expansion 36
- term 38
Moment 191
- equation 185, 189, 190, 194–196, 198, 205, 207, 208, 212, 216, 224, 234, 250, 261, 262
- function 188
Multi-component 97
Multi-scale 262
- expression 239
- representation 236, 261
Mutual coherence function 60

Non-optical range 49, 81, 90
Normalization 27, 28, 32, 35, 61, 206
Normalized intensity moment 209, 213, 214

Operator representation 2, 9
– solution 88
Optical condition 16, 51, 70
– expression 20, 77, *see also* transport equation 18
– range 52, 57, 82, 91
– relation 58, 60, 62, 63, 65, 73, 76, 81–83, 107, 109, 110, 119, 121, 143, 145, 148, 154, 157, 159, 163, 166, 168, 181
– transformation 20, 25, 150
Ordered equation 92
Ordering 96, 205, 209, 237
– symbol 91, 241
Orthogonality 40

Pair bond 213, 214
Phase screen 200, 202, 205, 239, 246, 253, 254
– stationary point 244
Power conservation 17
– equation 30
– flux vector 10, 18, 21, 29, 46, 51, 52, 86, 115
– spectrum density 84
P-ordered product 213

Random function 6
– quantity 1, 6
Randomly distributed bosses 72, 83
Random-volume scattering matrix 130
Reciprocity 49, 149, 154, 155, 166
Reference (boundary) plane 45, 66, 71, 73, 86, 87, 114
Reflection coefficient 48, 52, 57, 59, 67, 109
Reflection-transmission coefficient 89, 124, 164
ꞁice-Nakagami distribution 217, 219, 224
ɔtational invariance 32, 34

turation effect 234
ıttering amplitude 14, 15, 51
ıatrix 12, 14, 15, 24, 28, 49, 50, 70, ɔ, 76, *see also* coherent scattering matrix 16, 123
– matrix for the entire system 144
– phase shift 67
Schrödinger equation 185, 207
Shadow area 62, 64
– effect 62
Shadowing 64
– effect 180
– function 66
Short-range function 38, 41, 61, 78, 135, 174
Singular expansion 37, 38
Spectrum function of the refractive index 193
– (density) function 193, 201, 204, 245, 250, 251, 253, 260
Specular reflection 79
Spherical harmonics 35
State of wandering 224
Statistical Green's function 9, 30, 66, 100, 117, 171
– – – of first order 11, 100
– – – of second order 11, 14, 23, 66, 100, 111, 117
Structure constant 193
Surface Green's function 46, 66, 89, 91
– impedance 45, 52, 54, 84, 92, 114
– wave 48, 71, 76, 85, 115
Symbolical Green's function 11

Tangent plane 58, 62
Three-point intensity correlation 262
Threshold value 226, 228, 234, 235
Transfer matrix 131
Translational invariance 19, 68
Transmissible (two-sided) boundary 43
Transport equation 18, 21, 30, 40, 131, 138, 148, 157, 183
Transposed matrix 90, *see also* transposition 124, 136
Truncated intensity 234
– moment 235
Turbulent continuum 202, 246, 253, 255
Two frequency intensity correlation 236, 249, 251, 262

Two-point intensity correlation 235, 261
Two-scale expression 245, 255, 260
– integral 244, 255
– method 202, 235, 236, 242, 262
– representation 262

Universal 216, 229

Variable function of integration 236, 261
Von Karman spectrum function 193

Watson (integral) 217, 221
Watson-Sommerfeld transformation 216, 221
Weighting function 27, 63